Studies in the History of Civil Engineering

Volume 7

The Civil Engineering of Canals and Railways before 1850

STUDIES IN THE HISTORY OF CIVIL ENGINEERING

General Editor: Joyce Brown

with the assistance of M.M. Chrimes, A.W. Skempton, N.A.F. Smith and R.J.M. Sutherland

---

Twelve volumes are to appear from the Autumn of 1997:

1. **The Engineering of Medieval Cathedrals**
   Edited by LYNN T. COURTENAY, University of Wisconsin

2. **Masonry Bridges and Viaducts**
   Edited by TED RUDDOCK, University of Edinburgh

3. **Land Drainage and Irrigation**
   Edited by SALVATORE CIRIACONO, University of Padova

4. **Dams**
   Edited by DONALD C. JACKSON, Lafayette College

5. **Water Supply and Public Health Engineering**
   Edited by DENIS SMITH, formerly University of East London

6. **Port and Harbour Engineering**
   Edited by ADRIAN JARVIS, Merseyside Maritime Museum and Liverpool University

7. **The Civil Engineering of Canals and Railways before 1850**
   Edited by MIKE CHRIMES, Institution of Civil Engineers

8. **The Development of Timber as a Structural Material**
   Edited by DAVID T. YEOMANS, Liverpool University

9. **Structural Iron, 1750–1850**
   Edited by R.J.M. SUTHERLAND, Consultant, Harris and Sutherland, London

10. **Structural Iron and Steel, 1850–1900**
    Edited by ROBERT THORNE, Editor, *Construction History Journal*

11. **Early Reinforced Concrete**
    Edited by FRANK NEWBY, formerly F.J. Samuely and Partners, London

12. **Structural and Civil Engineering Design**
    Edited by WILLIAM ADDIS, University of Reading

Studies in the History of Civil Engineering
General Editor: Joyce Brown

Volume 7

The Civil Engineering of Canals and Railways before 1850

edited by
Mike Chrimes

Routledge
Taylor & Francis Group
LONDON AND NEW YORK

First published 1997 by Ashgate Publishing

Published 2016 by Routledge
2 Park Square, Milton Park, Abingdon, Oxon OX14 4RN
52 Vanderbilt Avenue, New York, NY 10017

Routledge is an imprint of the Taylor & Francis Group, an informa business

British Library CIP data
The Civil Engineering of Canals and Railways before 1850. –
p. cm. – (Studies in the History of Civil Engineering; v. 7).
1. Railroad engineering—History. 2. Canals—Design
and construction—History. 3. Civil Engineering—History.
I. Chrimes, Mike.
625.1'09032

US Library of Congress CIP data
The Civil Engineering of Canals and Railways before 1850/
edited by Mike Chrimes.
p. cm. – (Studies in the History of Civil Engineering; v. 7).
Includes bibliographical references.
1. Canals–Great Britain–Design and construction–History–
17th century. 2. Canals–Great Britain–Design and
construction–History–18th century. 3. Railroad engineering–
Great Britain–History–17th century. 4. Railroad engineering–
Great Britain–History–18th century.
I. Chrimes, Mike. II. Series.
TC657. C58 1998 98–3723
627' 13' 0941–dc21 CIP

ISBN 13: 978-0-86078-756-3 (vol 7, hbk)
ISBN 13: 978-0-754-60654-3 (set)

STUDIES IN THE HISTORY OF CIVIL ENGINEERING – VOL. 7

Contents

Acknowledgements

The chapters in this volume are taken from the sources listed below, for which the editors and publishers wish to thank their authors, original publishers or other copyright holders for permission to use their material as follows:

Chapter 1: A.W. Skempton, 'Canals and river navigations before 1750', *History of Technology*, 3 (Oxford, 1957), pp. 438–470. Copyright © 1957 by Oxford University Press. Reproduced by permission of the publisher.

Chapter 2: K.R. Fairclough, 'The Waltham pound lock', *History of Technology*, 4 (London, 1979), pp. 31–44. Copyright © 1979 by Mansell Publishing (a Cassell imprint), London. Reproduced by permission of the publisher. All rights reserved.

Chapter 3: Charles Hadfield, 'Rivers and canals', *John Smeaton, FRS*, ed. A.W. Skempton (London, 1981), pp. 103–129. Copyright © 1981 by A.W. Skempton. Reproduced by permission of the publisher Thomas Telford Limited, London.

Chapter 4: R.B. Schofield, 'The construction of the Huddersfield narrow canal, 1794–1811: with particular reference to Standedge tunnel', *Transactions of the Newcomen Society*, 53 (London, 1981–82), pp. 17–38. Copyright © 1981–82 by the Newcomen Society, London.

Chapter 5: S.R. Broadbridge, 'John Pinkerton and the Birmingham canals', *Transport History*, 4 (1971), pp. 33–49. Copyright © 1971 by David & Charles, Newton Abbot. Reproduced by permission of Cassell Academic, London.

Chapter 6: Alastair Penfold, 'Managerial organization on the Caledonian canal, 1803–1822', *Thomas Telford: Engineer*, Proceeding of a seminar held at Coalport China Works Museum, Ironbridge, April 1979, under the auspices of the Ironbridge Gorge Museum Trust and Telford Development Corporation (London, 1980), pp. 129–50. Copyright © 1980 by Ironbridge Gorge Museum Trust, Ironbridge, Telford, Shropshire.

Chapter 7: Pietro Redondi, 'Along the water: the genius and the theory. D'Alembert, Condorcet and Bossut and the Picardy canal controversy', *History and Technology*, 2, no. 1 (Amsterdam, 1985), pp. 77–110. Copyright © 1985 by Harwood Academic Publishers GmbH and OPA Ltd. Reprinted by permission of Gordon and Breach-Harwood Academic, Amsterdam.

Chapter 8: Poverty, distress and disease: labour and the construction of the Rideau canal', *Labour/Le Travailleur*, 11, no. 11 (St John's, NF, 1983), pp. 7–29. Copyright © 1983 by the Canadian Committee on Labour History. Reprinted by permission of the editor.

Chapter 9: Mike Chrimes, 'Hugh McIntosh (1768–1840), national contractor', *Transactions of the Newcomen Society*, 66 (1994–95), pp. 175–192. Copyright © 1994–95 by The Newcomen Society, London.

Chapter 10: Charles E. Lee, 'Some railway facts and fallacies', *Transactions of the Newcomen Society*, 33 (1960–61), pp. 1–16. Copyright © 1960–61 by The Newcomen Society, London.

Chapter 11: Frank A. Sharman, 'The influence of landowners on route selection', *Journal of the Railway and Canal Historical Society*, 26, no. 2 (1980), pp. 47–53. Copyright © 1980 by The Railway and Canal Historical Society, Gwermynynydd, Clwyd.

Chapter 12: Richard S. Smith, 'England's first rails: a reconsideration', *Renaissance and Modern Studies*, 4 (1960), pp. 119–134. Copyright © 1960 by *Renaissance and Modern Studies*, University of Nottingham.

Chapter 13: P.J. Riden, 'The Butterley Company and railway construction, 1790–1830', *Transport History*, 6, no. 1 (1973), pp. 30–52. Copyright © 1973 by David & Charles, Newton Abbot. Reproduced by permission of Cassell Academic, London.

Chapter 14: A.W. Skempton, 'Cast iron edge-rails at Walker Colliery, 1798', *Transactions of the Newcomen Society*, 48 (1976–77), pp. 119–122. Copyright © 1976–77 by The Newcomen Society, London.

Chapter 15: A.W. Skempton, 'Embankments and cuttings on the early railways', *Construction History*, 11 (1996), pp. 33–49. Copyright © 1996 by The Construction History Society. Reproduced by permission of The Chartered Institute of Building, Ascot, Berks.

Chapter 16: David Brooke, 'The railway navvy: a reassessment', *Construction History*, 5 (1989), pp. 35–45. Copyright © 1989 by The Construction History Society. Reproduced by permission of The Chartered Institute of Building, Ascot, Berks.

Chapter 17: Harold Pollins, 'Railway contractors and the finance of railway development in Britain', *Railways in the Victorian Economy: Studies in Finance and Economic Growth*, ed. M.C. Reed (Newton Abbot, 1996), pp. 212–228. Copyright © David & Charles, Newton Abbot, Devon.

Chapter 18: Darwin H. Stapleton, 'The origin of American railroad technology, 1825–1840', *Railroad History*, 139 (1978), pp. 65–77. Copyright © 1978 by The Railway and Locomotive Historical Society, Clemson, SC.

Chapter 19: John H. White, 'Tracks and timber', *IA: The Journal of the Society of Industrial Archeology*, 2 (1976), pp. 35–46. Copyright © 1976 by The Society for Industrial Archeology, Houghton, MI.

***

Chapters 4, 9, 10 and 14 are reproduced from the *Transactions of the Newcomen Society* by permission of the Society's Council. The Newcomen Society promotes, encourages and co-ordinates the study of the history of all types of engineering and technology from the earliest times to the present day. Details from the Executive Secretary, The Newcomen Society, The Science Museum, London SW7 2DD, Tel/Fax 0171–589–1793.

Every effort has been made to trace all the copyright holders, but if any have been inadvertently overlooked the publishers will be pleased to make the necessary arrangement at the first opportunity.

General Editor's Preface

Joyce Brown

Civil engineering has a long history. Although practitioners did not begin to describe themselves as 'civil engineers' until the eighteenth century, the origins of their work lie in the construction skills of the ancient world and in the works of military engineers. Shortly after British civil engineers had formed themselves into a professional body, the Institution of Civil Engineers in London, their role was described in 1828 by Thomas Tredgold in these words:

> Civil engineering is the art of directing the great sources of power in Nature for the use and convenience of man.[1]

He thus neatly encapsulated what the civil engineer's job is about – the provision of the means to give society what it wants for its survival.

As time has gone by, the civil engineer's skills have had to diffuse over an ever-widening field and at the same time become more specialised. In all periods he has had to be the ingenious 'fixer', but more than a mere artisan – a point observed in the Institution's motto, *scientia et ingenio*.

The history of civil engineering is a fascinating one. Certainly 'the great sources of power in Nature' are not always directed without a certain amount of opposition by them, and the provision of the basic requirements – habitation, water supply, main drainage, harbours bridges, places of worship, transport systems by land and water – has called for inventiveness and sometimes for daring and courage.

Yet interest in the history of civil engineering is not of long standing, and it seems fair to state that most of the important work has been done in the last forty years. With a few notable exceptions, most of the major books on the subject have appeared in that era; many specialist societies with their own transactions have been founded; and many of the journals in which articles might be published have come into existence. While archaeologists and architectural historians have long made their contribution to this subject, their ranks have been swelled by a large number of serious scholars – including engineers themselves – now interested in recording and evaluating the engineering achievements of earlier times. In particular, there is a lively interest in the study of our industrial past, at least in part because some of the evidence of it are already disappearing from the landscape.

Our intention in creating the series of which this volume forms part is to provide, through the reproduction of important contributions to this subject, an

---

[1] *Minutes of Council, Institution of Civil Engineers*, 4 January 1828.

invaluable reference collection for its study. The series encompasses many different branches of engineering from early times to the beginning of this century; its perspective is global; and the chosen articles have an international authorship, to the extent that this can be achieved for an essentially English-language series. The introduction to each volume has allowed the volume editor to set the pages selected in the context of the whole history of the subject and its historiography, while the provision of references and a bibliography will enable the reader to go further into the study of the topic.

This series will have succeeded if it gives civil engineers some unexpected insights into their own craft and other readers a new way of looking at engineering structures.

***

The volume editors were chosen for their particular knowledge and expertise in the field. They have worked with enthusiasm and good will to put together collections of articles intended to be both informative and stimulating, and I am grateful to them for their efforts. Where possible, they have incorporated the work of the best scholars, while at the same time giving preference to articles often obscurely published and not readily available. The learned journals are represented, as also are conference proceedings and essay chapters contributed to books.

In all of this, I have been much assisted by my Advisory Panel. Emeritus Professor A.W. Skempton (Imperial College) has made himself freely available to me for consultation on many aspects of this venture. Mike Chrimes (Librarian of the Institution of Civil Engineers), Dr. Norman Smith (Emeritus Reader, Imperial College) and James Sutherland (formerly a Senior Partner in Harris and Sutherland, London, consulting engineers) have all contributed valuable advice based on their extensive knowledge of particular fields. The idea for the series was developed in discussion with Dr John Smedley of Ashgate, and Publisher of the Variorum Collected Studies Series, who has given me much personal support and the benefit of his experience.

I am grateful also for the help given by the staff of several libraries, notably Kay Crooks and Susannah Parry in the library of the Department of Civil Engineering, Imperial College, and Mike Chrimes and staff of the library of the Institution of Civil Engineers, London. Many friends have given me encouragement and useful advice; in particular, I want to thank Paula Kahn and Mark Baldwin.

Formerly Department of Civil Engineering,
Imperial College, London

Introduction

Mike Chrimes

This volume is concerned with the civil engineering of canals and railways, more specifically the construction of transport systems which revolutionised communications in Europe and North America *c.*1600–1850, a revolution largely achieved by the use of manual labour. The significance of these civil engineering works can scarcely be understated. Robert Stephenson, writing in January 1856, described them graphically thus,

> At the end of 1854 ... the aggregate length of railways opened in Great Britain and Ireland – measured about 8,054 miles – about the diameter of the globe ... It will naturally be asked what amount of capital has been required for the construction of these vast works ... £286,000,000 has absolutely been raised ... It is more than four times the amount of the annual value of all the real property of Great Britain ... If you consider the extent of the earthworks ... they will measure 550,000,000 cubic yards ... Imagine a mountain half a mile in diameter at its base, and soaring into the clouds one mile and a half in height – that would be the size of the mountain of earth which these earthworks would form.[1]

By 1850 the British achievement was by no means unique.

It is, however, by no means easy to obtain an accurate impression of, never mind detailed information on, how these works were carried out. In 1798 John Rennie, arguably the leading British engineer of his generation, wrote, concerning advice on the superintendence of canal construction in America, 'few authors have written on the subject of canals who are in any respect versant in the practical execution of them'.[2] One can argue that this absence of 'primary', contemporary published sources has made it difficult for modern historians dealing with the history of canals to provide an adequate explanation of how they were built. Before 1820 there were virtually no textbooks on civil engineering in the English language, and such published literature as exists is largely in the form of engineering reports, while the articles in encyclopediae by Farey on canals and Telford on inland navigation remain the best sources on canal construction in the British Isles to this day.[3]

---

[1] R. Stephenson, 'Presidential address', *Minutes of the Proceedings of the Institution of Civil Engineers*, 15 (1856), 124–5.

[2] Quoted in T.C. Proctor, 'The Middlesex Canal: prototype for American canal building', *Canal History and Technology Proceedings*, 7 (1984), 154–5.

[3] Farey, 'Canals'; and Telford, 'Navigation inland'.

Railway historians do not have the same excuse; the volume of railway construction encouraged a publishing explosion in Britain, and there is an abundance of contemporary monographs on surveying, earthwork measurement, tunnelling and railway construction in general[4] which provide useful guidance on how railways were built. One might perhaps mention two works in particular, Simms' *The Public Works of Great Britain*[5], and Brees' *Railway Practice*,[6] which contained specifications and contract drawings which could serve as a model for any aspirant engineer. The advent of the railways also coincided with the growth of civil engineering periodicals, a form of literature familiar to researchers today, but largely absent for the canal age in Britain and elsewhere.[7] Despite these advantages the history of the civil engineering of railways has largely been neglected, and is arguably in a weaker state than that of canals.

The papers gathered here are designed to take the reader through the main developments in civil engineering associated with canals and railways between 1600 and 1850. The focus is on British developments as it was Britain which, in the eighteenth and early nineteenth centuries, led the world in its development of a transport network.

The papers have been subdivided into 'canals' and 'railways'. This can be justified on chronological grounds, as canals generally preceded railways, and also on technological grounds in that different engineering solutions were required for the design of waterways from those needed for iron railroads. However, as this introduction will demonstrate, much linked the two, and the engineering successes of the canal era laid the foundations for the rapid development of the railway network.

---

[4] P. Bruff, *A Treatise of Engineering Fieldwork* (London, 1838); F.W. Simms, *A Treatise of the Principles and Practice of Levelling* (London, 1837); J. Macneill, *Tables for Calculating the Cubic Quantity of Earthwork in the Cuttings and Embankments of Canals, Railways and Turnpike Roads* (London, 1833); F.W. Simms, *Practical Tunnelling* (London, 1844, and later editions); W.D. Haskoll, *Railway Construction*, 2 vols. (London, 1857–1864). Earlier published as *Assistant Engineer's Railway Guide* (London, 1846–1848); G.D. Dempsey, *The Practical Railway Engineer*, 4th edn. (London, 1855).

[5] F.W. Simms, *The Public Works of Great Britain* (London, 1838).

[6] S.C. Brees, *Railway Practice*, 4th series (London, 1837–1847). Similar works appeared on the continent such as A. Perdonnet and C. Polonceau, *Portefeuille de l'ingénieur des chemins de fer*, 2 vols. (Paris, 1843–1846).

[7] The earliest civil engineering periodical is arguably the *Annuaire des ponts et chaussées* (1804–1807). However, the *Journal des voies de communications* (1826–), a Russian journal modelled on the French *Journal des mines* is more deserving of the title in the modern sense of a periodical concerned with theory, research and case studies. It was closely followed by the *Journal du génie civil* (France, 1828–1831). *Annales des Ponts et Chaussées* (France, 1831–), Crelle's *Journal für die Baukunst* (Germany, 1829–1838), *American Railroad Journal* (1832–1886), and *Transactions of the Institution of Civil Engineers* (1836–1842).

It is intended to indicate not only some of the major technological developments in this period, but also to give an idea of how construction was organised, how finance was raised and how technology was transferred.

Civil engineering today is normally assumed to include the non-moving parts of transport systems – the railway track, its substructure, bridges and buildings, waterways, locks and aqueducts. But what then is meant by the 'civil engineering' of canals and railways prior to 1850? For much of the period the profession of civil engineering would, in Britain at least, have comprised all branches of engineering outside the military sphere, including millwork, steam engine and machine design. Increasing use of machinery and locomotives in the first half of the nineteenth century encouraged the development of a specialist branch – mechanical engineering – and anachronistically these branches of 'civil engineering' have been excluded from the scope of this collection except as incidental topics.

The 'civil engineering of canals and railways' is taken to comprise the design and execution of canals, artificial waterways and railways, and the attendant professional work of the civil engineers responsible. The construction of canals and railroads had much in common: identification and justification of a suitable route; surveys; land acquisition; and, frequently, legislative approval. All required finance, design and specification of structures, as well as the award and supervision of contracts.

The design process involved some understanding of the engineering principles underpinning the successful construction of structures. A significant body of literature now exists on the development of structural design in the period, associated with advances in theoretical understanding, particularly elastic theory.[8] While not specifically concerned with the civil engineering of canals and railways, these structural engineering developments were increasingly associated with the demands of the developments in transport, particularly the design of long span structures such as railway bridges and station roofs. These subjects are outside the scope of this volume, but, as subjects of historical research are amply represented by other volumes in this series,[9] as well as in the collected papers of Heyman[10] and Skempton.[11] Also outside the scope of this volume is an account of two important aspects of the civil engineering of canals: their water supply and the design of aqueducts to convey them across roads and valleys.[12]

---

[8] E. Benvenuto, *A History of Structural Mechanics* (Berlin, 1991) is the latest in a series of such works. S.P. Timoshenko's *History of Strength of Materials* (New York, 1953) is arguably the most readable.

[9] Pertinent volumes are 2, 8, 9, 10 and 12 (see list at front of this volume).

[10] J. Heyman, *Arches, Vaults and Buttresses* (Aldershot, 1996).

[11] A.W. Skempton, *Civil Engineers and Engineering in Britain, 1600–1830* (Aldershot, 1996).

[12] The engineering of water supply is closely related to dam design, covered by Jackson in volume 4 in this series. The most important monograph on the subject is that by Binnie, *Early*

With these notable omissions, this volume still has considerable scope: the organisation of preparatory work, design and construction of earthworks, contract organisation and management, the role of the engineer, client and contractor, navigation devices on canals, the development of rail technology on the railways, and, associated with all these, the spread of ideas, encompassed in the concept of technology transfer.

Historiographical Overview

The historical literature of canals and railways is as old as the enterprises it describes. Works like the Canal du Midi, Canal de Aragon and London-Birmingham Railway were recognised by contemporaries as major human achievements, and accounts of the pioneering works appeared almost as the last spike was driven.[13] While the authors of some of these accounts may have lacked technological understanding to describe accurately the civil engineering involved, accounts, such as that by Lecount,[14] are today outstanding sources of information on civil engineering of the time. As the number of canals and railways multiplied, more general accounts were published, such as those by Phillips and by Priestley.[15] While of value today in establishing a chronological and geographical framework for the subject, they are inadequate as accounts of the construction of engineering works. These descriptive accounts were followed, in the second half of the nineteenth century, by the era of engineering biography, led by Samuel Smiles.[16] This had been anticipated by the publication of Smeaton's reports, with a biographical account of his life, largely instigated by members of the Smeatonian Society, who were acutely aware of the dearth of readily accessible

---

Dam Builders... and relevant contributions by Legget (see Bibliography) and R.B. Schofield, 'Design and construction of the Cromford Canal', *Transactions of the Newcomen Society*, 57 (1985–86), 101–23. The design of aqueducts is also tied in with bridge design, covered in volume 2 of the series by Ted Ruddock, and in his book *Arch Bridges and their Builders* (Cambridge, 1979). There has been considerable controversy regarding the design of the Pontcysyllte Aqueduct, and the relative responsibility of Telford or Jessop for its design. This is related in Charles Hadfield's *Thomas Telford's Temptation* (Cleobury Mortimer, 1993), although few would accept all of Hadfield's conclusions, it being generally held that Telford was largely responsible for the aqueduct as built. Other material on cast iron aqueducts is to be found in Schofield (chapter 4 in this volume) and in his paper, 'Benjamin Outram, canal engineer...', *Proceedings of the Institution of Civil Engineers*, 66 (1979), 539–55, and A.M. Faulkner, 'The Wolverton aqueduct: a case study in canal engineering, *Transport History*, 2 (1969), 155–69.

[13] For example, Andreossy, *Histoire du Canal du Midi* (Paris, 1799); Conde de Sastago, *Descripción de los Canales Imperiales de Aragon*... (1796).

[14] Lecount, *The History of the Railway*... (1839).

[15] Phillips, *A General History*... (1792); Priestley, *Historical Account*... (1831).

[16] Smiles, *Lives of the Engineers* (1861).

information on civil engineering works in print,[17] and also by Thomas Telford's autobiography.[18] Some of these Victorian works, such as Jeaffreson's life of Robert Stephenson,[19] contain useful information on civil engineering, but these biographical accounts, and works such as Francis' *History of the English Railway*[20] lack any critical historical perspective.

As the nineteenth century progressed, the existence of well-established railway companies encouraged the publication of 'company' histories. One of the earliest, and in terms of its coverage of civil engineering topics, best, was Williams' account of the Midland Railway.[21] In the early twentieth century several outstanding company histories were published: Tomlinson on the North Eastern Railway, McDermott on the Great Western, and Dendy Marshall on the Southern and Liverpool-Manchester.[22] The latter coincided with the centenary of the Liverpool-Manchester Railway, and such anniversaries have continued to inspire works, often of indifferent historical accuracy. There has been, fortunately, a continuing corpus of good company histories and Dow's work on the Great Central Railway,[23] and Dilts' on the Baltimore and Ohio contain much on the early construction of these lines.[24]

On the continent, official histories of the railway systems of Sweden, Austria-Hungary and latterly Germany appeared from the late nineteenth century; these provide useful and detailed information on the development of their railways.[25] Caron's more recent work on the Compagnie du Nord in France is in the highest traditions of economic history, but definitely is not an engineering history.[26] Generally, the works mentioned here are exceptional in their detail, and in their relatively extended coverage of the construction of the lines. Many histories are slight in every sense, and rarely get beyond a chronology of opening and closing dates, and details of locomotive-working. The literature of the civil engineering of railways is overshadowed by the vast amount published on the history of

---

[17] Smeaton, *Reports of the Late John Smeaton* (1797–1812).

[18] Rickman, *Life of Thomas Telford* (1838).

[19] Jeaffreson, *The Life of Robert Stephenson* (1864).

[20] Francis, *History of the English Railway* (1851).

[21] Williams, *The Midland Railway...* (1876).

[22] Tomlinson, *North Eastern Railway* (1915); MacDermott, *History of the Great Western Railway* (1927–31); Dendy Marshall, *Centenary History...* (1930).

[23] Dow, *Great Central Railway.*

[24] Dilts, *The Great Road...*

[25] Jarnvagsstyrelse, *Statens Jarnvagar...* (1906); Zuffer, *Geschichte der Eisenbahnen* (1989); Deutsche Reichsbahn Hundert Jahre (1936). A comprehensive geographical and conventional historical approach to the last is provided in A.M. Mayer's *Geschichte und Geographie der Deutschen Eisenbahnen von 1835 bis 1890*, 4 vols. (1894, reprinted Steiger, 1984).

[26] F. Caron, *Histoire de l'Exploitation d'un Grand Réseau: la Compagnie du Chemin de Fer du Nord, 1846–1937* (Paris, 1973).

locomotives, and nothing comparable to Ahrons' work on locomotives has appeared on civil engineering.[27]

As for engineering biography, this can be said to have emerged in Britain in the period after the First World War, led by members of the Newcomen Society, notably H.W. Dickinson. He assessed the engineering achievements of figures such as James Watt, moving beyond the chronological approach of conventional biographies.[28] While Dickinson was chiefly concerned with 'mechanical' engineers, as Skempton has noted,[29] there was a contemporary renaissance of interest in early civil engineering, including figures like Thomas Steers and Henry Berry who were responsible for some of the early eighteenth century inland navigation schemes.[30] While Gibb's biography of Telford, and Rolt's post-war biographies of Telford, Stephenson and Brunel do much to illuminate their engineering achievements,[31] the basic chronological approach inevitable in any conventional biography tends to fragment discussion of achievements in specific civil engineering disciplines, such as canal design. Gibb highlighted Telford's contribution to the development of the civil engineering profession in terms of contract organisation, but the scale of Telford's works made it difficult to assess his engineering achievement outside key works such as the Menai Bridge. In a sense, Boucher attempted to deal with the limitations of the biographical approach in his work on John Rennie, by looking at Rennie's approach to bridge design outside the narrative framework.[32] However, such is the breadth of Rennie's, and other leading engineers' achievements, that it is difficult to deal with them all in one work. More recently, a different approach was tried in books on Brunel, Telford, and Smeaton.[33] Here subject specialists reviewed particular aspects of the engineer's career, and generally a more satisfactory feel for the engineering achievements of the engineers was obtained.[34] Two of the papers included, by Penfold (chapter 6) and Hadfield (chapter 3), come from these works.

---

[27] Ahrons, *The British Steam Railway Locomotive...*

[28] H.W. Dickinson and Rhys Jenkins, *James Watt and the Steam Engine* (Oxford, 1927) is probably the finest example of these engineering biographical studies.

[29] A.W. Skempton, 'Introduction' to *Civil Engineers and Engineering* (see note 11).

[30] H. Peet, 'Thomas Steers, the engineer of Liverpool's first dock', *Transactions of the Historical Society of Lancashire and Cheshire*, 82 (1930), 163–242; S.A. Harris, 'Henry Berry, Liverpool's second dock engineer', *Transactions of the Historical Society of Lancashire and Cheshire*, 89 (1937), 91–116.

[31] Gibb, *The Story of Telford*; Rolt, *Isambard Kingdom Brunel*; idem, *George and Robert Stephenson*; idem, *Thomas Telford.*

[32] Boucher, *John Rennie...*

[33] Pugsley, *The Works of Isambard Kingdom Brunel*; Penfold, *Thomas Telford...*, Skempton, *John Smeaton.*

[34] This had been tried earlier by Jeaffreson in *The Life of Robert Stephenson* with a specialist contribution by William Pole.

The focus remains on the great engineers, and coverage of others is at best patchy. The lack of a biography of a major figure such as Jessop has only recently been remedied.[35] Overseas, the picture is similar. In France there have been recent studies of Gauthey, and specialist contributions on Seguin.[36] In the USA there is an excellent recent work on Jervis, and comprehensive treatment of Latrobe Snr.,[37] but many of the other pioneers are still awaiting serious study. In Spain there has been a recent study of the early canal engineer Carlos Lemaur.[38]

Generally, engineering biography has been poorly served by historians, in comparison with the treatment of other professions. Relatively few are found in national biographical dictionaries, whose 'whole life' approach makes difficult an engineering assessment. The American Society of Civil Engineers has for a number of years been attempting to remedy this by a series of volumes of biographical dictionaries;[39] a standard work has existed for many years on French engineers.[40]

More recently, the Institution of Civil Engineers has embarked on a similar project, dealing initially with engineers active before 1830. At present many of the individuals are known merely as names in books such as Hadfield's regional canal histories. No engineering assessment is possible without an overview of these individuals' careers, and a proper understanding of the influence of their training on their designs and the like.

The Making of the Canals and Railways

The origins of canal engineering can be traced from attempts by man to improve natural river navigation; they are as old as civilisation itself. Water always provided convenient transport corridors even when it was a barrier to land transport, and as

[35] Hadfield and Skempton, *William Jessop, Engineer.*

[36] Presse de l'Ecole National des Ponts et Chaussées, *Un Ingénieur des Lumières: Emiland-Marie Gauthey* (Paris, 1993); for Seguin, see M. Cotte and I. Kobayashi, 'The first French railways of St Etienne (1823–1833)', *Historical Studies in Civil Engineering*, 16 (1996), 105–16; M. Cotte, 'La genèse de la chaudière tubulaire chez Marc Seguin (1925–1830)', *Revue d'Histoire des Chemins de Fer*, 14 (1996), 23–41.

[37] F.D. Larkin, *John B. Jervis: An American Engineering Pioneer* (Iowa, 1990); D.H. Stapleton, ed., *The Engineering Drawings of Benjamin Henry Latrobe* (New Haven, 1980) is best for his engineering career. A full biography is available: T. Hamlin, *Benjamin Henry Latrobe* (New York, 1955).

[38] T. Sanchez La Zaro, *Carlos Lemaur y el Canal de Guadarrama* (Madrid, 1995).

[39] American Society of Civil Engineers, *A Biographical Dictionary of American Civil Engineers*, 2 vols. (New York, 1972–1991). A dictionary approach was earlier adopted by C.B. Stuart in his *Lives and Works of Civil and Military Engineers of America* (New York, 1871).

[40] F.P.H. Tarbe de St Hardouin, *Notices Biographiques sur les ingénieurs des Ponts et Chaussées depuis la création du Corps, en 1716, jusqu'à nos jours* (Paris, 1884).

a means of carrying heavy loads must always have provided a welcome relief from land haulage. By the use of rivers it was possible to extend sea-borne trade many miles inland. Even today the great river arteries of the world – the Rhine, Danube, Volga, Yangtze, Mississippi and Amazon – are of major importance in world trade, far more so than the more famous artificial canals at Panama and Suez.[41]

The early development of river navigations, and the canals which followed, has been traced by Skempton (chapter 1), whose paper relies largely on contemporary published accounts. This demonstrates that the major early engineering developments and achievements took place outside the British Isles. Nevertheless, in the medieval period, improvements were made to river navigations in Britain, with flash locks known to have been used on the Thames from the thirteenth century,[42] and by 1600 there were around 700 miles of navigable rivers in England. The Elizabethan period saw the introduction of pound locks to Britain, with their use on the Exeter Canal, Trent, and Lea Navigation.[43] Although first used in China, and developed, independently, in Europe from the fourteenth century, these British developments can be viewed as the beginning of modern civil engineering in the United Kingdom. While the great navigation achievements of the Renaissance and early modern period were in Northern Italy[44] and France, the forerunners of the British civil engineering profession made considerable improvements to inland navigation, and by 1760 had added 600 miles, or almost double the length available 150 years before. This achievement has been explored by Willan,[45] and the engineering aspects by Skempton in more detail in two papers.[46] With regard to

---

[41] For a summary of these developments, and more recent events, see Charles Hadfield's *World Canals* (Newton Abbot, 1986).

[42] An impression of their extent can be obtained from a major industrial archaeological survey: M.J. Lewis and others, 'Flashlocks on English waterways', *Industrial Archaeology*, 6 (1969), 209–53. Other articles on their use include A.W. Skempton, 'Engineering on the Thames navigation, 1770–1845', *Transactions of the Newcomen Society*, 55 (1983–84), 153–76, and K. Fairclough, 'The River Lea before 1767: an adequate flash lock navigation', *Journal of Transport History*, 3rd series, 10, no. 2 (Sept. 1989), 128–44.

[43] De la Garde, 'Memoir of the canal of Exeter...'; C.R. Salisbury, 'Documentary evidence for an Elizabethan pound lock on the River Trent in Nottinghamshire', *Transactions of the Thoroton Society*, 86 (1982), 115–17; see Fairclough, chapter 2 in this volume.

[44] More information on these developments can be found in W.B. Parsons, *Engineers and Engineering in the Renaissance* (Baltimore, 1939), especially 323–477. More recent works include José García Diego, 'Giovanni Francesco Sitoni, an hydraulic engineer of the Renaissance', *History of Technology*, 9 (1984), 103–25, reprinted in volume 3 of this series.

[45] Willan, *River Navigation in England*... Some French historians appear ignorant of these British developments; see Abel Poitrineau, 'The construction and operation of the Briare Canal' [in French], *Revue d'Auvergne*, 102, no. 3 (1988), 197–224.

[46] A.W. Skempton, 'Engineers of the English river navigations, 1620–1760', *Transactions of the Newcomen Society*, 29 (1953–55), 25–54; *idem.*, 'Engineering on the English river navigations to 1760', in M. Baldwin and A. Burton, ed., *Canals: A New Look* (Chichester, 1984), 23–44.

the papers reprinted here, Fairclough's (chapter 2) is one of several he has written describing the engineering innovations which characterised the improvements in navigation in the early modern period.[47] The Waltham Pound Lock can be regarded almost as a defining moment in the emergence of modern canal engineering in the British Isles.

Hadfield's paper (chapter 3) describes the achievements of John Smeaton in this field. Smeaton is justly regarded as the founder of the civil engineering profession in Britain. His professional practice served as a model for canal and railway engineers, and his methods of project organisation, developed for the Forth-Clyde Canal, have remained in use to this day. His career encompassed the river navigation and canal age, and included, at the start of his career, work on the Calder-Hebble navigation, and subsequently the Forth-Clyde Canal, one of the longest canals.

Hadfield was the doyen of canal historians. His career, spanning fifty years, raised the standard of scholarship in this field to a level rarely encountered in civil engineering history.[48] He described his approach in a paper in 1956,[49] and his insistence on studying all aspects of a canal's history – the general social and economic background, the interests of the promoters, and the engineering – ensured that his regional histories of canals always contain engineering information. The regional histories were associated with a growing interest in many of the individual canals and these, often produced under Hadfield's overall editorship, have generally been associated with a similar level of scholarship. Some, of which Household's history of the Thames and Severn Canal is perhaps the best example, provide general background on canal construction at the time.[50] They have been underpinned by a growing body of work by local historians, mostly focusing on economic rather than engineering history.[51]

---

[47] K. Fairclough, 'Navigation devices along the River Lea, 1600–1767', *Transactions of the Newcomen Society*, 64 (1992–1993), 21–40; *idem.*, 'A survey of the River Lea by Christopher Wren', *Railway and Canal Historical Society Journal*, 31, no. 1 (1993), 10–17; 31, no. 2, 107–111; *idem.*, 'A successful Elizabethan project: the River Lea improvement scheme', *Journal of Transport History*, 11, no. 2 (1990), 54–65.

[48] Hadfield's first paper was published in 1942: E.C.R. Hadfield, 'Canals between the English and Bristol Channels', *Economic History Review*, 12 (1942), 59–67. Hadfield, *The Canals of Southern England* (1955) was the first in his series of regional histories of canals. A complete Bibliography of Hadfield's work on all subjects appears in Baldwin and Burton, *Canals: A New Look*, 4–8, a book produced in his honour (see note 46).

[49] C. Hadfield, 'An approach to canal research', *Railway and Canal Historical Society Journal*, 1, no. 3 (1956), 23–6.

[50] Household, *The Thames and Severn Canal.*

[51] Examples include G.G. Hopkinson, 'The development of inland navigation in South Yorkshire and north Derbyshire, 1697–1850', *Transactions of the Hunter Archaeological Society*, 7 (1951), 229–51; A. Temple Patterson, 'The making of the Leicestershire canals, 1766–1814', *Transactions of the Leicestershire Archaeological and Historical Society*, 27 (1951), 67–99; C.R.

The best consideration of engineering issues has been the work of engineers themselves, most notably papers in the Newcomen Society by Skempton, Schofield and Legget. Despite their work, one still has to turn to contemporary sources for discussion of basic information, such as that on manufacture and use of puddle clay for lining canals.

The great age of canal building was between about 1760 and 1830. Canals were not immediately superceded by railways, but serious development of railways took place only from about 1830. The origins of the railway were, of course, earlier, and the introduction of steam locomotives in the British Isles in the early nineteenth century acted as a catalyst for rapid growth. Consequently, the overwhelming bulk of literature on the railways is concerned with developments since 1800. In reality, the rapid exploitation of steam traction was dependent on experience built up over centuries of railroad construction, most recently the use of iron rails to form part of the permanent way, an innovation better suited to sustaining the heavy loads of a locomotive than the timber more generally employed until the late eighteenth century.

The origins of the modern railway lie in the primitive wooden guides developed for use with mine trucks in central European mining areas in the fifteenth century. The existence of such prototypes is well attested in early mining literature such as the writings of Agricola, reprinted earlier this century, and widely reproduced.[52] The first British historian seriously to investigate these early lines was Charles E. Lee in *The Evolution of Railways*, published in 1937. His paper (chapter 10) here identifies many of the key issues in the development of railways and their historical treatment. The post-war division of Europe made it difficult for western researchers to investigate East (Central) European locations first hand, and language barriers made it difficult to access work being pursued in old mining regions in Hungary and Slovakia. Despite these problems, a definitive international study of early wooden railways was produced by M.J.T. Lewis in 1970.[53] Smith (chapter 12) had earlier established the introduction of

---

Clinker and C. Hadfield, 'The Ashby de la Zouch Canal and its railways, *Transactions of the Leicestershire Archaeological and Historical Society*, 34 (1958), 53–76; J. Tann, 'The Yorkshire Foss Navigation', *Transport History*, 3 (1970), 80–90; Peter L. Smith, 'The Sheffield and South Yorkshire Navigation, *Transport History*, 11, no. 3 (1980), 228–43; *idem.*, 'The Calder and Hebble Navigation, *Transport History*, 12 (1981), 7–10; G.N. Crawford Thomas, 'Telford and the Gloucester and Berkeley Canal, *Industrial Archaeology Review*, 11, no. 2 (1989), 155–70; E.A. Shearing, 'The Birmingham and Liverpool Junction Canal: planning and construction in the Norbury District', *Railway and Canal Historical Society Journal*, 30, no. 2 (1990), 73–85; 30, no. 3 (1990), 139–46; P. Jones, 'Vermuyden's navigation works on the River Don', *Railway and Canal Historical Society Journal*, 31 (1994), 248–59.

[52] Agricola, *De re metallica*.

[53] Lewis, *Early Wooden Railways*. The title 'wooden' is a misnomer, as the introduction of iron is discussed in detail.

railroads in this country by documentary research; to go further will require archaeological finds.[54] The significance of Smith's paper is that the 'railways' it describes mark the start of the development of the modern railway system in Britain. Henceforth Britain increasingly became the leader in railway technology, taking it away from the pit head into a relatively well-developed freight transport system by the late eighteenth century. Our knowledge of this system has been expanded by the work of Mott and others.[55]

Of great interest is the introduction of iron rails, more specifically the use of edge rails – the basis of modern rails – rather than the flanged plateways which preceded them.[56] The claims of the Walker Colliery are detailed by Skempton and Andrews (chapter 14) and it remains the best documented early example. The exploitation of the iron for rails was encouraged by and fostered the growth of specialist suppliers such as the Butterley concern described by Riden (chapter 13). In the early years a vast variety of rail profiles were developed, described by Whishaw and others.[57] The so-called Vignoles or flat-bottomed rail, much used on the continent, is generally believed to have been developed in the United State.[58] By mid-century, British producers were losing out to continental works such as Cockerills due to poor quality control.[59] This was in part dealt with by

---

[54] That such discoveries may be made is clear from those made recently regarding old plateways: P. Burgess, 'The use of plate rails in the Godstone firestone quarries – some recent discoveries', *Proceedings of the Croydon Natural History and Scientific Society*, 18, no. 1 (1994), 102–7; R.W. Rennison, 'A rare find: a wooden waggon way unearthed', *Panel of Historical Engineering Works* (Institution of Civil Engineers), *Newsletter*, 73 (Spring 1997), 1. Recent finds in the Ironbridge area are summarised in C. Clark, *Ironbridge Gorge* (London, 1993), 95–107. The evolution of plateways on the Kilmarnock and Troon Railway is discussed in detail in R.A. Paxton, 'Conservation of the 1811 railway viaduct at Laigh Milton, Scotland', *Historical Studies in Civil Engineering* (Japan Society of Civil Engineers), 16 (1996), 1.

[55] R.A. Mott, 'English waggonways of the eighteenth century', *Transactions of the Newcomen Society*, 37 (1964), 1–33; *idem.*, 'Tramroads of the eighteenth century and their originator: John Curr', *Transactions of the Newcomen Society*, 42 (1969–70), 1–23; B. Baxter, 'Early railways in Derbyshire', *Transactions of the Newcomen Society*, 26 (1947–49), 185–97; but see P.J. Riden, 'Tramroads in North East Derbyshire', *Industrial Archaeology*, 7, no. 4 (1970), 373–96, for a corrective to this.

[56] Latest, as yet unpublished, research by Fitzgerald, suggests that the Middleton Railway, Leeds, may be regarded as the 'home' of the modern rail.

[57] Francis Whishaw's *Railways of Great Britain and Ireland* (London, 1840; 2nd edn. 1842) illustrates many of the early rail profiles; A Hertwig, 'Die Eisenbahn und das Bauwesen', *Tecnikgeschichte*, 3 (1936), 29–37. Plates 9 and 10 illustrate the early German rail profiles.

[58] J.E. Watkins, 'The development of the American railroad track as illustrated by the collection in the U.S. National Museum', *Report of the National Museum* (Washington, DC, 1888–1889).

[59] R. Fremdling, 'Belgische und britische Schienenexporte für die Eisenbahn von Genua nach Turin-Qualitatsmangel eines britischen Exportproduktes um 1850', *Scripta Mercaturae*, 27 (1993), 165–82.

introducing steel rails, a process discussed by Brooke, in a paper reproduced in volume 10 of this series.[60]

Construction: Organisation and Building

Turning now to the way in which construction of canals and railways was achieved, we can see that development of a country's infrastructure was frequently the result of central government planning and initiatives. While France provides the best known example of this approach, with its national road network, and state involvement in canals from the sixteenth century,[61] it was true of most other European states, both in planning, design and some form of financial support. A joint stock/shared risk approach had been developed for land drainage schemes in the Netherlands and was exported to France and England by the early seventeenth century. Many of the British canals and railways developed from this model. Although early railways or tramroads were built by a way-leave approach (i.e. the grant of permission to transport coal etc. across property), this was impractical for navigations and canals as they affected the interests of riparian owners, and Acts of Parliament were involved. This procedure, by private bill, with most schemes privately financed, reached its maturity in the railway age. The methods of raising finance, which was generally possible using local resources for relatively low capital canal schemes, developed rapidly in the early railway age. Pollins (chapter 17) demonstrates how the system developed into one where those with the greatest interest in the realisation of these projects, the contractors, came to play an ever more important role.

In Britain, the successful passage of any scheme through the Houses of Parliament involved surveying and route selection, taking into account the route's economic viability, the specialist property interests along it, and a skilled engineer capable of arguing the case in the Houses of Parliament.[62] Sharman's paper

---

[60] D. Brooke, 'The advent of the steel rail, 1857–1914', *Journal of Transport History*, 3rd series, 7 (March 1986), 18–31.

[61] R.G. Geiger, *Planning the French Canals: Bureaucracy, Politics and Enterprise under the Restoration* (Newark, 1994). Although this focuses on Becquey's proposals, it does provide background on the early development of canals in France. Problems with realising French government policy on railways are discussed by Ribeill, *La Révolution Ferroviare...*, and Barrie M. Ratcliffe, 'The origins (building) of the Paris Saint Germain Railway. Some entrepreneurial and financial problems of railways in France in the 1830s (and 1840s)', *Journal of Transport History*, 1, no. 4 (1972), 197–219; 2, no. 1, 20–40.

[62] For geographical and property factors, see D. Whomsley, 'A landed estate and the railway: Huddersfield', *Journal of Transport History*, 2, no. 4 (1974), 189–213; P.S. Richards, 'Geographical analysis of some of the surveys made for the London and Birmingham Railway line', *Transactions of the Birmingham Archaeological Society*, 80 (1965), 17–25; John H. Farrington, 'The Leeds and

(chapter 11) effectively summarises the issue of property influence on canal and railway schemes.

In the USA, while there is much evidence of state and federal interest in developing the infrastructure, most notably in the Gallatin plan of 1808, one also sees many examples of private enterprise taking the lead.[63]

The scale of major canal projects like the Canal du Midi and the Gotha Canal, and the main line railways, demanded a sophisticated degree of project organisation which in many countries had only been seen before in a military context. It is instructive to note that the Gotha Canal in fact relied on the Swedish army to a large extent for its execution,[64] many of the early American railways were surveyed, laid out and even supervised by West Point trained engineers,[65] and the Ponts et Chaussées administration which drew up the plans for the French canal and railway network had many of the hallmarks of a military command structure.[66]

In Britain, a system of delegated authority can be seen in seventeenth-century projects such as the Bedford Level Drainage, but is generally felt to have emerged following Smeaton's example on the Forth-Clyde Canal, with a resident engineer acting in the client's interest to supervise the execution of the designs of the (consulting) engineer, and assistant engineers under him to supervise the contractors at work.[67] This system can seem to have reached maturity under Telford on works such as the Caledonian Canal, and Penfold (chapter 6) describes a system which would be familiar to most British engineers working today. The superior authority of the engineer was by no means clear before then, and even

Liverpool Canal: a study in route selection', *Transport History*, 3 (1970), 52–79; Schofield discusses the engineering evidence to Parliament in his paper on 'The design and construction of the Cromford Canal...' (see note 12). The biographies of Brunel and the Stephensons by Rolt discuss these aspects of their work with particular reference to the Great Western, London and Birmingham, and Liverpool and Manchester Railways.

63 Vance, *The North American Railroad*. It can be unwise to generalise. C.A. Dunlavy, *Politics and Industrialisation: Early Railroads in the United States and Prussia* (Princeton, 1994) demonstrates that, despite America's reputation for a *laissez-faire* approach to development, there is much evidence for government involvement in American railroads, while, in contrast, Prussia, the archetypal German authoritarian state, often stood back from railway development. Despite the preference of some early Prussian proponents of railways for the cheap American construction model, there were some critics of what some saw as a short-sighted approach. For Federal involvement in waterways, see T. Shallat, 'Building waterways, 1802–1861: science and the United States Army in early public works', *Technology and Culture*, 31, no. 1 (1990), 18–50.

64 Stromback, *Balzar von Platen...*

65 Vance, *The North American Railroad.*

66 See note 61.

67 Skempton, *John Smeaton*. Professor Roland Paxton provided more documentation on this in 'Smeaton in Scotland', *Institution of Civil Engineers' Smeaton Lecture: Introductory Notes* (London, 1992).

on the Forth-Clyde canal it is evident that members of the committee played an important role in engineering decisions on a regular basis.[68]

One aspect of construction common to both canals and railways was the building of earthworks in the form of embankments and cuttings. It seems incredible that until Skempton's recent (1996) paper (chapter 15) nobody had attempted a review of how such work was carried out. This work was undoubtedly facilitated by the use of locomotives to move spoil, but the techniques were clearly developed in Britain in the canal age using horses and wagons. It is interesting to observe that the contractors' tipping wagons were as studiously noted by foreign visitors, such as Lame and Melnikov, as were rail profiles and more obvious aspects of railway technology.[69] There is evidence of how such work was carried out on canals, but as is the case with railways until recently, no overall review exists. In overseas literature the picture is much the same.

A regular feature of American literature is reference to the cheaper methods of construction favoured there. The paper by White (chapter 19) points to a stark contrast between the American and British model. At its extreme, even earthworks were dispensed with in the United States, the track being laid on various forms of timber sub-structures. The 'lighter' forms of construction were initially preferred in Germany to the English model,[70] but in France, once the Ponts et Chaussées engineers had assumed full control after the English invasion of the early 1840s, durability of construction rather than short-term cost savings became a standard feature.[71] However, more reviews in the manner of Skempton are required to obtain a true international picture of earthworks construction, and how attendant problems of slope stability were dealt with.

One of the most challenging aspects of 'earthwork' construction is tunnelling. While tunnelling techniques were hardly new – the links with mining are obvious, given the involvement of coal owners in the early British canals, and the support of soft ground was a necessary aspect of military mining – transport tunnels presented new problems. They had to be permanent structures, rather than temporary access routes to minerals or military targets, and in the case of railway

---

68 J. Jones, 'James Hutton and the Forth and Clyde canal', *Annals of Science*, 39, no. 3 (1982), 255–63.

69 Melnikov and Lame manuscripts in the Library of the Institute of Railway Engineers in St Petersburg: G. Lame, *Observations relatifs à l'art de l'ingénieur recueilliés durant le voyage en Angleterrre fait d'après l'ordre de...Nicholas Ier* (1830); P.P. Melnikov, *Angliya, Schotlandiya, Irlandiya, 1837–38* (1839). Telford had drawings prepared of contractors' equipment on the Caledonian Canal in the first decade of the nineteenth century (Telford MSS, Library Institution of Civil Engineers, London).

70 See note 63.

71 Ribeill, *La Révolution Ferroviare...* For the English-built lines, see V. Maréchal, 'La construction des lignes de chemins de fer du Paris à Rouen et de Rouen au Havre', *Revue d'Histoire des Chemins de Fer*, 14 (1996), 64–89.

tunnels in particular, they had to be reasonably dry. There are valuable contemporary accounts of tunnelling techniques by Simms, Drinker and Rziha, and Sandstrom provides a useful modern overview of the major engineering advances and achievements.[72] Junca Ubierna gives an excellent account of early tunnelling, but his volume on recent techniques is not yet published.[73] West has described the most important technological innovations, while Stack provides detailed descriptions of most of the machinery.[74] Drilling machinery, however, is a development outside the scope of this period, first being applied at Mount Cenis in 1861. Shield tunnelling was introduced on the Thames Tunnel in the 1820s by (Sir) Marc Isambard Brunel for a road tunnel, but not successfully applied to railway construction until the City and South London Railway in the 1880s. Compressed air likewise was only at an experimental stage before 1850.[75]

What all the general sources reveal are the problems presented by tunnelling. It remains one of the difficult civil engineering arts, in part because complex ground conditions may be unexpectedly encountered. These problems, present today, were exacerbated by the primitive site investigation methods employed at the time, which could frustrate even great engineers like the Brunels,[76] while accurate methods of aligning tunnels were lacking for much of the canal period, with a transit instrument being used for the first time by William Tierney Clark on the Thames-Medway Canal.[77] Some ideas of what was involved, and how problems could be overcome is revealed by Schofield (chapter 4). A major problem was the cost of tunnelling work, which could be a source of financial embarrassment to both client and contractor. Brooke devotes a chapter to tunnel construction in his book on navvies.[78] Many of the Pinkertons' problems (see chapter 5) stemmed from tunnel contracts.

This was an area where contractors early began to specialise. The firm of Pritchard and Hoof carried out seven tunnel contracts. Despite the great problems

---

[72] Simms, *Practical Tunnelling...* (see note 4). H.S. Drinker, *Tunnelling, Explosive Compounds and Rockdrills* (New York, 1878 and later editions); F. Rziha, *Lehrbuch der gesammten Tunnelbaukunst*, 2nd edn., 2 vols. (Berlin, 1874); K.E. Sandstrom, *The History of Tunnelling* (London, 1963).

[73] J.A. Junca Ubierna, *El Túnel: Historia y Mito* (Madrid, 1990).

[74] G. West, *Innovation and the Rise of the Tunnelling Industry* (Cambridge, 1988); B. Stack, *Encyclopaedia of Tunnelling, Mining and Drilling Equipment*, 3 vols. (Hobart, 1995). Volume 1 deals with the history of tunnelling equipment. This work updates Stack's earlier *Handbook of Mining and Tunnelling Machinery* (New York, 1982).

[75] R. Glossop, 'The invention and early use of compressed air to exclude water from shafts and tunnels during construction', *Géotechnique*, 26 (1976), 253–80.

[76] A.W. Skempton and M.M. Chrimes, 'Thames tunnel: geology, site investigation and geotechnical problems', *Géotechnique*, 44 (June 1994), 191–216.

[77] D. Smith, 'The works of William Tierney Clark (1783–1852), civil engineer of Hammersmith', *Transactions of the Newcomen Society*, 63 (1991–1992), 181–207.

[78] Brooke, *The Railway Navvy*, 50–72.

on the London-Birmingham railway contract at the Kilsby Tunnel, of which Robert Stephenson took direct control, the contractors, the Nowell family, also specialised in tunnel work, with Jonathan Willans Nowell describing himself as the 'Great Borer'. Josiah Clowes established a reputation as a specialist canal tunnel engineer.[79] One reason why Mackenzie and Brassey were invited to tender for work on the Paris-Rouen Railway, the origins of the firm's overseas work, was the anxiety of Locke, the English engineer, about the lack of experience of French contractors of tunnelling work; Mackenzie and Brassey were therefore invited to tender for the Rolleboise Tunnel.[80]

P.K. Roberts in his doctoral thesis has carried out a thorough investigation of British canal tunnels; unfortunately only his study of canal tunnels associated with the exploitation of mineral wealth has been published.[81] Of those he investigates, the construction of Butterley canal tunnel has also been described by Schofield.[82] While Harding gives an introduction to Brunel's tunnel work,[83] railway tunnels have not really been studied as a group, and the construction of the great British tunnels, such as Kilsby and the Severn, require modern critical assessments.[84]

Although the earliest canal tunnel was built in France, relatively few transport tunnels had been built outside Britain before 1850 and hence the absence of a large body of relevant literature here. The paper by Redondi (chapter 7), which takes as its starting-point the proposal for a major canal tunnel in Picardy, seems to illustrate many of the differences between British and French engineering. In Britain, it was practical problems of money and geology, combined occasionally with incompetence, which tended to hamper the successful completion of tunnels. In France, the Picardy project became enmeshed in bureaucratic politics and attendant scientific disputes, which held back the work for half a century.

So much for the works themselves. Turning now to the men who built the canals and railways, perhaps the most serious gap in our knowledge of the history of civil engineering concerns the civil engineering contractor. In Britain, early

---

[79] C. Lewis, 'Josiah Clowes (1735–1794)', *Transactions of the Newcomen Society*, 50 (1978–79), 155–8.

[80] M. Murphy, 'William Mackenzie' in D. Smith, ed., *Perceptions of Great Engineers* (London, 1994).

[81] P.K. Roberts, 'British Canal Tunnels (a Geographical Study)' (Doctoral thesis, University of Salford, 1977); *idem.*, 'Canals associated with mineral exploitation', *Industrial Archaeology Review*, 5, no. 1 (1980), 5–14.

[82] Schofield, 'The design and construction of the Cromford Canal...' (see note 12).

[83] H. Harding, 'Tunnels' in Pugsley, *The Works of Isambard Kingdom Brunel*, 25–50.

[84] G.Y. Hemingway, 'Kilsby Tunnel', *Railway and Canal Historical Society Journal*, 26, no. 1 (1980), 17–26, is a narrative; T.A. Walker, *The Severn Tunnel: Its Construction and Difficulties* (London, 1891) needs reassessment, particularly as regards the role of Charles Richardson, the original engineer.

medieval, large scale 'civil' engineering schemes such as the canalisation of river access to Rhuddlan Castle would have been carried out by what would now be considered direct labour, albeit 'forced' in some cases. Small construction schemes such as bridges might have been undertaken by local carpenters or masons, but larger structures would have required the co-operation of several such tradesmen. Modern histories of British contractors, of which Joby's book[85] is the best example, although both Brooke and Coleman[86] provide insight, gloss over the early development of contracting, focusing on the 'mature' period of railway construction from the mid-1840s onwards, where several major and many medium-sized contractors, employing large numbers of men and reasonable capital resources, were in existence.

One reason for the concentration on the Victorian period is the existence of contemporary biographical sources on some of the leading mid-nineteenth century contractors – Brassey, Peto, Jackson, and their Edwardian successors, Pauling and Cowdray. Perhaps a dozen such works exist,[87] but they do little to shed light on earlier developments, and the absence of archives for the firms themselves has made critical examination of the biographies difficult.

It is surprising, given the scale of the mid-Victorian enterprises, that few have ventured to ask what happened before. As the volume of civil engineering work was sufficient in eighteenth century in Britain for individuals to take up civil engineering as a profession, this would suggest there might be sufficient work for a contractor. One can imagine that early contractors would have operated on a small scale, excavating a relatively short section of a canal, but this would be a function of the capital resources available – capital would be required to pay the workforce and obtain the equipment – but clearly if capital was available from a wealthy partner, then more work could be undertaken in return for profit-sharing. Sadly, our knowledge of the extent to which this took place is very limited.

The first serious study of early contractors was that by Dickinson on Jollife and Banks. This relied heavily on a contemporary account by Wells.[88] The canal histories by Hadfield and his colleagues do mention the names of some contractors, but it is unclear whether names are omitted because they are unknown or were considered of minor significance by the volume authors. The picture

---

85 Joby, *The Railway Builders*.

86 Brooke, *The Railway Navvy*; Coleman, *The Railway Navvies*.

87 For example, Helps, *The Life and Labour of Mr Brassey*; H. Peto, *Sir Morton Peto: A Memorial Sketch* (London , 1892); T. Jackson, *Industry Illustrated: A Memoir* (London, 1884); G. Pauling, Chronicles of a Contractor (London, 1926); J.A. Spender, *Weetman Pearson, first Viscount Cowdray* (London, 1930).

88 Dickinson, 'Joliffe and Banks...'; S. Wells, *History of the Drainage of the Great Level of the Fens* (London, 1830), vol.1, 773–9.

which emerges is one of small-scale contractors, often stricken by financial and technical problems. In several cases, engineers acted as contractors and vice versa, as circumstances dictated. One name which crops up with frequency is that of the Pinkerton family, and these are really the only early contractors to have been studied in the post-war period. Broadbridge's paper (chapter 5) stems from his studies on the Birmingham Canal, and cannot be regarded as the definitive work on the family, as their most successful work appears to have been associated with fen drainage.[89]

Another family worthy of serious study were the McIntoshs (chapter 9), whose work embraced canal and railway construction, on a massive scale. Since this paper was published, more information has come to light on Hugh McIntosh's entrepreneurial activities in the London gas industry. McIntosh, along with Jollife and Banks, developed from a relatively small-scale enterprise to a major firm from about 1790 to 1830. One cannot be sure that there were no others; in this context table 1 (chapter 9) can only be regarded as tentative.

During this study I benefited from discussions with Dr David Brooke, whose own recently published work on the McIntoshs reveals how difficult it was for even major contractors to operate without the co-operation of the client and engineer,[90] the litigation between the McIntoshs and the Great Western Railway being literally the stuff of novels.

If one turns to work on the later contractors, this can generally all said to be flawed, as not enough work has yet been done on the many firms operating at mid-century and later. In the last ten years it has at least been possible to identify the names of the leading firms, thanks to the work of Popplewell.[91] It is going to be difficult to go beyond this in the absence of business records. That it needs to be done cannot seriously be disputed. The economic significance of some of these enterprises can scarcely be overstated. Thomas Brassey is estimated to have built 5 per cent of the world's railways between 1830 and 1871. Apart from Brassey, of the mid-Victorian contractors Edward Mackenzie and George Wythes died millionaires, and others like Sir Samuel Morton Peto would have done so but for the cyclical nature of their industry. Their role in the financing of the railways is revealed by Pollins (chapter 17).

Although some further work has been done on developments in the 1860s by Cottrell,[92] in the light of Popplewell's work more needs to be done to see

---

[89] M.M. Chrimes, 'The Pinkertons'. Draft for *Biographical Dictionary of Civil Engineers* (in preparation, Library, Institution of Civil Engineers, London, 1996).

[90] D. Brooke, 'The equity suit of McIntosh v. the Great Western Railway', *Journal of Transport History*, 3rd series, 17, no. 2 (1996), 133–49.

[91] Popplewell, *A Gazeteer... of Railway Contractors...*

[92] P.L. Cottrell, 'Railway finance and the crisis of 1866: contractors' bills of exchange and the finance companies', *Journal of Transport History*, new series, 3 (1975–76), 20–40.

what entrepreneurial activities were being undertaken outside railway construction pure and simple by some of the less well-known contractors. While the later period is likely to be more fruitful, the scope of McIntosh's activities (chapter 9) suggests contractors' activities were of some significance earlier than this. Indeed, we have some evidence of this in the Mackenzie archive at the Institution of Civil Engineers – the only known British archive of an early British contractor which survives in anything like completeness. Some idea of its scope and importance can be obtained from the Institution of Civil Engineers' exhibition catalogue of 1994.[93] The collection reveals that Mackenzie was investing his surplus capital in various speculative ventures from the beginning (i.e., the 1830s), and was a major figure in European railway finance in the mid-1840s, well before Pollins (chapter 17) and Cottrell suggest. At present Dr Brooke is working on an edited version of the diaries; their publication will hopefully encourage further research, not least into the activities of Mackenzie's contemporaries. Much can be discovered about the human and financial management techniques employed by the contractors, and the extent to which better techniques led to success or failure.

Overseas, in general, there seems to have been little written beyond corporate histories of firms with origins in the late nineteenth century. In part this may be the result of the influence of Brassey and his colleagues in the early period of railway construction and the role the state had to play in mobilising resources in the absence of private undertakers – most obvious in the case of the Gotha canal in Sweden, built by the army.

Outside Britain, research has been done by Way on the origins of American contracting.[94] Some contractors, such as John Hay, had worked on Waterloo Bridge, London, in the 1820s, but most were American entrepreneurs, who faced similar problems to their European equivalents, caused by low margins and undercapitalised transport companies as well as the contractors' personal inadequacies. Mechtler has explored the origins of the contractors (and engineers) active in the Austrian Empire in the early railway period, revealing the role of the Klein Brothers and Felix Tallachini, who had experience of road construction.[95]

Surprisingly, while little serious study has been made of the contractors themselves, their workforce, the main resource with which civil engineering work was carried out before the 1880s, has been the subject of some excellent studies. No doubt this is in part due to the growth of interest in socio-economic history. One could perhaps qualify this by stating that in Britain it is the railway navvy

---

93 M.M. Chrimes and others, *Mackenzie: Giant of the Railways* (London, 1994).

94 Way, *Common Labour...*

95 P. Mechtler, 'Bauunternehmer und Arbeiter in der ersten Staatsbahnperiode Osterreichs (1842–1858), *Osterreich in Geschichte und Literatur*, 12, no. 6 (1968), 317–30.

rather than the canal labourer, or 'navigator' from which the term 'navvy' is derived, who has been studied in earnest. This is no doubt due to a large degree to the availability of a growing amount of information through the release of census records. This resource has been combined with a critical re-examination of contemporary published sources.

The first work of importance was that of Coleman, a masterly journalistic account of the railway navvy, taking fully advantage of contemporary published resources, most notably the *Select Committee on Railway Labourers* (1846).[96] Many myths surround the navvy – the drunken Irishman marauding across the countryside, sleeping in turf huts, exploited by the contractors, and so on. Since Coleman's work, Brooke in particular has made use of unpublished and contemporary sources to attempt to find the truth behind some of the myths. The subject is vast and the availability of data variable; Brooke's paper of 1989 (chapter 16) is a masterly state-of-the-art review. Apart from Brooke's own work,[97] there have been two other monographs since Coleman's on the British navvy. Both Handley's and Sullivan's studies are of value as they do not confine themselves to railway construction, but deal with the origins of the navvy in the late eighteenth century.[98] The geographical focus of Handley on Scotland meant that less data had to be researched, but one could not generalise about the navvy over the whole of the British Isles from the Scottish experience, because of different geographical, economic and social circumstances. It is nonetheless a valuable indication both of where the navvy came from and where he was going.

Outside Britain, historians have pursued similar lines, demystifying myths and mastering archival data.[99] Although Kerr's work extends beyond the chronological period under review, it is concerned with the early period of railway construction in India;[100] he has revealed the relatively intense use of labour on Indian lines, perhaps twice as many navvies on average as were used

---

[96] Coleman, *The Railway Navvies; Select Committee on Railway Labourers*, Parliamentary Papers, 1846, XIII. The background to this is discussed in R.A. Lewis, 'Edwin Chadwick and the railway labourers', *Economic History Review*, new series, 3, no. 1 (1950), 107–18.

[97] Brooke, *The Railway Navvy*; *idem.*, 'Railway navvies on the Pennines', *Journal of Transport History*, new series, 3, no. 1 (1975), 41–53; *idem.*, 'The railway navvy of the 1881 census', *Quarterly Journal of Social Affairs*, 2, no. 4 (1986), 363–77; *idem.*, 'The lawless navvy: a study of the crime associated with railway building', *Journal of Transport History*, 3rd series, 10, no. 2 (Sept. 1989), 145–65; *idem.*, 'The great commotion at Mickleton Tunnel', *Railway and Canal Historical Society Journal*, 30, no. 2 (1990), 63–70; *idem.*, 'The criminal navvy in the West Riding of Yorkshire', *Railway and Canal History Society Journal*, 30, no. 7 (1992), 365–72.

[98] J.E. Handley, *The Navvy in Scotland* (Cork, 1970); D. Sullivan, *Navvyman* (London, 1983).

[99] See, for example, Dennis Rowe, 'The robust navvy: the railway construction worker in northern New South Wales, 1854–1894', *Labour History* (Australia), 39 (1980), 28–46.

[100] Kerr, *Building the Railways*...; I.J. Kerr, 'Constructing railways in India – an estimate of the numbers involved, 1. 1850–80', *Indian Economic and Social History Review*, 20, no. 3 (1983), 317–39.

in Britain. Kerr is now investigating the history of the Great Indian Peninsula Railway in detail. A paper by Way provides an American insight into peripatetic Irish workforce, and its interface with other labour groups, which has been broadened into a more general study.[101] Although ultimately some American navvies may have been better paid than their European counterparts, earlier American working practices were unknown in western Europe – most obviously the use of slaves, but also the use of bonded labour. 'Common labour' was frowned upon by the skilled urban working class, and for the unskilled labourers existence was marginalised, and characterised by fighting over perceived social differences and concerns over wage levels. In Canada a good deal of work has been done on the canal labour force, and Wylie (chapter 8) describes the working conditions on the Rideau Canal, one of the key early Canadian civil engineering works.[102]

The Spread of Ideas

An important aspect of the history of any scientific or technological subject is the study of the spread of ideas and techniques between people and around the globe. An obvious agency for the spread of ideas is the published word, and there can be no doubt that by the early nineteenth century scientific and technical publishing was an established means of communication. Before the 1830s, however, there was not an abundant civil engineering literature, and personal contact, travel and movement of people formed an important, perhaps the most important, method of technology transfer. Movement of capital encouraged technology transfer, and export of machinery would often necessitate the export of people capable of operating it.[103] Engineers and advisors would move in response to bribes, or in search of new opportunities. Engineers visited other countries or travelled within their own country in search of ideas: the engineers' version of the architects' Grand Tour. One feature of the literature of engineering history has been the reprinting or publishing of such 'journals', which provide a useful insight into construction methods of the time.[104]

---

[101] P. Way, 'Shovel and shamrock: Irish workers and labor violence in the digging of the Chesapeake and Ohio Canal', *Labor History*, 30 (1989), 489–517; Way, *Common Labour*.

[102] These canals have been discussed in engineering detail by Legget in a series of books and articles. See Bibliography and R.F. Legget, 'The Ottawa canals and portage railways', *Transactions of the Newcomen Society*, 40 (1967–68), 61–73. More information on the working conditions is provided by R. Bleasdale, 'Class conflict on the canals of upper Canada in the 1840s', *Labour Le Travailleur*, 7 (1981), 9–39.

[103] D.J. Jeremy, 'Damming the flood: British Government efforts to check the outflow of technicians and machinery, 1780–1843', *Business History Review*, 51, no. 1 (1977), 1–34. For a more general review, see W.O. Henderson, *Britain and Industrial Europe*, 3rd edn. (Leicester, 1972).

The importance of these visits is revealed by Stapleton's study (chapter 18) of early American railway engineers who had visited Britain, and the role they played in developing the early American railways. The theme of technology transfer has been taken up by Stapleton more fully elsewhere.[105] Similarly, Cotte's work, particularly with relevance to the Seguin brothers, demonstrates how railroad technology was transferred to France in the period.[106] British influence on the Gotha Canal has been discussed by Stromback.[107] Of contemporary biographies, it is perhaps that of Vignoles which gives the fullest flavour of the peripatetic nature of civil engineering at the time.[108]

Such technology transfer was voluntary. On other projects it was the result of colonial government policy – the canals built by the British military in Canada[109] and Indian Railways being obvious examples.[110] More work needs to be done to distinguish the relative importance of foreign influence and indigenous engineering solutions.

***

It is intended that the papers selected here provide an insight into the civil engineering of canals and railways, and also the themes of historical inquiry in the field. Railway history too often can be caricatured as sophisticated train-spotting, and debates over the histories of canals and railways sometimes concentrate on two apparently irreconcilable issues – the need for a total approach considering technical, social, economic and political issues, and those who claim engineering history can only be written by people who understand engineering, i.e., engineers.[111] While often the classic examples of the total approach, such

[104] Examples include A. Cross, 'A Russian engineer in eighteenth-century Britain: the journal of N.I. Korsakov, 1776–1777', *The Slavonic and East European Review*, 55, no. 1 (1977), 1–20; 'Diary of Horatio Allen', *Railway and Locomotive Historical Society Bulletin* (1953), 97–138; E.S. Ferguson, ed., *Early Engineering Reminiscences (1815–1840) of G.E. Sellers* (Washington, 1965); C. von Oeynhausen and H. von Dechen, *Railways in England, 1826 and 1827* (Cambridge, 1971).

[105] Stapleton, *The Transfer of Early Industrial Technologies...*

[106] M. Cotte (see note 36) and 'Circulation de l'information technique et innovation: quelques exemples (France debut XIXe)', *Le Culture della Tecnica*, 2 (1994), 23–60.

[107] Stromback, *Balzar von Platen.*

[108] O.J. Vignoles, *Life of Charles Blacker Vignoles* (London, 1889).

[109] Legget, *Ottawa River Canals...*; *idem., Rideau Waterway.*

[110] Kerr, *Building the Railways...*

[111] See C. Hadfield, 'An approach to canal research', *Journal of the Railway and Canal Historical Society*, 1, no. 3 (1955), 23–6; M. Robbins, 'What kind of railway history do we want?", *Journal of Transport History*, 3, no. 2 (1957), 65–75; R.C. Post, 'Railroad history – what's the object?', *Railroad History: Bulletin*, 167 (1992), 5–8 and Discussion in *Bulletin*, 168 (1992), 76–80; Terry Gourvish, 'What kind of railway history did we get?: forty years of research', *Journal of Transport History*, 14, no. 2 (1993), 111–25.

as Hadfield's canal histories or Jack Simmons' *Victorian Railway*, pay due account to civil engineering achievements, there remain great gaps in our knowledge of how civil engineering was carried out before 1850. Skempton's work on earthworks is unique, and that is eloquent testimony to the need for much more work before the civil engineering history of railways can be considered a mature discipline. That much has been done is revealed by Biddle and Simmons' *Oxford Companion to Railway History* (1997) and hopefully, in the articles reproduced here.

Bibliography

Canals

[Asterisked items refer to both railway and canals].

Andreossy, F., *Histoire du Canal du Midi* (Paris, 1799).

Barker, T. C., 'The Sankey navigation', *Transactions of the Historical Society of Lancashire and Cheshire*, 100 (1948), 121–55.

Binnie, G.M., *Early Dam Builders in Britain* (London, 1987).

Boucher, C.T.G., *John Rennie, 1761–1821* (Manchester, 1963).

—, *James Brindley, Engineer* (Norwich, 1968).

Broadbridge, S.R., *The Birmingham Canal Navigation*, Vol. 1 (Newton Abbot, 1974).

Conde de Sastago, *Descripción de los Canales Imperiales de Aragon y Real de Tauste* (Zaragoza, 1796).

De la Garde, P.C., 'Memoir of the canal of Exeter, from 1563 to 1724', *Minutes of the Proceedings of the Institution of Civil Engineers*, 4 (1845), 90–102.

de La Lande, J., *Des Canaux de Navigation, et spécialement du Canal de Languedoc* (Paris, 1778).

de Maré, Eric, 'Telford and the Gotha Canal', *Architectural Review* (August, 1956).

Dickinson, H.W., 'Jolliffe and Banks. Contractors', *Transactions of the Newcomen Society*, 12 (1931), 1–6.

*Farey, John, 'Canals' in *Rees' Cyclopedia* (London, 1819–1820).

Fernandez Ordonnez, J.A., *Catálogo de Treinta Canales Españoles anteriores a 1900* (Madrid, 1986).

Gibb, A., *The Story of Telford* (London, 1935).

Hadfield, E.C.R., *The Canals of Southern England* (London, 1955).

Hadfield, C., and A.W. Skempton, *William Jessop, Engineer* (Newton Abbot, 1979).

The History of Inland Navigations. Particularly those of the Duke of Bridgwater, in Lancashire and Cheshire (London: Lowndes, 1766).

Hogrewe, J.L., *Beschreibung der in England seit 1759 Angelegten, und jetzt grusstentheils vollendeten schriffbaren Kanale* (Hanover, 1780).
Household, H., *The Thames and Severn Canal* (Newton Abbot, 1969).
Jackman, W.T., *The Development of Transportation in Modern England* (Cambridge, 1916).
Leech, Sir Bosdin, *History of the Manchester Ship Canal*, 2 vols. (Manchester, 1907).
Legget, R.F., 'The Jones Fall dam on the Rideau canal, Ontario, Canada', *Transactions of the Newcomen Society*, 31 (1957–59), 205–18.
—, *Rideau Waterway* (Toronto, 1967).
—, *Ottawa River Canals and the Defence of British North America* (Toronto, 1988).
Paget-Tomlinson, E., *Complete Book of Canal and River Navigations* (Albrighton, 1978).
Penfold, A., ed., *Thomas Telford, Engineer* (London, 1980).
Phillips, J., *A General History of Inland Navigation, Foreign and Domestic: Containing a Complete Account of the Canals Already Executed in England* (London, 1792 and subsequent editions).
*Pratt, E., *A History of Inland Transport and Communications in England* (London, 1912).
*Priestley, J., *Historical Account of the Navigable Rivers, Canals and Railways throughout Great Britain* (London, 1831).
*Rickman, J., ed., *Life of Thomas Telford, Civil Engineer* (London, 1838).
Roberts, C., *The Middlesex Canal, 1793–1860* (Cambridge, Mass., 1938).
Rolt, L.T.C., *Thomas Telford* (London, 1958).
Sanderton, W.S., *The Great National Project: a History of the Chesapeake and Ohio Canal* (Baltimore, 1946).
Shaw, R.E., *Canals for a Nation: The Canal Era in the United States, 1790–1860* (Lexington, 1996).
Skempton, A.W., ed., *John Smeaton, FRS* (London, 1981).
Smeaton, J., *Reports of the late John Smeaton* vol. 1 (1797); vols. 2 and 3 (1812).
*Smiles, S., *Lives of the Engineers*, 5 vols. (London, 1861).
Stapleton, D.H., *The Transfer of Early Industrial Technologies to America* (Philadelphia, 1987).
Stromback, L., *Baltzar von Platen, Thomas Telford och Gota kanal* (Stockholm, 1993).
*Taylor, G.R., *The Transportation Revolution, 1815 to 1860* (New York, 1951).
Telford, T., 'Navigation inland', *Edinburgh Encyclopedia* (Edinburgh [1812], 1830).
Thacker, F.S., *The Thames Highway*, 2 vols. *General History: Locks and Weirs* (London, 1914 and 1920).

Way, P., *Common Labour: Workers and the Digging of North American Canals, 1780–1860* (Cambridge, 1993).
Whitford, N.E., *History of the Canal System of the State of New York, with Brief Histories of the Canals of the United States and Canada*, 2 vols. (Albany, 1906).
Willan, T.S., *River Navigation in England, 1600–1750* (Oxford, 1936).

Railways

Agricola, G., *De re metallica* (1556, reprinted 1912).
Ahrons, E.L., *The British Steam Railway Locomotive*, 1825–1925 (London, 1927).
Berridge, P.S.A., *Couplings to the Khyber* (Newton Abbot, 1969).
Biddle, G., and J. Simmons, *Oxford Companion to Railway History* (Oxford, 1997).
Brooke, D., *The Railway Navvy* (Newton Abbot, 1983).
Coleman, T., *The Railway Navvies* (London, 1965).
Dendy Marshall, C.F., *Centenary History of the Liverpool and Manchester Railway* (London, 1930).
Deutsche Reichsbahn, *Hundert Jahre deutsche Eisenbahnen* (Berlin, 1936).
Dilts, J.D., *The Great Road: The Building of the Baltimore and Ohio, the Nation's First Railroad, 1828–1853* (Stanford, 1993).
Dow, G., *Great Central Railway*, 3 vols. (London, 1959).
Francis, J., *A History of the English Railway*, 2 vols. (London, 1820–1845).
Helps, A., *The Life and Labours of Mr Brassey* (London, 1872).
Henderson, W.O., *Britain and Industrial Europe* (Liverpool, 1954).
Hungerford, E., *The Story of the Baltimore and Ohio Railroad, 1827–1927* (New York, 1928).
Hunter, M., and Robert Thorne, ed., *Change at King's Cross: from 1800 to the Present* (London, 1990).
Jarnvagsstyrelse, *Statens Jarnvagar, 1856–1906*, 4 vols. (Stockholm, 1906).
Jeaffreson, J.C., *The Life of Robert Stephenson*, 2 vols. (London, 1864).
Jenks, L.H., *The Migration of British Capital to 1875* (New York, 1928).
Joby, R.S., *The Railway Builders* (Newton Abbot, 1983).
Kerr, I.J., *Building the Railways of the Raj* (Delhi, 1995).
Lecount, Peter, *The History of the Railway Connecting London and Birmingham* (London, 1839).
Lee, Charles E., *The Evolution of Railways* (London, 1937).
Lewis, M., *Early Wooden Railways* (London, 1970).
MacDermott, E.T., *History of the Great Western Railway* (London, 1927–1931. Enlarged edn., Ian Allen, 1964).
Meek, C.L.V., *The Railway Station* (Yale, 1957).

Popplewell, L., *A Gazetteer of...Railway Contractors...*, 11 parts (Bournemouth, 1982–1992).
Pugsley, Sir Alfred, ed., *The Works of Isambard Kingdom Brunel* (London and Bristol, 1976).
Reed, M.C., ed., *Railways in the Victorian Economy* (Newton Abbot, 1969).
Ribeill, G., *La Révolution Ferroviare: la Formation des Compagnies de Chemins de Fer en France (1823–1870)* (Paris, 1993).
Rolt, L.T.C., *Isambard Kingdom Brunel* (London, 1958).
—, *George and Robert Stephenson* (London, 1960).
Sandstrom, K.E., *The History of Tunnelling* (London, 1962).
Simmons, J., *The Victorian Railway* (New York, 1991).
Tomlinson, W.W., *North Eastern Railway* (Newcastle-upon-Tyne, 1915).
Vance, J.E., *The North American Railroad* (Baltimore, 1995).
Williams, F.S., *Our Iron Roads* (London, 1852).
—, *The Midland Railway: its Rise and Progress* (London, 1876).
Zuffer, J., *Geschichte der Eisenbahnen der Oesterrisch-Ungarischen*, 3 vols. (Vienna, 1898).

1
Canals and river navigations before 1750

A.W. Skempton

I. INTRODUCTION

It is well known that the English canals of the eighteenth and early nineteenth centuries served as the arteries of the industrial revolution, but even in pre-industrial times the rivers and canals of Europe and Asia played a significant part in trade and commerce. One reason for this is that rivers (and canals are but extensions of river-systems) form the most natural inland routes to and from sea-ports. Moreover, so long as men and horses provided the only practical source of motive power for inland transport, heavy and bulky goods could be carried more economically and efficiently on water-ways than by any other means.

To illustrate this point quantitatively, the loads that can be carried or drawn by a single horse are set out in the table below. The gain in mechanical efficiency was partially offset by the greater capital expenditure required for the engineering works of the canal and river navigations; nevertheless the average cost of carriage by water in eighteenth-century England, for example, was rarely more than half and often only a quarter of the cost by road [1]. And the roads were frequently impassable for commercial traffic in winter.

Typical loads carried or drawn by a single horse

Pack-horse	$\frac{1}{8}$ ton
Wagon on 'soft' roads	$\frac{5}{8}$ ton
Wagon on macadam roads	2 tons
Wagon on iron rails	8 tons
Barge on river	30 tons
Barge on canal	50 tons

II. TRANSPORT CANALS IN CHINA

The earliest major civilizations were in the valleys of the Euphrates, Nile, Indus, and Huang-Ho, and these great rivers provided a ready means of transport in addition to their life-giving supply of water for irrigation. The Phoenicians and the Greeks were maritime peoples, while the Romans chiefly made use of the many naturally navigable rivers in their empire. Apart from a few notable canals such as that from the Nile to the Red Sea, cut by order of Darius *c* 510 B.C., and several in France, Lombardy, and the Netherlands made during Roman times, the first sustained effort in canal-construction was made by the Chinese [2]. Among the more important of their works are the Ling Ch'u canal in Kuangsi (215 B.C.), the 90-mile-long canal from the Han capital Ch'ang-an to the Yellow River (133 B.C.), the Pien canal in Honan (A.D. 70), the Shanyang canal in Chiangsu (A.D. 350), and the first sections of the Grand Canal completed in 610. This had a total length of 600 miles, and along its banks ran an Imperial road planted with elms and willows. It served to transport grain from the lower Yangtze and the Huai to Kaifeng and Loyang. Under the T'ang dynasty in the eighth century the traffic on the canal was known to exceed 2 million tons annually.

In most cases the land traversed by these canals had a small gradient, and water-levels could readily be controlled by single gates separated from each other by considerable distances. For instance, on the Pien canal in A.D. 70 the engineer Wang Ching built the gates typically some 3 miles apart. They consisted of stone or timber abutments, each with a vertical groove into which squared logs of timber could be lowered or raised by ropes attached to their ends (plate 18). These simple stop-log gates evidently derived from the sluices used on irrigation canals. On the smaller transport canals, and especially in places where an appreciable difference in land-level had to be overcome, a double slipway was usually built, over which the barges were hauled. References to this device occur in Chinese literature at least as early as A.D. 348.

Occasionally a more elaborate gate was adopted, in which a solid door could be raised or lowered by a windlass. Two of these were built by Ch'iao Wei-Yo in 984 to replace a double slipway, during the remodelling of a section of the Grand Canal, and by placing the gates only 250 ft apart he created the first known example of the canal-lock: a device of fundamental technological importance. With single gates widely spaced along a canal or river considerable delays and great losses of water are involved in waiting for the levels to equalize after any particular gate has been opened, unless the difference in level is slight. With the lock, however, only the comparatively small volume within the lock-basin has to be filled or emptied, and the water-level in the long reaches or pounds between two locks is never altered.

It is rather curious, therefore, to find that little use was made of the lock in China. Its later development was entirely due to western engineers. Yet, if the Chinese remained content, in general, with their stop-log gates and slipways, they nevertheless achieved canal-works on a most impressive scale. Outstanding among these was the construction, between the years 1280 and 1293, of the northern branch of the Grand Canal from Huaian to Peking, having a length of 700 miles. Parts of this utilized existing rivers, and other sections were 'lateral' canals; but the section crossing the Shantung foothills, completed in 1283, was the earliest example of a 'summit-level' canal. A lateral canal has a continuous fall in one direction and, with intakes from the river alongside which it runs, there are few problems in water-supply. The conception of a lateral canal is relatively simple, since it is essentially an improvement of an existing river. In contrast, the idea of taking a canal over the summit of a watershed dividing two

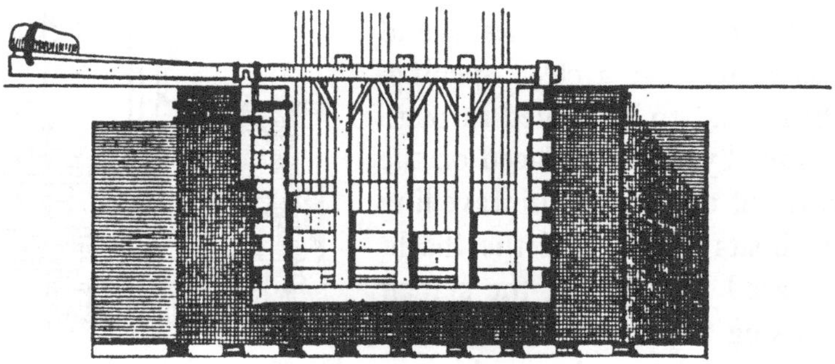

FIGURE 277—*Section of a typical stanch or flash-lock. From Belidor, 1753.*

rivers requires bold imagination and considerable technical skill in providing an adequate water-supply at the summit. The Shantung section of the Grand Canal linked the Yellow River with a group of lakes, situated some 100 miles to the south at approximately the same elevation as the river. But over a short length the intervening land rose to a height of about 50 ft above the river and the lakes. The upper part of the canal was taken in a cutting with a maximum depth of 30 ft, yet this still left a fall of 20 ft to the north and south. To provide for the inevitable losses of water occasioned by operating the gates, two small rivers situated to the east, higher up the foothills, were partially diverted to flow into the summit-level. The engineers of this notable undertaking were Li Yueh and Lu Ch'ih.

III. MEDIEVAL CANALS AND RIVER WORKS

At the time when the Grand Canal of China was completed, water-transport in Europe was still in a primitive state. Few canals had been constructed, and rivers were chiefly used as a source of power for water-mills. On many rivers each mill had its weir, to provide an adequate head of water for the mill-wheel, and these weirs were a serious obstacle to navigation. In the later Middle Ages, however, important developments took place in the Netherlands, as we shall see, while throughout the more commercially active countries of Europe improvements were made in the rivers by building stanches in the weirs and also at intervals along the river, between the mills, to reduce the gradient and increase the depth of water in the shallow places [3].

A typical stanch (also called a flash-lock or navigation-weir) is shown in figure 277. When a boat wished to pass, the wooden boards or 'paddles', with their long handles, were lifted out, and after the rush of water had somewhat abated the balance-beam, with the vertical posts which had supported the paddles, could be turned aside to leave a clear opening. Often the boats had to be hauled through against the flow, by a winch placed on the bank upstream of the stanch; in travelling downstream boats would find the 'flash' of water released by opening the stanch a help in crossing any shoals below it.

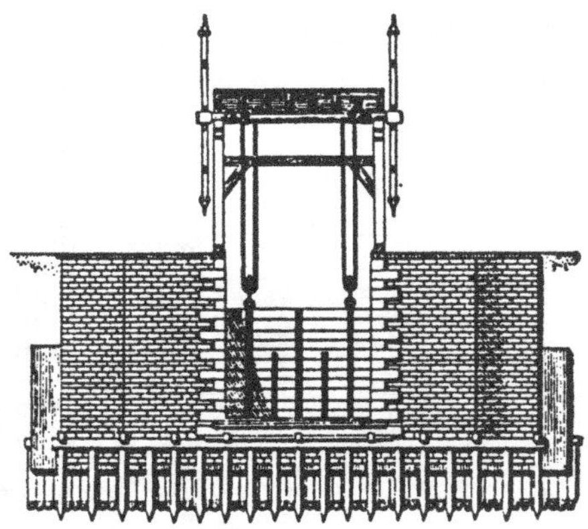

FIGURE 278—*Section of a typical 'portcullis' sluice. From Belidor, 1750.*

The early history of stanches is obscure, but it is practically certain that they were in existence on a number of rivers in Flanders,[1] Germany, England, France, and Italy before the end of the thirteenth century. A reference to the winch for a stanch on the Thames at Marlow occurs in 1306. The oldest complete account of the Thames navigation between Oxford and Maidenhead, in 1585, shows twenty-three stanches; all but four were situated in the weirs of the mills on this 62-mile stretch of river [4]. Many of these stanches were still in place in the mid-eighteenth century, while as recently as the early nineteenth century twenty-two stanches were to be seen on the Marne between Châlons and Paris.

The very existence of Holland depends upon the dikes and drainage canals, and it is not surprising that some of the canals were enlarged and suitably equipped for transport at an early date. Originally the drainage canals had outlets through the dikes controlled by sluices. In some cases goods had to be transhipped over the dike, and in other places boats were hauled over a double slipway similar to those in China.[2] Perhaps the first examples were at Het Gein and Otterspoor, built in 1148 on the Nieuwe Rijn canal near Utrecht. Where

[1] Possibly as early as 1116 on the river Scarpe.

[2] The double slipway was used also in Italy, and one was built in 1437 at Fusina where the river Brenta enters the lagoon south of Venice. This slipway, as reconstructed in the sixteenth century, is illustrated by Zonca [5]. The slipway at Fintelle, in Flanders, remained in use until 1824.

hydraulic conditions permitted, an obvious improvement was to make the sluice-gate large enough for the passage of boats. These navigation sluices in Holland were of the lifting-gate or portcullis type shown in figure 278. When the tide in the estuary or river was at the same level as the water in the canal, the gate was raised by a windlass and boats could pass through. At other states of the tide a difference in water-level existed across the gate. To safeguard the sluice from under-seepage the foundations and abutment-walls were extended, typically 20 or 30 ft beyond the gate and also well into the body of the dike. Figure 278 represents a sluice built in 1708; it will be seen that the danger of under-seepage is further prevented by sheet-piling. This was a characteristic feature of construction in the sixteenth century and later, but seems to have been unknown to medieval engineers. A *magna slusa* at Nieuport, mentioned in 1184, may have been of the portcullis type, as also the sluice at Governolo on the river Mincio, built in 1188–98 by Alberto Pitentino, and that at Gouda of *c* 1210.

The next step was of vital importance. It involved building two sluice-gates, one behind the other, enclosing a basin or chamber and thereby forming a lock. The first example that can be dated with certainty was built in 1373 at Vreeswijk, where a canal from Utrecht enters the river Lek [6]. From regulations of 1378 and 1412 we learn that the Vreeswijk lock was operated three times weekly, at 2 o'clock in the afternoon [7]. First the outer gate was wound up, and the boats from the river-harbour entered the large basin of the lock. The outer gate was then closed and the inner one raised, and when a level had been made the boats were hauled through into the canal. The inner gate was then closed again and the outer gate opened, and in due course the boats that had been waiting in the basin could pass into the river. This leisurely procedure was not without advantage to the townspeople of Vreeswijk since it provided the boatmen with ample time for shopping and gossip. It is possible that a similar lock had been built somewhat earlier at Spaarndam, but it was only in the later fourteenth century that this form of construction became widely adopted. Examples include basins at Delfshaven (1389) and Schiedam (1395). At Gouda in 1413 a lock was built by Jan van Rhijnsburch with a triple set of gates.

A large basin was typical of the early Dutch tide-locks and is essential for intermittent working, but in 1394–6 a lock was built at Damme, near Bruges, in which the chamber formed by the two gates and the masonry side-walls was 100 ft long and 34 ft wide. This suggests that the lock at Damme was operated as often as any boat arrived there, and it is the earliest lock with what may be regarded as modern proportions.

By 1400, therefore, important advances had taken place in the construction of locks. But those made in the Netherlands were intended only to overcome differences in water-level, and the first locks to overcome differences in land- as

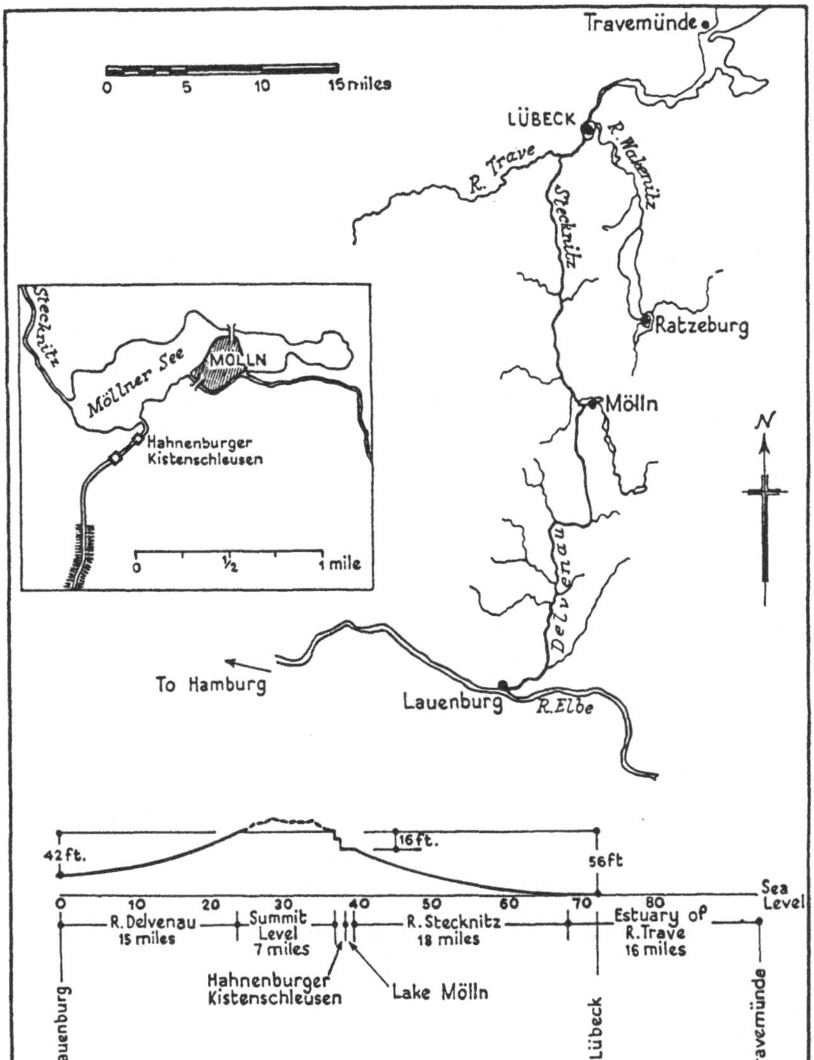

FIGURE 279—*Map and longitudinal section of the Stecknitz canal, 1391–8.*

well as water-level were built on the Stecknitz canal, 1391–8 [8]. This is also the oldest summit-level canal in Europe, and therefore a work of considerable historical interest.

In the early fourteenth century the river Stecknitz had been made navigable from Lake Mölln down to Lübeck, a distance of 21 miles and a fall of 40 ft, with four stanches. In 1391 an extension of this waterway southwards to Lauenburg on the Elbe was begun, to establish a link between the Baltic and the North Sea (figure 279). From the lake a canal was dug, rising 16 ft in a distance of less than half a mile, then running horizontally for 7 miles in a cutting with a maximum depth of about 12 ft to meet the river Delvenau. This river falls 42 ft in a distance of 15 miles to Lauenburg. It was rendered navigable by eight stanches. The sections south of the summit (formed by the Delvenau) and north of Lake Mölln (formed by the Stecknitz) naturally had an adequate water-supply. The steep section between the summit-level and the lake, however, was largely dependent for its supply upon water seeping from the sandy soil in which the cutting had been excavated, and with single-gate sluices or stanches the loss of water would have been excessive. In order to overcome this difficulty two locks were built, known as the Hahnenburger Kistenschleusen. Each could contain ten small barges (35 ft long and of 11-ft beam) and, like the tide-lock at Vreeswijk, they were operated every second or third day. The men responsible for this scheme are not known, but the close connexion between Lübeck and the Netherlands, through the Hanseatic League, may be significant.

The next development of importance took place in Lombardy soon after 1450, but we must first note the work carried out there in the early fifteenth century [9]. Between 1179 and 1209 a canal was constructed with an intake on the river Ticino, whence its course ran south to Abbiate and then east to Milan. It fell 110 ft in 31 miles (figure 280). Initially the canal was designed for irrigation, but in 1269 its cross-section was enlarged and sluices or stanches were built in the various weirs along its length. The canal was then known as the Naviglio Grande. It did not enter the city, but ended in a basin just outside the western wall.

In 1387 plans were being prepared for the new cathedral at Milan, and when construction began the chief building-material was marble from quarries near Lake Maggiore. This was brought down the Ticino and along the Naviglio Grande. How the marble was at first carried from the canal-basin to the cathedral is not clear, but early in the fifteenth century a canal was made linking the Naviglio Grande with the old moat, which had surrounded the city at an earlier date when its area was much smaller. This moat passed quite close to the cathedral. By means of a single-gate sluice in the link canal, situated alongside the Via Arena, boats could pass from the Grande to the mason's yard. But the water-level in the moat, which was fed by the Seveso running in from the north, was several feet above that of the canal. Consequently, whenever the sluice was opened the water-level in the moat had to fall to that of the canal; when the boats had passed the sluice, the moat had to fill again before they could proceed along it. In 1438 this inconvenient arrangement was improved by the construction of a second gate in the link canal along the Via Arena, thus forming the first pound-lock in Italy. The engineers responsible for this work were Filippo da

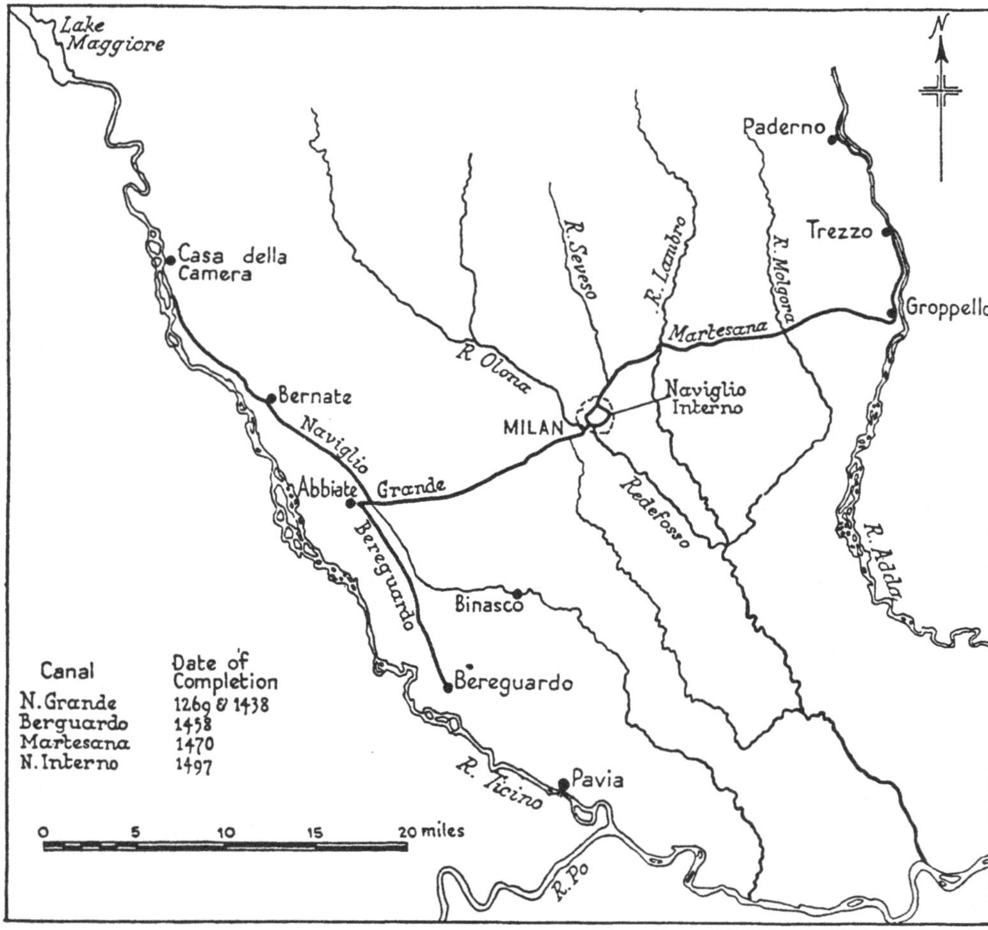

FIGURE 280—*Map of canals near Milan before 1500.*

Modena and Fioravante da Bologna, and it appears that before 1445 they had built a second lock near Sant' Ambrogio on the old moat which, by now, had been enlarged and was known as the Naviglio Interno.

IV. CANALS OF THE ITALIAN RENAISSANCE

The canal-works in Lombardy of the later fifteenth century mark a new epoch in canal-construction. In 1451 Bertola da Novate (*c* 1410–75) was appointed ducal engineer of Milan, and one of his first tasks was to consider the possibilities of building a canal from Milan to Pavia on the river Po [10]. In 1359 a small irrigation canal had been constructed from Milan as far as Binasco, about half-way towards Pavia. The question of enlarging and extending this canal was naturally investigated, but Bertola concluded that it would be too difficult an operation and recommended instead that a canal be cut from Abbiate, on the Naviglio Grande, southwards to the village of Bereguardo, with a land-portage to the banks of the Ticino (figure 280). Work began in 1452. By 1458 the canal was completed, with a length of 12 miles and a fall of 80 ft taken in 18 locks; it was the earliest canal in which a considerable gradient was controlled entirely by pound-locks.

FIGURE 281—*Sketch of a lock with portcullis gates, c 1470.*

While the Bereguardo canal was under construction, Bertola was consulted on the construction of five locks near Parma. These were built between 1456 and 1459 under the supervision of an assistant, Bertola visiting the site from time to time. Meanwhile in 1457 he prepared plans for a canal joining Milan to the river Adda, east of the city. This canal, constructed between 1462 and 1470 and known as the Martesana, had an intake on the Adda at Trezzo, where a large weir was constructed. The canal then proceeded alongside the river for a distance of some 5 miles until a point was reached at which it could turn westward over the Lombardy plain to Milan. The great masonry wall separating the canal from the river in the first 5 miles remains today much as it was originally built. The canal was carried over the river Molgora on a small three-arch masonry aqueduct, the earliest known example, and the stream of the Lambro was carried under the canal in a culvert of the type illustrated in figure 294. By careful planning Bertola was able to lay out this 24-mile-long canal with only two locks. Boats entered a basin or dock near San Marco in Milan.

By this time knowledge of the pound-lock was becoming general in Italy. In his book *De re aedificatoria*, completed during the 1460s and published in 1485, Leone Battista Alberti (1404–72) gives an account of a lock with a pair of gates separated by a distance equal to the length of a boat [11]; it is possible that he was in fact describing the locks built by Bertola, unquestionably the leading canal engineer of his time. In 1481 the brothers Domenico da Viterbo built a lock at Stra on the canal joining Padua to the river Brenta [12], and in 1491–3 several timber locks were built by 'an engineer of Milan' on the canal near Bologna [13]. Few details are known of these early Italian locks, and the only contemporary illustration so far discovered is a sketch in the *Codice Laurenziano*, dated between 1460 and 1490 [14]. The drawing (vol II, figure 626)

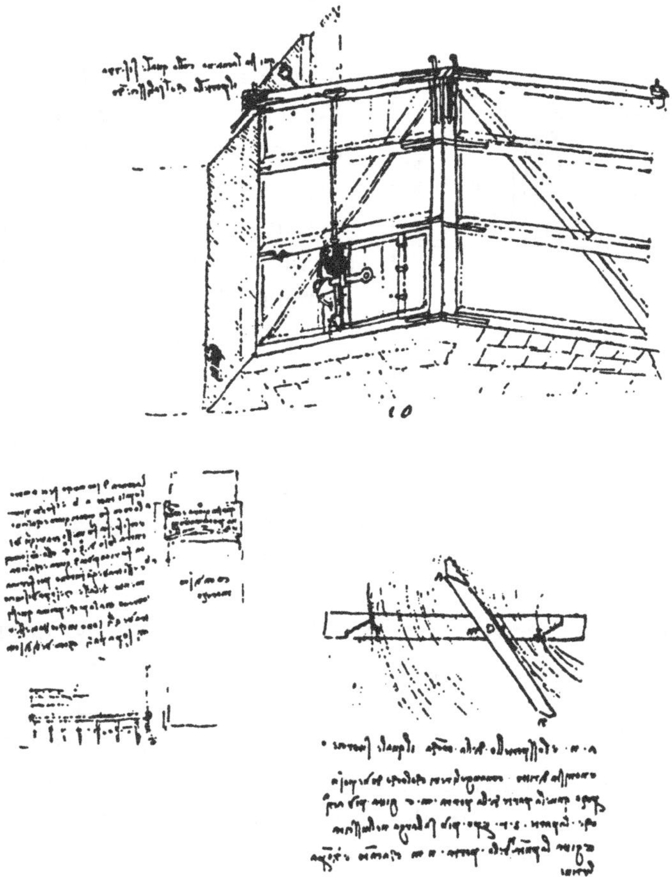

FIGURE 282—*Sketch by Leonardo da Vinci of mitre-gates for the San Marco lock, Milan.* c *1495.*

as a whole is difficult to interpret, but a pair of the vertical-lift gates represented may be taken as constituting a lock similar to that described by Alberti (figure 281).

The portcullis type of gate was not ideally suited to navigational purposes. It was superseded in the last decade of the fifteenth century by Leonardo's wonderful conception of the mitre-gate. Leonardo da Vinci (1452–1519) was appointed ducal engineer of Milan in 1482; some ten years later he turned his attention to hydraulics and, in particular, to the construction of six new locks on the Naviglio Interno. These were completed in 1497, and Leonardo's drawing for the gates of the lock at San Marco, situated just below the terminal basin of the Martesana canal, is shown in figure 282. In another drawing (figure 283) Leonardo shows a rectangular masonry lock 95 ft long between the gates and 18 ft wide, having a pile cut-off wall and mitre-gates, in which there are small sluice-doors, identical with those at San Marco. These drawings, which give the first complete design of a modern form of lock, are of the highest importance in the history of canal-construction.

In 1503 Leonardo went to Florence, where he was engaged upon very

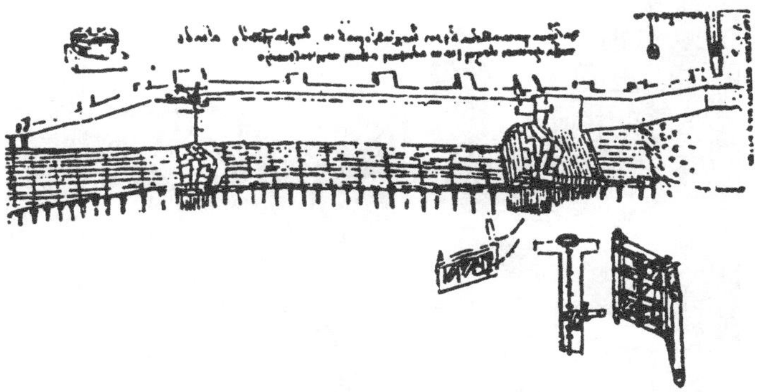

FIGURE 283—*Leonardo's sketch of a longitudinal section of a lock with mitre-gates.*

ambitious plans for a canal from that city to the river Arno near Vicopisano. This was to have been a summit-level canal, and Leonardo gave careful thought to the problem of an adequate water-suppy at the summit [14]. His project, however, was in advance of its day, and the first summit-level canal of real importance was not built until the early seventeenth century, in France.

The Martesana canal had successfully linked Milan with the river Adda, but upstream of Trezzo the river was impassable and it was clearly desirable to by-pass this section, to enable navigation to continue up to Lake Como. In 1518 Benedetto da Missaglia designed a lateral canal for this purpose with headworks near Paderno, where there was to be a diversion-weir 14 ft high at a location chosen after borings had been made in the river bed. The canal would then run in an excavation on the right bank, to re-enter the Adda at a point situated 3000 yd downstream and 90 ft lower than the intake. Ten locks were proposed and the amount of water admitted to the canal was to be regulated, all the surplus above a predetermined maximum flowing back to the river over spill-ways in the canal bank. Work began in 1519 and it seemed that this fine example of river navigation would be successfully completed. In 1515 Milan had come under the rule of Francis I of France and it was by his order that the canal had been started. In 1522, however, Francis lost Milan to the Emperor Charles V; the main driving force was thus removed, and work on the canal ceased.

In 1584 plans were prepared for a canal between Milan and Pavia, but in the following year a serious flood damaged the intake of the Naviglio Grande.

FIGURE 284—*Lock at Brandenburg, 1548–50.*

Attention had therefore to be directed immediately towards the reconstruction of this important component of the Lombardy canal-system. The work was entrusted to Giuseppe Meda (*c* 1540–99), who took the opportunity of completely remodelling the old intake, still little changed from its medieval form. He built a weir and control-works similar in principle to those designed by Missaglia on the Adda. Meda's later career was tragic. In 1591 he prepared a new scheme for the Paderno canal, using an entirely novel form of lock, of the type later known as a shaft-lock [15]. After various mishaps and interference from the authorities in Milan his plans were only partially carried out before his death. Meanwhile in 1595 he had drawn up proposals, complete in all details including the locks, culverts, aqueducts, and bridges, for the canal between Milan and Pavia. A year later these proposals were accepted, but work had barely started before the engineer died and this scheme too was abandoned.

V. THE POUND-LOCK IN THE SIXTEENTH CENTURY

The increasing importance of canal and river navigations in the sixteenth century is reflected in the widespread adoption of the mitre-gate pound-lock. The first examples, as we have seen, were completed by Leonardo in 1497 on the Milan canal. These were followed in 1518 by Missaglia's designs for the locks on the Paderno canal, and by three locks near Bologna built in 1548 by Giacomo da Vignola (1507–73). The latter were oval in plan, 100 ft long, 25 ft in maximum width, and 12 ft wide at the entrance [13].

The earliest examples in France that can be securely dated are those designed in 1550 in connexion with improvements of the rivers in the neighbourhood of Bourges. Detailed specifications of these locks still exist in the archives at Bourges [14]. In all, fifteen locks were required, two of masonry on the river Yèvre, four similar ones on the river Cher, and nine of timber construction on the river Auron. The locks were rectangular with a length of 90 ft between the gates and a width of 13 ft. The floor and side walls extended 15 ft beyond each gate, beneath which sheet-pile cut-off walls were constructed.

FIGURE 285—*Sketch of a lock on the Brussels canal*, c *1560*.

In the same period improvements were being made on the Havel and Spree in the Mark Brandenburg [16]. Work began on two locks at the towns of Rathenow and Brandenburg in 1548, at Spaarndam in 1572, and in Berlin in 1578. The lock at Brandenburg, with its large octagonal basin and timber walls, still exists (figure 284), but the type of gate originally used is not known. A sketch by Tilemann Stella (1524–89) of the locks that he built on the Mecklenburg canal between 1572 and 1582 shows mitre-gates and a rectangular chamber 18 ft wide and 90 ft long. He had visited the Netherlands in 1561 to study hydraulic engineering and would have seen the Brussels canal, completed in that year.

This important canal, begun in 1550, is described in the next section (p 452). Here it may be noted that the works included four locks of octagonal shape, 200 ft long and 70 ft wide, with falls varying from 6 ft to 10 ft. One of these locks is shown in figure 285. It can be seen that there were mitre-gates and that the chamber was emptied and filled by means of culverts in the walls. This is the earliest known example of the ground-sluice, a device enabling the basin to be operated more rapidly and with less disturbance than with the usual sluice-doors in the lock-gates.

In England the first pound-locks were constructed in 1564–7 by John Trew on a lateral cut beside the Exe, known as the Exeter canal [17]. Like the locks on the Brussels canal, on the Havel, and the old tide-locks in Holland, those at Exeter had a basin of sufficient size to hold several boats. They were 189 ft long by 23 ft wide. A manuscript sketch shows three vertical-lifting sluice-paddles in each leaf of the mitre-gate at the upstream end of the lock [18]. The lower end, however, was closed by a single gate. This arrangement also occurs in the sixteenth-century Italian lock on the Brenta, illustrated by Zonca (vol II, figure 625).[1] Both upper and lower mitre-gates, however, were provided in the lock built 1571–4 on the Lea, near London, at Waltham Abbey, where

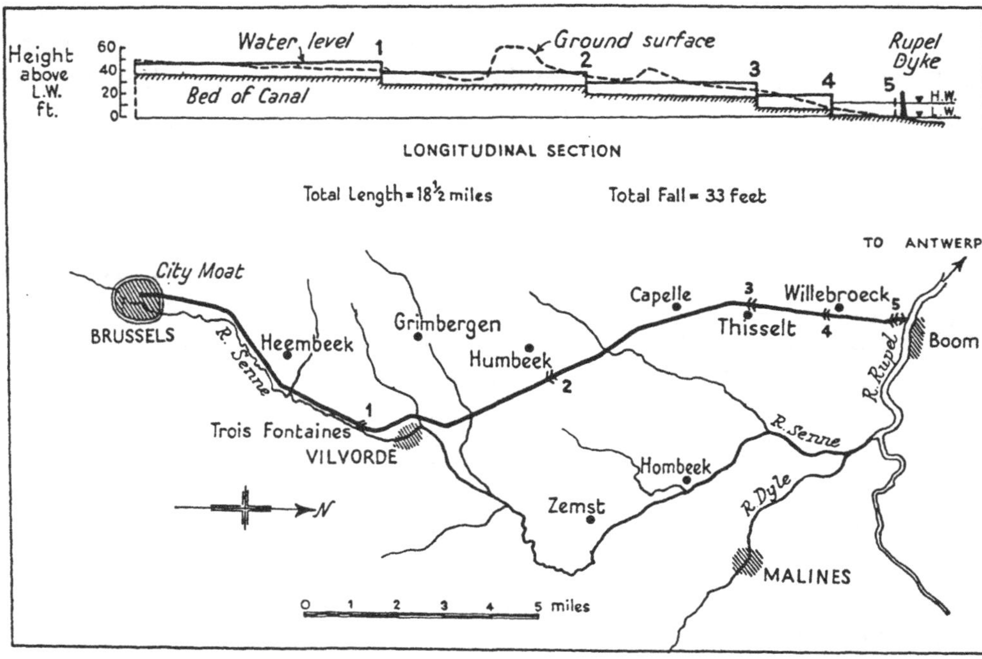

FIGURE 286—*Map and longitudinal section of the Brussels canal, 1550–60.*

> . . . a rare device they see,
> But newly made, a waterworke; the locke
> Through which the boates of Ware doe passe with malt.
> This locke contains two double doores of wood,
> Within the same a cesterne all of Plancke,

[1] Vittorio Zonca (1568–1602), an architect-engineer of Padua, does not give the location of this lock in his book [5], but says that gates of the type shown are to be found in the locks at Padua and Stra (i.e. on the Brenta canal). There is no reason to assume that the gates built a century earlier in the Stra lock, by the brothers Domenico, were similar to those depicted by Zonca.

Which onely fils when boates come there to passe
By opening of these mightie dores.
(Vallans, 'Tale of Two Swannes', 1577.)

The first mitre-gates in Holland appear to be the triple set built in 1567, and described by Andries Vierlingh (1507–79), at Spaarndam in a new tide-lock (*Grote Haerlemmer Sluys*) 25 ft wide and 122 ft long [19]. There may have been earlier examples, however, for Simon Stevin (1548–1620), writing at Leiden in 1617, says that the mitre-gate pound-lock 'has been in use for a long time' [20].

Finally, it is interesting to note that in Sweden eleven locks were built between 1603 and 1610 as part of the canalization of the river Eskilstuna between Lakes Hjälmar and Mälar (p 455) [21]. Thus, by the beginning of the seventeenth century knowledge of the canal-lock had spread practically throughout Europe.

VI. CANALS IN FLANDERS

Of the canals built in the sixteenth and seventeenth centuries, the Brussels canal was the first of importance to be constructed outside Italy [22]. Navigation had long existed between Brussels and Antwerp along the rivers Senne, Rupel, and Schelde. From time to time during the Middle Ages improvements were made in the Senne navigation, but in 1531 it was decided to cut a canal from Brussels to Willebroeck on the Rupel (figure 286). This would have a length of $18\frac{1}{2}$ miles, which was almost half the distance that boats had previously to travel. Some of the towns on the old route opposed the plan, fearing a loss of trade, but by 1550 agreement had been reached and work was started under the direction of Jean de Locquenghien (1518–74). The fall of 34 ft between Brussels and high-water mark in the Rupel was taken in four locks (p 450). These locks were large enough to hold twelve of the small coasting-vessels using the canal, and the depth of water provided was 5 ft. Seven streams were taken beneath the canal in culverts, four road-bridges were built, and one section of the canal ran in a cutting 2 miles long with a maximum depth of 30 ft. For a distance of about $1\frac{1}{2}$ miles between its entrance on the Rupel and the lowest lock the canal was contained between dikes raised some 10 or 12 ft above the low-lying marshland, and as completed in 1561 the canal was tidal in this reach. But excessive silting occurred, and in 1570 Locquenghien constructed at the debouchment of the canal a tide-lock with three pairs of mitre-gates.

The organization of the undertaking is known in some detail. Locquenghien as director of works received a regular salary during the period 1550–63. Adrien van Bogaerden, the surveyor and principal assistant, was also regularly employed throughout. The day-to-day supervision was carried out by two divisional resident engineers, and there was a treasurer. From time to time expert advice was sought. In this way Willem Maertense came from Holland in 1554 as a consultant on the opening of the Rupel dike at Willebroeck, and two years earlier Gilbert Van Schoonbeke of Antwerp was called in to give advice on the construction of the locks. The canal was not completed without difficulty. The contractor had trouble with the lock at Willebroeck during its construction; the lock at Humbeek failed in 1562, owing to seepage beneath the foundations of one of the gates, and a slip occurred in the canal dike. Remedial measures were successfully carried out in each case. The largest of the culvert aqueducts (all of

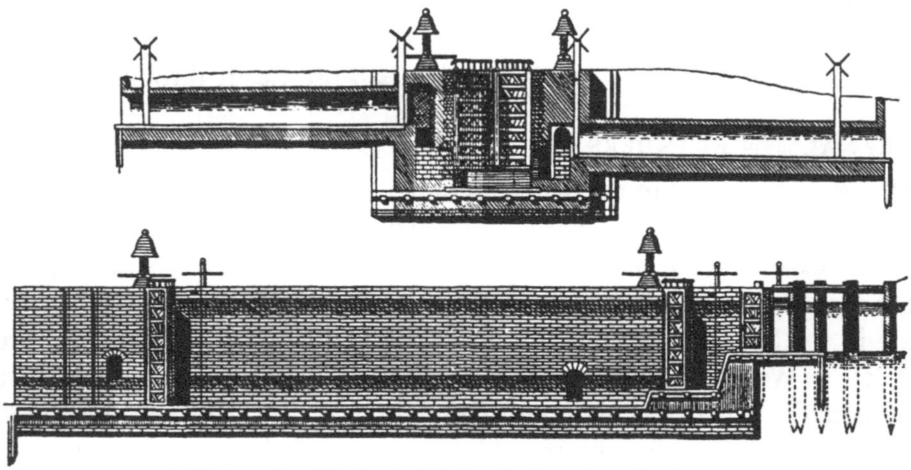

FIGURE 287—*The Boesinghe lock on the Brussels canal, 1643–4.*

which had been built in timber) was not sufficiently strong; it was rebuilt in masonry in 1569 by Georges Rinaldi—a construction said to have been greatly admired by Peter the Great when he travelled on the canal.

The Brussels canal was completed only just before the outbreak of war with Spain. When peace was restored in the early seventeenth century the canal system of Flanders expanded greatly [23]. The more notable works included a 44-mile canal linking Bruges, Passchendaele, Nieuport, and Dunkirk *c* 1622, the enlargement of this canal in 1641–61, and an extension in 1666 to Ostend, where a very fine tide-lock was constructed in 1669. In 1670 Dunkirk was linked to the river Aa, at the mouth of which, before the end of the century, another great tide-lock had been built at Gravelines. By 1692 the old river navigations near Lille had been improved and extended with a canal forming a connexion between the Lys and Scarpe.

The most interesting technical achievement of the period, in Flanders, was the celebrated lock at Boesinghe built by Maître Dubie. In 1643–6 navigation between Ypres and the river Yser was improved by the construction of a lateral canal, 4 miles long, beside the river from Ypres to Boesinghe. From this point navigation continued along the river to its junction with the Yser. The fall in this 4-mile reach amounted to 20 ft and, instead of building three locks of normal dimensions, the whole of the fall was taken in one great lock at Boesinghe (figure 287). This is remarkable not only for its size but for the use of side-ponds, introduced here for the first time, as a means of reducing the loss of water in operating the lock. Each side-pond takes one-third of the water as the lock is being emptied (the remaining third flowing down the canal), and this volume is available when the lock is to be filled again. Ground-sluices were provided.

VII. CANALS IN GERMANY AND SWEDEN

It has been mentioned that in Germany a number of locks were built in the mid-sixteenth century. As early as 1540, at the suggestion of the Elector Joachim II of Brandenburg, plans were drawn up for a canal linking the rivers Havel and Oder. In 1548 he proposed another, more practicable, scheme for linking the river Spree with the Oder upstream of Frankfurt [24]. Work began on this latter canal in 1558, but it appears that considerable difficulties were encountered and eventually by 1563 operations came to a stop. After many political delays new plans were prepared by order of Frederick William the Great Elector, after whom the canal was named. In 1662, under the direction of the Italian engineer Philippe de Chiese, the earthwork was begun, while construction of the locks and bridges was entrusted to Michael Schmidts, a Dutch engineer who had built a new lock in Berlin five years earlier. The Friedrich-Wilhelm canal, completed in 1669, had a total length of 15 miles and was the third summit-level canal to be built in Europe. As with the early Stecknitz canal the summit was supplied by ground-water from the sandy soil through which the upper part of the canal was carried in a cutting. On the eastern side the canal rose from Neuhaus on the Spree for a height of 10 ft in two locks, and from the western end of the summit-level it fell 65 ft to Brieskow on the Oder.

The construction of the Finow canal, the earlier of Joachim II's schemes, was begun in 1605. After an interruption between 1609 and 1617, for lack of funds, it was completed in 1620. The first reach, 8 miles long, was level with the Havel. The canal then descended the valley of the Finow for 17 miles to Liepe, on a short tributary of the Oder, only eleven locks being provided in the fall of 120 ft. This canal proved to be very unsatisfactory, for the Havel was able to flood down it, and owing to lack of maintenance during the Thirty Years War the works fell into complete decay. By the beginning of the eighteenth century scarcely a trace of them existed. The canal was rebuilt between 1744 and 1751 on a different plan, having two locks up from the Havel to a summit-level, a feeder-channel with an intake several miles upstream on this river, and fourteen locks down to Liepe. The system of water-ways in this region of Germany was completed by the Plaue canal, built in 1743–6, from the Elbe to the Plaue lake and hence to the Havel.

In Sweden, when Telford designed the Göta canal (1808–10), providing water-transport between Stockholm and Lake Väner and thence by the river Göta to Göteborg, he was following in broad principle the scheme first put forward nearly 300 years before in 1526 by Gustavus I, another Renaissance prince with an ambition for canals on a grand scale. It was characteristic of these schemes that they were beyond the technical and financial possibilities of their day. Yet the objective was never forgotten, and in 1596 work began on a much smaller project, which, nevertheless, by canalizing the river Eskilstuna between the Hjälmar and Mälar lakes formed the first step of a route between Stockholm and Göteborg [25]. This was completed in 1610, with eleven timber locks built by Petter von Lübeck, with a Dutch engineer acting as consultant. It was not altogether successful, however, and in 1628 surveys were made by Andreas Bureus for a canal linking Lake Hjälmar and the river Arboga which flows into Lake Mälar. This canal was constructed 1629–39 and entirely superseded the older Eskilstuna route. The fall of 75 ft between Arboga and Hjälmar was taken in ten masonry locks. The works, under the direction of Carl Bonde, included a cutting half a mile long in rock. Trade was considerable and before the end of the century the canal had to be enlarged. The Dutch engineer Tilleman de Moll was in charge of these improvements, which were carried out in 1691–1701. The old locks were replaced by eight new ones, each 24 ft wide and 100 ft long with falls of up to 13 ft, and the depth was increased to 8 ft. Meanwhile in 1635 surveys were made of the country west of Lake Vatter along the course of what was later to be the summit of the Göta canal, and as early as 1607 a lateral cut and lock had been built at Lilla Edet on the Göta river. The subsequent stages in the construction of this great waterway linking the Baltic and the North Sea, associated with Polhem, Telford, and Count von Platen, lie outside the scope of the present chapter [26].

VIII. ENGLISH RIVER NAVIGATIONS

In the British Isles the first summit-level canal was built as late as 1737–45, by Thomas Steers, from Newry in Northern Ireland to Lough Neagh, to bring coal from the Tyrone collieries over the canal and thence by sea to Dublin.

Indeed, canal-construction was not undertaken on a large scale until the days of the industrial revolution, though in the preceding 150 years a great deal had been accomplished in extending and improving the river navigations [27]. To give a measure of this progress, it may be mentioned that by the end of the eighteenth century some 2000 miles of navigable water existed in England, of which approximately one-third was in the form of canals built between 1760 and 1800; one-third was in the form of 'open' rivers which were naturally navigable; and the remaining third had been created as a result of the work of engineers, chiefly between about 1600 and 1760.

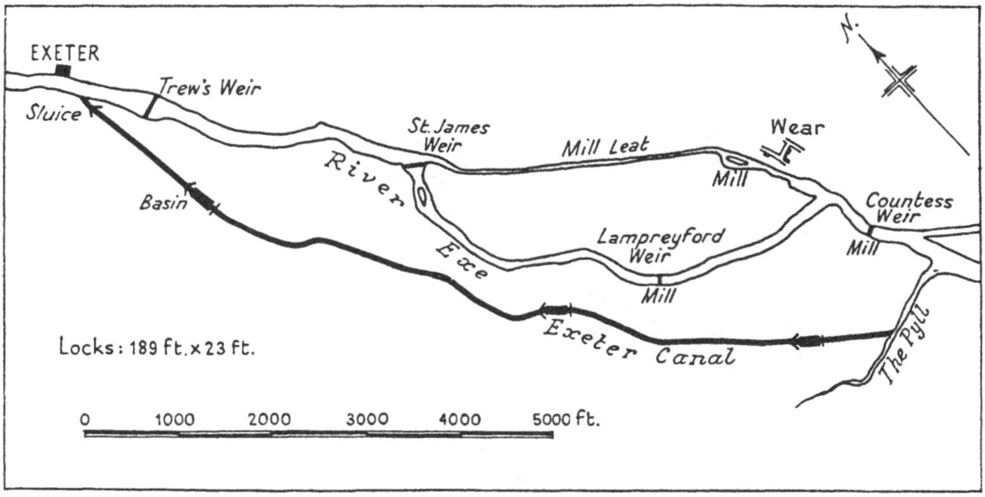

FIGURE 288—*Map of the Exeter canal, 1564–7.*

It will be recalled that during the Middle Ages stanches had been built, usually in mill-weirs. These were works consequent upon the existence of the mills, rather than works designed from the outset to improve navigation. And for small medieval craft navigation was possible on many rivers that would be considered impracticable by modern standards. In the course of time, however, the medieval river system was increasingly felt to be inadequate, and the works at Exeter in 1564–7 (p 451 and figure 288) represent the beginning of a new outlook on water-transport in England.

Shortly afterwards, improvements were undertaken on the Thames and its tributaries. Reference has been made to the lock completed in 1574 on the Lea (p 451), while between 1624 and 1635 three locks were built on the Thames at Iffley, Sandford, and Abingdon. The Wey was made navigable in 1651–3 for a length of 15 miles, with a fall of 86 ft, from Guildford to Weybridge (figure 289).

This work was carried out by Sir Richard Weston (1591–1652) and involved the construction of 7 miles of new cut and ten locks, together with twelve bridges and wharves at Guildford and Weybridge. Seventy years later the Kennet was made navigable from Reading to Newbury. John Hore (*c* 1690–1762) was the

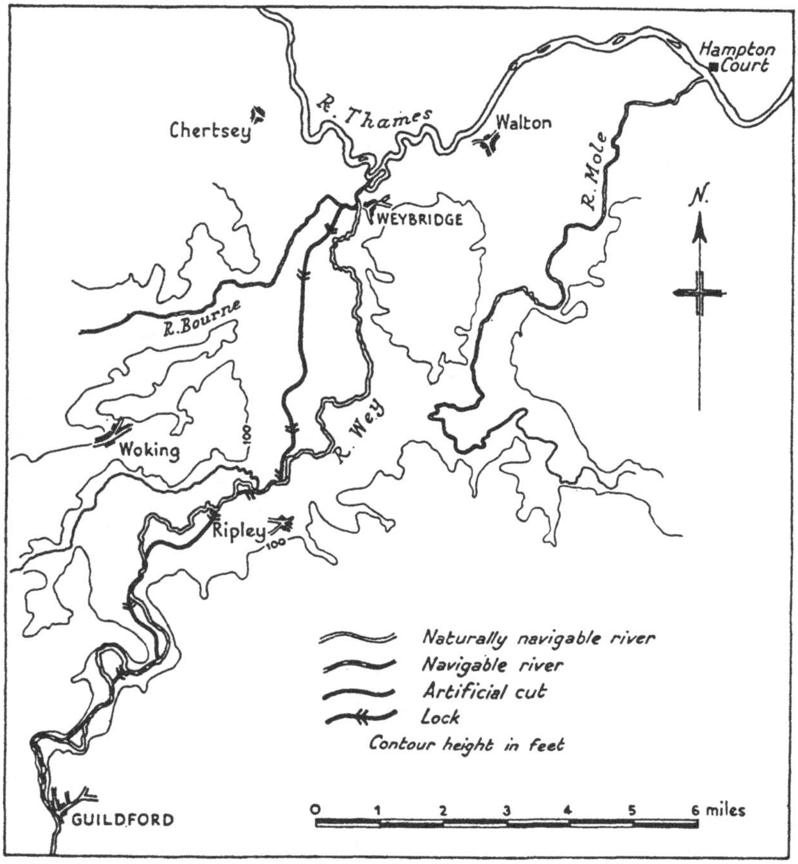

FIGURE 289—*Map of the Wey canal, 1651–3.*

engineer for this work, and for the Avon navigation from Bristol to Bath. On the Kennet, between 1718 and 1723, he built 11½ miles of new cuts 54 ft wide, and in the total length of 18½ miles, with a fall of 138 ft, he provided eighteen locks each having a length of 122 ft and a width of 19 ft. In the west country the Severn was the main artery of transport, the river being an open navigation up to Shrewsbury. It was evident that great advantages would be gained by making some of the larger tributaries of the Severn navigable. Work on the Warwick Avon from Tewkesbury upstream for a distance of 32 miles to Bidford was undertaken by William Sandys in 1636–9. This navigation was extended to Stratford-on-Avon in 1675–7 by Andrew Yarranton (1616–84). In 1662 Sandys attempted to make the Wye navigable from Chepstow to Monmouth, but the technical difficulties proved too great. In the same year, however, Yarranton started work on the Worcestershire Stour and brought the navigation up to Stourbridge.

The first engineering works of importance on the rivers of Yorkshire were carried out by John Hadley on the Aire and Calder (1699–1703), from the Ouse up to Leeds and Wakefield (figure 290). Between 1726 and 1729 William Palmer

FIGURE 290—*View of a lock on the river Aire Navigation near Leeds*, c 1702.

extended the Don navigation up to Sheffield, and he also supervised improvements on the Ouse in 1727–32. Meanwhile, in the north-west, Thomas Steers (1672–1750), who had completed Liverpool's first dock in 1715, made the rivers Mersey and Irwell navigable to Manchester in 1722–5. By 1732, after three years' work, the Weaver navigation was constructed by Thomas Robinson, with eleven locks in a length of 20 miles and a fall of 42 ft from Winsford to the Mersey estuary.

This summary of a few of the English river navigations indicates the scope of the works. Their economic importance is shown by the fact that on the Weaver, for example, over 50 000 tons were carried annually in the period 1750–60. Their construction involved weirs, flood-gates, bridges, and wharves as well as locks and, although individually their technical importance cannot be compared with that of the contemporary canals in France, they represent altogether a considerable engineering achievement.

IX. PROJECTS OF THE FRENCH RENAISSANCE

Inland navigation in France followed a pattern in marked contrast to that in England [28]. It is true that some attention was paid to river improvement, but the principal feature of French work in the seventeenth century was the construction of two great summit-level canals which are so important in the history of engineering that they will be considered separately. Indeed the Languedoc canal (1666–81) was the greatest feat of civil engineering in Europe between Roman times and the nineteenth century; it was largely based upon experience gained from the Briare canal, on which work began in 1604.

The Languedoc canal was conceived in the ambitious and imaginative mind of Francis I (r 1515–47) who clearly foresaw great advantages both to trade and to the prestige of his country if an inland waterway could be established between the Mediterranean and the Atlantic. When Francis returned home from Milan in 1516, accompanied by Leonardo da Vinci, they discussed together the means for effecting this notable enterprise of constructing a *canal des deux mers*. Two possible routes were found. That in the south would link the rivers Garonne and Aude (the Languedoc canal or Canal du Midi); the other, in central France, would join the rivers Loire and Saône (the Charolais or Canal du Centre). The latter was economically more attractive but technically more difficult, and the first detailed studies were made for the southern canal. The survey was carried out by Nicolas Bachelier (1485–1572), who reported in 1539 that the best route would be up the Aude to Carcassonne, then by canal by way of Villefranche to the Garonne just above Toulouse (figure 292). This was essentially the route of the Languedoc canal as built 125 years later but, though locks were mentioned, Bachelier produced no working plans.

After twenty years Adam de Crapponne (1526–76) again investigated this route, and made reconnaissance surveys for the Charolais canal: When the Wars of Religion put a stop to public works further progress had to await the peace of 1598. Then, under Henri IV and his great minister the Duc de Sully, bridges and water-supply schemes were started, land-drainage was initiated on a considerable scale, and a re-examination of the southern canal was ordered. Humphrey Bradley, the dike-master of Henri IV, who in 1584 had acted as a consultant on Dover harbour and in 1589 had prepared the earliest comprehensive proposals for draining the Fens (pp 316, 318), was engaged on this task. A little later he was concerned with a scheme for a canal linking the Seine and Saône, the origin of the Canal de Bourgogne.

The three great Renaissance projects, the Languedoc, Charolais, and Bourgogne canals, were all eventually accomplished, but they were beyond the resources of the early seventeenth century and Sully very wisely decided upon a more practical scheme for a canal linking the Loire and Seine. This was an attractive commercial proposition; it presented few serious technical difficulties; and it would form a branch of the Atlantic to Mediterranean waterway, leading to the Seine valley and the capital.

X. THE BRIARE AND ORLEANS CANALS

Initial plans for a canal linking the Loire and Seine were drawn up in 1603 [29]. The small river Trezée was to be made navigable from its junction with the Loire at Briare for a distance of some 10 miles up to the village of Breteau; then a cutting with a maximum depth of 75 ft was to be excavated in the high land between the Trezée and the Loing, and this river rendered navigable for a distance of 24 miles down to Montargis (figure 291). The rivers and the canal at the summit were to have a minimum depth of 4 ft and a width of 40 ft, and it was thought that 48 locks would be required. These were to be 90 ft long and 16 ft wide, with mitre-gates and a fall varying from 3 to 5 ft. Bids for the contract were made in January and again in February 1604, when Hugues Cosnier was appointed contractor.

Cosnier then examined the site in detail, and found several important errors in the original scheme, which must have been no more than a sketch-plan. For, instead of the rivers Trezée and Loing being at the same level at the points where they were to be joined by the canal in the summit cutting, there was a difference in elevation of more than 40 ft. Provision had been made neither for supplying water to the summit, nor to secure the works against damage from floods. Cosnier was therefore compelled to produce a new design. He proposed to follow the valley of the Trezée only 7 miles up from the Loire and then, striking northwards, to take the canal over a plateau, crossing this with a summit-level $3\frac{3}{4}$ miles long, from which the canal would be taken down the steep side of the Loing valley to meet this river at Rogny. The new route was nearly 3 miles shorter than the original scheme, it avoided the deep cutting, and it made possible the supply of water to the summit. Cosnier also proposed to construct a true canal along the whole length from Briare to Montargis, realizing that this would prove far more satisfactory than merely making the rivers Trezée and Loing navigable. At the same time he suggested that the locks be enlarged and built of masonry.

These radical alterations were examined on behalf of the king by Jean Fontaine, who found himself in complete agreement with Cosnier's proposals, and the new design was accepted by the royal council in December 1604. Sully drafted 6000 troops to provide the labour force, and himself inspected the works from time to time; in 1608 Henri IV, accompanied by the queen, paid Cosnier the honour of visiting the site. By 1610 about three-quarters of the work had been

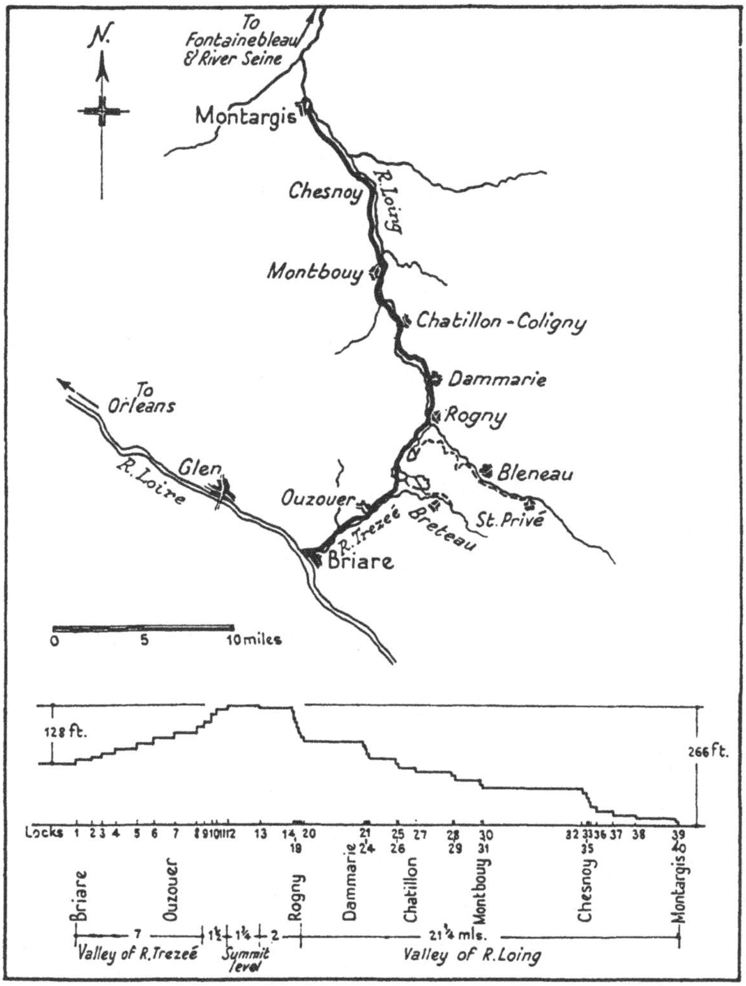

FIGURE 291—*Map and longitudinal section of the Briare canal, 1604–42.*

completed, but in May of that year the king was assassinated and the canal lost its royal patronage. Work nevertheless continued, but in 1611 Sully was forced to resign from the government and under the new régime a commission was appointed to investigate the state of the canal and the cost of its completion. The commission was headed by the Marquis de Rhoissy, who had with him four experts, including Humphrey Bradley (p 459). A careful survey of the works showed that Cosnier had completed 26 of the required 34 miles and that he had built 35 locks, together with many ancillary works such as 15 discharge-weirs and 9 bridges. The locks had masonry walls 6 ft thick, the distance between the gates was 105 ft, and boats of 15-ft beam could be accommodated. The fall of the locks was usually 10 ft, but in some cases was as great as 14 ft. Cosnier had visited the Brussels canal, and adopted the ground-sluices that he saw in the locks there. These were so efficient that even the locks with a 14-ft fall could be operated in ten minutes. In descending from the plateau to the Loing at Rogny, Cosnier had constructed a 'staircase' of six locks with a total fall of 65 ft, a remarkable achievement for the period. He firmly believed in the principle of providing a series of levels each as long as possible, and to this end he built staircase locks not only at Rogny but also near Dammarie and Chesnoy, as well as a number of double locks. For the summit-supply Cosnier proposed, but had not yet built, a feeder-channel 3¼ miles long with an intake near the head waters of the Trezée running down to a small lake, the Étang de la Gazonne, which acted as a reservoir and from which the water was taken directly into the canal. To provide further storage a lock had been built without a step but capable of impounding the water on one side to a depth of 8 or 9 ft. By this means a length of 1¾ miles of the canal at the summit could act as a reservoir, since the water in this reach could be lowered by 4 or 5 ft without interfering in any way with navigation. At intervals along the canal in the valleys of the Trezée and the Loing a number of intakes from these rivers had also been provided.

Thus between the years 1604 and 1611 Hugues Cosnier had designed and almost completed a summit-level canal, rising 128 ft from the Loire and falling 266 ft to the Loing at Montargis. He had built 35 locks, and had designed the summit-supply as well as looking after the supply of water at other parts of the canal and safeguarding it against floods by means of the discharge-weirs.

All these works are mentioned in the report of the Commission of 1611, in which de Rhoissy wrote '*que ce serait un grand abus de laisser et abandonner une si louable entreprise, qui a esté fort bien entendue, conduite et presque parfaicte.*' Nevertheless, for political and financial reasons the scheme was abandoned. Cosnier made proposals to complete the work partly at his own expense, in return for the right to collect tolls from the canal during the first six years of its operation; but with wars and with the weak government of the Regency nothing was done for seventeen years. During this interval Cosnier naturally was engaged elsewhere, notably in connexion with the Arcueil water-supply to Paris. In 1628 a re-examination of the Briare canal was made by Francini and Le Mercier, who strongly recommended the completion of the work but pointed to the need for additional water-supply at the summit to compensate for the large losses of water involved in operating the six locks at Rogny. For this purpose they proposed a second feeder-channel with an intake on the Loing at St-Privé, leading to a small lake, which would act as a reservoir, at a level just above that of the summit. In order to forward the work to which he had already devoted so much thought and time, Cosnier offered to construct this feeder-channel provided he was paid a part of the money still owed to him since 1611. This offer was accepted, but he was now an elderly man in ill health and he died in the last month of 1629. *Saluons, au passage, la mémoire de ce grand ingénieur!* [30].

Another eight years passed before, in 1638, Guillaume Boutheroue and Jacques Guyon obtained letters patent from Louis XIII to complete the work and to pay the land-compensations still outstanding, in return for ownership of the canal. They formed a company for this purpose and, under the direction of Boutheroue's brother François, the damage of twenty-seven years of neglect was made good, the last five locks down to Montargis were built, and the canal was put into operation in 1642, exactly as conceived by Cosnier. The proposal of 1628 to provide a feeder-channel from St-Privé proved, however, to be well founded, and this channel was duly constructed in 1646. It is notable for having a fall of only 5½ ft in a total length of 13 miles (5 in per mile).

The undertaking was a satisfactory investment and during the latter half of the seventeenth century the company received an annual return of 13 per cent on its capital. The annual trade on the canal in this period was about 200 000 tons. This figure was maintained throughout the eighteenth century, doubled during the nineteenth, and is trebled at the present day.

The goods transported on the Briare canal consisted chiefly of coal and of wine brought down the Allier and the upper Loire to Paris, but part of the trade came up the Loire from Nantes, Tours, and Orleans. To shorten the journey of this latter commerce and to facilitate the export of timber from the forest of Orleans a canal was built in 1682–92 with a length of 46 miles, rising 98 ft from the Loire at Orleans to the summit in 11 timber-locks and descending 132 ft to the Loing at Montargis, in a distance of 18 miles, in 17 locks [31]. The chief technical problem in the Orleans canal was the summit-water supply. The canal crosses an extensive and very flat plateau, and the principal feeder-channel, known as the Rigole de Courpalet, has a length of 20 miles with a fall of only 4 ft (2½ in per mile). The flow in this channel was therefore very slow, and the system would have been ineffective but for the use of the summit-level as a storage reservoir, in the manner already devised by Cosnier on the Briare. With a length of 11 miles, the summit-level of the Orleans canal had sufficient capacity to store the water flowing from the feeder-channels during the night; this was used to replace the loss of water in operating the canal the following day. The consultant for the Orleans canal was Sébastien Truchet (1657–1729), a well-known *savant hydraulicien* and pupil of the physicist Mariotte [32]. The high degree of accuracy required in surveying the Rigole de Courpalet is remarkable, and calls to mind the achievement of Edward Wright in his survey for the New River water-supply for London as early as 1609 [33]. Here the fall was 18 ft in a length of 39 miles, or 5½ in per mile (almost the same slope as that of the Rigole de St-Privé). The Courpalet feeder is also memorable as having provided a valuable subject for investigation by Antoine Chézy in 1769 during his classic research on flow in rivers and open channels [34].

The Loing between Montargis and the Seine which, after 1692, took the traffic both of the Orleans and the Briare canals, was navigable only by virtue of twenty-six stanches situated in the weirs of the various mills along the river. With the two canals in operation the delays caused by operating these stanches became intolerable, and between 1719 and 1723 a lateral canal alongside the Loing was built by Jean-Baptiste de Règemortes, with twenty-one locks in a length of 32 miles [35]. In 1726 his son Noël rebuilt, in masonry, the locks on the Orleans canal, and thus the canal system between the Loire and Seine was perfected.

XI. THE LANGUEDOC CANAL

It is not surprising that progress was delayed on the Languedoc canal in the south of France since, after the death of Henri IV, the government was reluctant even to complete the Briare canal. It was not until 1662 that the requisite combination of an able administrator in Colbert and an engineer of genius in Pierre-Paul Riquet (1604–80) was found [36]. In the previous year Riquet, with the assistance of François Andreossy (1633–88), had for the first time worked out a scheme for the supply of an adequate amount of water to the summit. For this purpose he proposed a feeder-channel 26 miles long with an intake on the river Sor near Revel. Its flow was to be increased by a second channel taking water from three mountain streams, the Alzau, Vernassonne, and Lampy (figure 292). For the route of the canal, he adopted essentially the line proposed by Bachelier more than a century earlier. The river Lers [now l'Hers] was to be made navigable from Toulouse to a point east of Villefranche (Haute-Garonne). A canal was then to be cut rising over the summit and joining the river Fresquel near Castelnaudary. Below this point the Fresquel would be made navigable to its junction with the

Aude near Carcassonne, and this latter river would in its turn also be made navigable down to its mouth on the Mediterranean coast.

In November 1662 Riquet communicated his ideas to Colbert. Colbert had little difficulty in arousing the interest of Louis XIV, who preferred that even the practical undertakings of his reign should exhibit the quality of grandeur. The king appointed a royal commission to investigate the plans, and during 1663 and 1664 Riquet was busy working out his scheme in more detail, paying visits to Colbert in Paris, and to Hector Boutheroue, the younger brother of

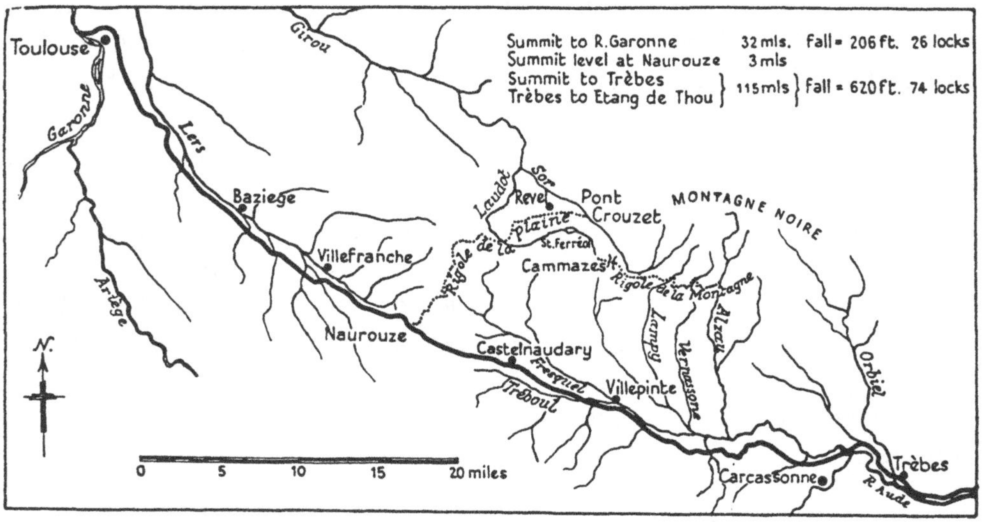

FIGURE 292—*Map of the western part of the Languedoc canal, 1666–81.*

François and one of the directors of the Briare canal, who had been appointed as the principal expert on the commission. In November 1664 Riquet presented his plan formally to the commissioners, who called upon the assistance of Andreossy and Jean Cavalier, Geographer Royal, and two other surveyors. During a period of seven weeks the commission made a detailed investigation of the terrain, approving Riquet's plans in general, but proposing that instead of making the rivers Lers, Fresquel, and Aude navigable, a canal be made along the entire length of the route from Toulouse to the Mediterranean. They also proposed that the canal should end not at the mouth of the Aude but in the Étang de Thau, with a port constructed at Sète.

The commission's report was approved by Colbert and the king, but as a practical demonstration of the validity of the plans for the summit-supply it was suggested that a pilot channel be dug from the Sor to the summit at Naurouze. This work was carried out by Riquet between May and October 1665 with complete success. Meanwhile de Clerville, chief royal engineer, prepared the contract documents for the first section of the canal between Toulouse and Trèbes. In October 1666 Riquet was appointed contractor, and by January of the following year 2000 men were employed, a number doubled by March. In 1668 specifications were drawn up for the second part of the canal from Trèbes to the Mediterranean, and in 1669 more than 8000 men were at work. Riquet had established an admirable organization: the canal was divided into twelve sections each under an Inspecteur-Général, one of whom was Andreossy. Under the Inspecteurs there were men with more local responsibilities, each having under his control the requisite number of foremen and labourers. There were

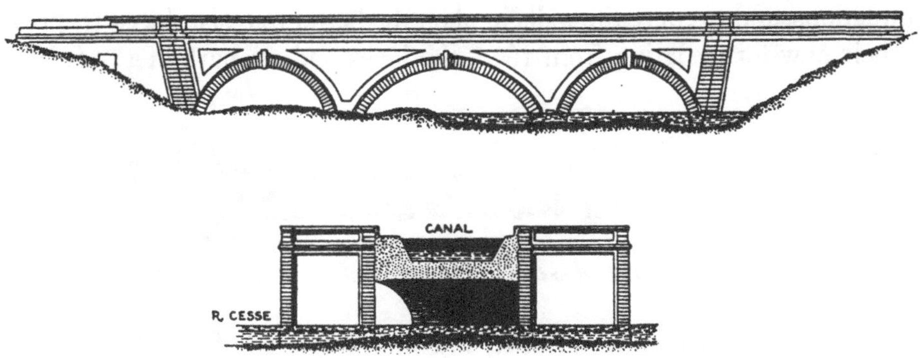

FIGURE 293—*Aqueduct over the river Cesse on the Languedoc canal*, c *1688*.

also seven surveyors permanently engaged. By 1680 the work was rapidly approaching completion, but in October Riquet died and so was denied the satisfaction of seeing the opening of the canal seven months later, in May 1681. Riquet was succeeded as director of the works by his son Jean-Mathias, under whom the work continued for a number of years: much had still to be done before the canal was perfected in 1692.

The Languedoc canal excited the admiration of the world. Travellers came to see it under construction and poets celebrated its engineer, Colbert, and the king. Voltaire, in his *Siècle de Louis XIV*, having mentioned the Louvre, Versailles, and other building works of the *Roi Soleil*, said: '*mais le monument le plus glorieux par son utilité, par sa grandeur, et par ses difficultés, fut ce canal de Languedoc qui joint les deux mers.*' Just as the Briare canal had served as the model for the Languedoc in France, so the Languedoc was the prototype for succeeding great canal undertakings in Europe.

From the Garonne at Toulouse the canal rises 206 ft to the summit, in a length of 32 miles and with 26 locks. The summit-level is 3 miles long, and the canal then descends 620 ft to the Mediterranean, with 74 locks in a distance of 115 miles. The whole canal from Toulouse to the Étang de Thau is therefore 150 miles long, with 100 locks. The contract plans called for a channel 50 ft wide at the water-surface, a depth of 8½ ft, and side-slopes of 1:1. These were too steep, and after several slips had occurred the section was modified to a top width of 64 ft, a depth of 6½ ft, and side-slopes of 2½:1, the original base-width of 32 ft being maintained.

The walls of one of the locks collapsed soon after construction, and to prevent any further trouble of this kind the existing locks were rebuilt and the whole lay-out revised to reduce the height of the walls by one-third, to provide stronger foundations, and to curve the walls in plan, the better to resist the earth-pressure. In their new form, dating from 1670, the locks had a length of 115 ft between

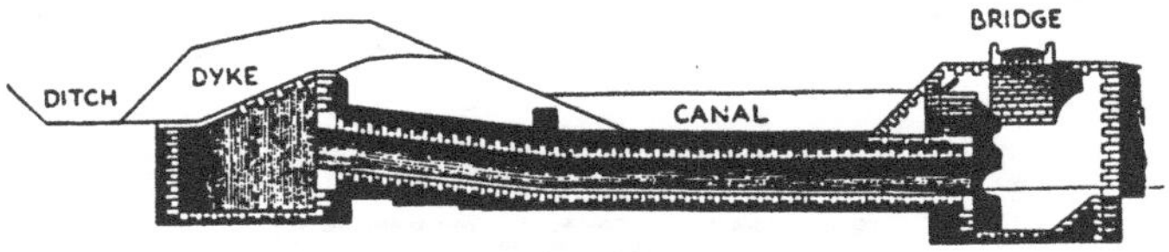

FIGURE 294—*Culvert under the Languedoc canal near Villepinte*, c *1680*.

the gates, an entrance-width of 21 ft, and an average fall of 8 ft. The oval shape, though of great solidity, leads to a greater loss of water than from the usual rectangular chamber. Many of the locks were grouped together, the most notable examples being near Béziers, where Riquet built a staircase of 8 locks having a fall of 70 ft. At a distance of about 6 miles upstream of these locks the canal passed through the Malpas tunnel, 180 yds long. There were three major aqueducts. One over the river Repudre with a single arch of 30-ft span was built before 1680. The other two over the rivers Orbiel and Cesse (figure 293) were designed in 1686 by the celebrated engineer Sébastien Vauban (1633–1707, p 372) and built by Antoine Niquet (1639–1724). The countless streams crossing the line of the canal were taken under it in culverts (figure 294). Numerous diversion-weirs, spill-ways, and road-bridges were constructed, as well as the new port of Sète.

This was civil engineering on a grand scale, but perhaps the chief technical interest lies in the scheme for the summit-water supply. The furthermost intake is on the river Alzau (figure 292), situated high up the southern flanks of the Montagne Noire. From this point a feeder-channel, known as the Rigole de la Montagne, runs for a distance of 12 miles, crossing two other streams; then it divides, one short branch leading to the river Sor and the other continuing for 3 miles and passing through a tunnel at Les Cammazes to supplement the head waters of the Laudot.[1] The Rigole de la Plaine, with an intake on the Sor at Pont Crouzet, flows for 8 miles to meet the Laudot and then, carrying the whole supply in a channel 20 ft wide and 9 ft deep, runs for a further 19 miles to the Naurouze summit of the canal. In the valley of the Laudot, at St-Ferréol, Riquet constructed between 1667 and 1671 a large earthen dam with a masonry core wall, 105 ft in height (figure 295), to form a storage reservoir of 250 million

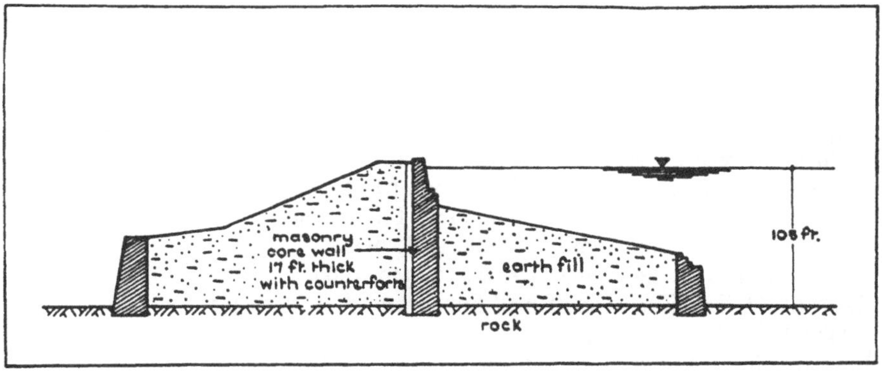

FIGURE 295—*Cross-section at maximum thickness of the St-Ferréol dam, Languedoc canal, built 1667–71. The capacity of the reservoir was 250 million cu ft.*

cu ft capacity [37]. During the summer the flow of the Rigole de la Montagne was directed entirely into the Sor, but in the winter the greatly increased flow in this channel was diverted partially into the St-Ferréol reservoir, the water of which could be used during the following summer. This brilliantly conceived plan proved to be wholly satisfactory, and it was not until 1777–81 that a second reservoir was constructed on the Lampy, with a masonry dam 54 ft high, to provide an additional source of water required by a branch canal to Narbonne.

With the perfection of the Languedoc canal in 1692 civil engineering had reached maturity in this field, and the technical basis had been provided for the vast expansion of the canal system of Europe which took place in the succeeding period of industrial development.

REFERENCES

[1] For extensive data on costs of transport in England see W. T. JACKMAN. 'The Development of Transportation in Modern England.' University Press, Cambridge. 1916.

[2] A general history of Chinese water-ways is given by CH'AO-TING CHI, 'Key Economic Areas in Chinese History, as revealed in the Development of Public Works for Water-control' (Allen and Unwin, London, 1936). Interesting descriptions will be found in LOUIS LECOMTE, 'Memoirs and Observations . . . made in a late journey through the Empire

[1] Although part of the original plan, this tunnel was built in 1686–8 by Niquet.

of China . . . translated from the Paris edition' (London, 1697), and SIR GEORGE L. STAUNTON, 'An Authentic Account of an Embassy to China, chiefly from the papers of Lord Macartney and Sir E. Gower' (3 vols, London, 1797). Details of the lock of A.D. 984 and the Shantung summit-canal have been elucidated in discussions with Dr Joseph Needham.

[3] For medieval locks in the Netherlands see G. DOORMAN. 'Techniek en Octrooiwezen in hun Aanvang.' Nijhoff, The Hague. 1953.

[4] THACKER, F. S. 'The Thames Highway: a History of the Inland Navigation.' Thacker, London. 1914.
Idem. 'The Thames Highway: a History of the Locks and Weirs.' Thacker, London. 1920.

[5] ZONCA, VITTORIO. 'Novo Teatro di Machine et Edificii.' Padua. 1607.

[6] TELFORD, T., article on "Inland Navigation" in 'The Edinburgh Encyclopaedia', vol. 15, pp. 209–315, writes: 'The lock at Vreeswyck dates from 1373 and is perhaps one of the oldest on record.' He also recognized that the Stecknitz canal should be 'reckoned amongst the first in which the invention of locks was employed'.
BREWSTER, D. *et al.* (Eds). 'The Edinburgh Encyclopaedia', vol 15. Edinburgh. 1830.

[7] Private communication from DR FOCKEMA ANDREAE.

[8] REHDER, P. *Z. Ver. Lübeckische Gesch.*, **11**, 339, 1909.
TEUBERT, O. 'Die Binnenschiffahrt.' Engelmann, Leipzig. 1912.
WREDEN, R. *V. D. I. Jb.*, **9**, 130, 1919.
ECKOLDT, M. *Wasserwirtschaft*, **40**, 255–60, 290–5, 1950.

[9] The standard works on the Lombardy canals are:
BRUSCHETTI, G. 'Storia dei Progetti e delle Opere per la Navigazione interna del Milanese.' Milan. 1821.
LOMBARDINI, E. 'Dell' Origine e del Progresso della Scienza Idraulica nel Milanese.' Milan. 1872.
LECCHI, A. 'Trattato de' Canali Navigabili.' Milan. 1776, 1824.
BASCAPE, G. C. 'Itinerarii della Nostalgia: il Naviglio di Milano e gli antichi Canali lombardi. Tavole di Giannino Grossi.' Delfino, Milan. 1950.

[10] For the work of Bertola da Novate see:
PARSONS, W. B. 'Engineers and Engineering in the Renaissance.' Johns Hopkins Press, Baltimore. 1939.
and also BRUSCHETTI, G. See ref. [9].

[11] ALBERTI, L. B. *De re aedificatoria.* Florence. 1485.

[12] ZENDRINI, B. 'Leggi e Fenomeni, Regolazioni ed Usi delle Acque correnti.' Venice. 1741.

[13] M——, G. B. 'Notizie storiche . . . del Canale naviglio di Bologna.' Bologna. 1825.

[14] PARSONS, W. B. See ref. [10].

[15] For a description of the modern shaft-lock see MCGAREY, D. G. *J. Jr Instn Engrs*, **49**, 1–28, 1938.

[16] KLEHMET. *Wschr. Arch. Ver., Berlin*, **3**, 177–81, 191–5, 1908.
ECKOLDT, M. See ref. [8].

[17] DE LA GARDE, P. C. *Min. Proc. Instn civ. Engrs*, **4**, 90–102, 1845.

[18] TWYNE-LANGBAINE MSS 1, fol. 76 (*c* 1620). Bodleian Library, Oxford.

[19] VIERLINGH, A. 'Tractaat van Dyckagie' (2 vols), ed. by J. DE HULLU and A. G. VERHOEVEN. Rijks Geschiedkundige Publicatiën, The Hague. 1920.

[20] STEVIN, S. 'Nieuwe Maniere van Sterctebou door Spilsluysen.' Rotterdam. 1617.

[21] NERMAN, G. 'Hjalmare Kanals Historia.' Uppsala. 1910.

[22] ENGELS, H. *Ann. Trav. publ. Belg.*, **1**, 120–73, 1843.

WAUTERS, A. 'Documents concernant le Canal de Bruxelles à Willebroeck.' Brussels. 1882.
Personal communication from DR O. GORTEMAN of Brussels.

[23] VIFQUAIN, J.-B. 'Des voies navigables en Belgique.' Brussels. 1842.
RIVE, B. L. DE. 'Précis historique et statistique des canaux et rivières navigables de la Belgique.' Brussels. 1835.

[24] The most convenient general reference for German canals is O. TEUBERT (see ref. [8]). More specialist studies include KLEHMET (see ref. [16]), R. WREDEN (see ref. [8]), and M. ECKOLDT (see ref. [8]).

[25] A detailed history of the Hjälmar canals is given by G. NERMAN (see ref. [21]).

[26] For the Gota canal and the navigation of the Gota river, including the Trollhatten canal, see T. TELFORD (see ref. [6]) and
SKJOLDEBRAND, A. F. 'Description des cataractes et du Canal de Trollhatta.' Stockholm. 1804.

[27] WILLAN, T. S. 'River Navigation in England 1600–1750.' University Press, Oxford. 1936.
SKEMPTON, A. W. *Trans. Newcomen Soc.*, **29**, 1953 (in the press).

[28] Standard works on the French canals of the seventeenth and eighteenth centuries are:
LALANDE, J. J. LE F. DE. 'Des canaux de navigation.' Paris. 1778.
POMMEUSE, H. DE. 'Des canaux navigables . . . avec des recherches comparatives sur la navigation intérieure de la France et celle de l'Angleterre.' Paris. 1822.
DUTENS, J. 'Histoire de la navigation intérieure de la France.' Paris. 1829.
BELIDOR, B. FOREST DE. 'Architecture hydraulique' (Vol. 1, Pt II, Paris, 1750; Vol. 2, Pt II, Paris, 1753) has many valuable drawings of canal works.
An important manuscript, 'Traité de plusieurs sortes de navigations' (Bibliothèque Nationale, Paris, MS fr. 18.954, pp. 47–71), was written by HUGUES COSNIER in 1628.

[29] The history of the Briare canal was first accurately worked out by LÈBE GIGUN, "Cosnier et les origines du Canal de Briare" (*Ann. Ponts Chauss.*, **18**, 509–57, 1889). His work has been amplified in the definitive monograph of P. PINSSEAU, 'Le Canal Henri IV ou Canal de Briare.' Houzé, Orléans. 1943.
Earlier books, notably H. DE POMMEUSE (see ref. [28]), contain valuable technical details.

[30] PINSSEAU, P. See ref. [29].

[31] POMMEUSE, H. DE. See ref. [28].

[32] TARBÉ DE ST. HARDOUIN, F. P. H. 'Notices biographiques sur les ingénieurs des ponts et chaussées.' Paris. 1884.
See also "Éloge du P. Sébastien Truchet." *Hist. Acad. R. Sci.*, 93–101, 1729.

[33] DAVIDSON, SIR JONATHAN. Presidential address, *J. Instn civ. Engrs*, **31**, 1, 1948.

[34] CHÉZY, A. 'Mémoire sur la vitesse de l'eau conduite dans une rigole.' (MS 1775). Publ. in *Ann. Ponts Chauss.*, **60**, 241, 1921.

[35] POMMEUSE, H. DE. See ref. [28]; for Regèmorte see TARBÉ DE ST. HARDOUIN, see ref. [32].

[36] Illustrated descriptions of the Languedoc canal are to be found in J. J. LE F. DE LALANDE (see ref. [28]), M. DE LA ROCHE, 'Atlas et description du Canal Royal de Languedoc' (Paris, 1787), and A. F. ANDREOSSY, 'Histoire du Canal du Midi' (Paris, 1804). RIQUET DE BONREPOS, 'Histoire du Canal de Languedoc' (Paris, 1805), gives a useful historical account based on original documents. Interesting details of the works, as seen by eye-witnesses, are contained in *Phil. Trans.*, **3**, 1123–8, 1670, and LOUIS DE FROIDOUR, 'La relation et la description des travaux qui se sont en Languedoc, pour la communication des deux mers' (Toulouse, 1672).

[37] DEVIC, C. and VAISSETE, J. J. 'Histoire générale de Languedoc', Vol. 14, p. 1088. Toulouse. 1876.

2

The Waltham pound lock

K.R. Fairclough

In 1571 an Act of Parliament was passed 'for the bryngïng of the Ryver of Lee to the Northside of ye Citie of London'.[1] This act, sponsored by the city authorities, gave details of plans proposed by the aldermen for building a new cut from the River Lee, through Hackney and Shoreditch, to terminate just outside the city walls near Moorgate. Once this canal had been completed, the aldermen further proposed that improvements be carried out along the existing river between Ware and the mouth of their new cut.

These ambitious plans, however, were never carried out. During the committee stage of the bill's passage through the House of Commons, several important additions and alterations were made to the original draft, the most important of which was the insertion of a veto on the right to collect tolls for using either the improved river or the new cut. Such a veto meant that the City was unable to finance the canal and although the scheme was not shelved immediately it became obvious within a couple of years that the project could not go ahead.

Interest in improving the Lee had been awakened, however, and on 27 September 1575 a Commission of Sewers was appointed to improve the navigation on the river[2] in order that it might become an important artery along which grain, meal and malt could be brought to the capital. John Norden ascribes this later initiative to 'the instant suyte of the inhabitants of Hartfordshire'[3] who saw that an improved navigation would allow them to capture an important share of a trade which had previously been dominated by land carriers known as 'badgers'.

There is no evidence to suggest that the Commissioners ever seriously considered building the new cut to Moorgate; instead they concentrated on improving the existing navigable channel. They scoured and cleaned this channel, ordered the removal of all fishing weirs and impediments to navigation, raised all bridges over the river to allow the barges more headroom, laid out a rough towpath which included towing bridges, and came to a series of differing compromises with the many millers in the valley whereby they were to take water out of the river without the use of any flash locks which would hold up the passage of the barges.[4] Though these improvements were much less ambitious than those originally proposed by the City, they were nevertheless extremely successful. Within four or five years of the Commissioners' appointment the river had been substantially improved and an expanding barge traffic was already arousing the opposition of the badgers.

The most ambitious task undertaken by the Commissioners was the construction of a pound lock at Waltham and it remains their best-known work for it was the first pound lock in England to be equipped with mitre gates at both ends. The only earlier pound locks known in England were those built between 1564 and 1567 by John Trew along the River Exe. These were constructed as large pools in which several boats could lie at once. Mitre gates were used at one end only; at the other end were single guillotine gates.[5]

The Waltham lock is described in a poem written by William Vallans some time during the 1580s.[6] In it two swans make a journey down the River Lee and one of the many sights they marvel at is:

> But newly made, a waterwourke: the locke
> Through which the boates of Ware doe passe with malt.
> This locke containes two double doores of wood,
> Within the same a Cesterne all of Plancke,
> Which onely fils when boates come there to passe
> By opening of these mightie dores with sleight,
> And strange devise, but now decayed sore.

Before the Commissioners began their work barges navigated the river above Waltham by means of a flash lock which stood across the main stream about three-quarters of a mile above Waltham High Bridge. When shut, this flash lock diverted water out of the river into the head stream of Waltham Mill. Since few barges navigated the river at this time the flash lock was usually shut and consequently, it was later claimed, 'the auntient Channell did decay for lacke of Continuall corse of Water and soe did become unpassable for Boates'.[7]

The Commissioners, however, chose not to scour and cleanse this traditional channel, but rather to open a new route. They ordered that 'the passage of the Boates shold be directed to passe by the sayde Mill-streame And for that purpose that a newe Cutt shold be made from the sayde Millstreame somewhat distaunte from the sayde Mill towardes the North unto the old River towardes the West', and that along this new cut 'a newe devised Lock to Cawse the water to swell upp wherby Boates may passe and repasse betwixt the sayd river of Ley and the water belonginge to the Mill' be constructed.[8]

To complete the arrangements for this new route the Commissioners further decreed that the old flash lock be pulled up and replaced by a 'loweshare of three foote and a half highe from the bottome of the river for the forcinge of the water to his (Edwards Denny's) sayde Milles and yett not be suffered to be anie higher for that all white and superfluous waters may passe that way for the drayninge of the groundes adioyninge and keepinge open of the olde channell'.[9] As a precaution the Commissioners added that the bargemen were to have the right of pulling up this loweshare and using the traditional channel, if, for any reason, they were unable to use the newly opened route. This particular proviso was to assume great importance later when (in 1592) disputes arose and riots ensued over the rights of navigation through Waltham.

In October 1576 the Commissioners were still debating which route to take through Waltham.[10] Yet by November 1576 they were meeting to discuss the completion of their work in the area. Thus the pound lock must have been built some time during the autumn of 1576 and not earlier as is usually stated.

Besides discussions about the route the Commissioners seem to have considered at least two alternative plans for the construction of the lock. A surviving estimate of costs,[11] reproduced in full in the Appendix, provides a rather inadequate comparison between the cost of building the lock entirely of wood, and an alternative proposal whereby the walls of the pound would be built of stone. Vallans' poem suggests that the Commissioners chose to build entirely with wood but no firm evidence remains to throw light on the reasons for such a choice.

The estimate itemizes the different sections of the lock and calculates the amount of timber necessary for each section. Although the arithmetic is not quite correct a total of 44 loades of timber is estimated and it was reckoned that the cost of this timber and the associated carpentry work would be £40 5s. 0d. No allowance, however, is made for any mechanism to open the doors nor for any paddles or other devices to let water into the lock when the gates were shut. The estimate for the lock built with stone walls is even more incomplete, although it does seem likely that this would have been the more expensive piece of work.

if it be don	Tymb xliiij lodes at x^{s}	xxii^{li}		
with timber	Sawinge + Carpenters woorke	xviij^{li}	v^{s}	
		xl^{li}	v^{s}	
the same with	Tymber xxiiij lodes	xii^{li}		
stone	Lyme viij lodes at 16.8	vi^{li}	xiii^{s}	iiij^{d}
	Sand xvj lodes at ii^{s}		xxxii^{s}	
	The walling	viij^{li}		
	Carpenters woork			
	Stone			
		xxviij^{li} [*sic*]		

In addition it was estimated that the digging of a new cut, 36 poles in length,[12] would cost a further £48, while heightening Waltham High Bridge by two feet would cost £4. These estimates, however, turned out to be much too optimistic. Though the total estimated cost of the work was given as £88 5s. 0d. beforehand, the actual cost was reported to be £271 18s. 0½.[13]

Assuming that the wooden lock was built according to the specifications laid down in the estimate, the reconstruction shown in Figures 1a and 1b indicates how the frame of the lock was constructed.[14] First, timbers known as 'grounsylles' and 'gysses' (joists) were laid in the bed of the cut to provide a foundation upon which to fix the frame of the

The Waltham Pound Lock

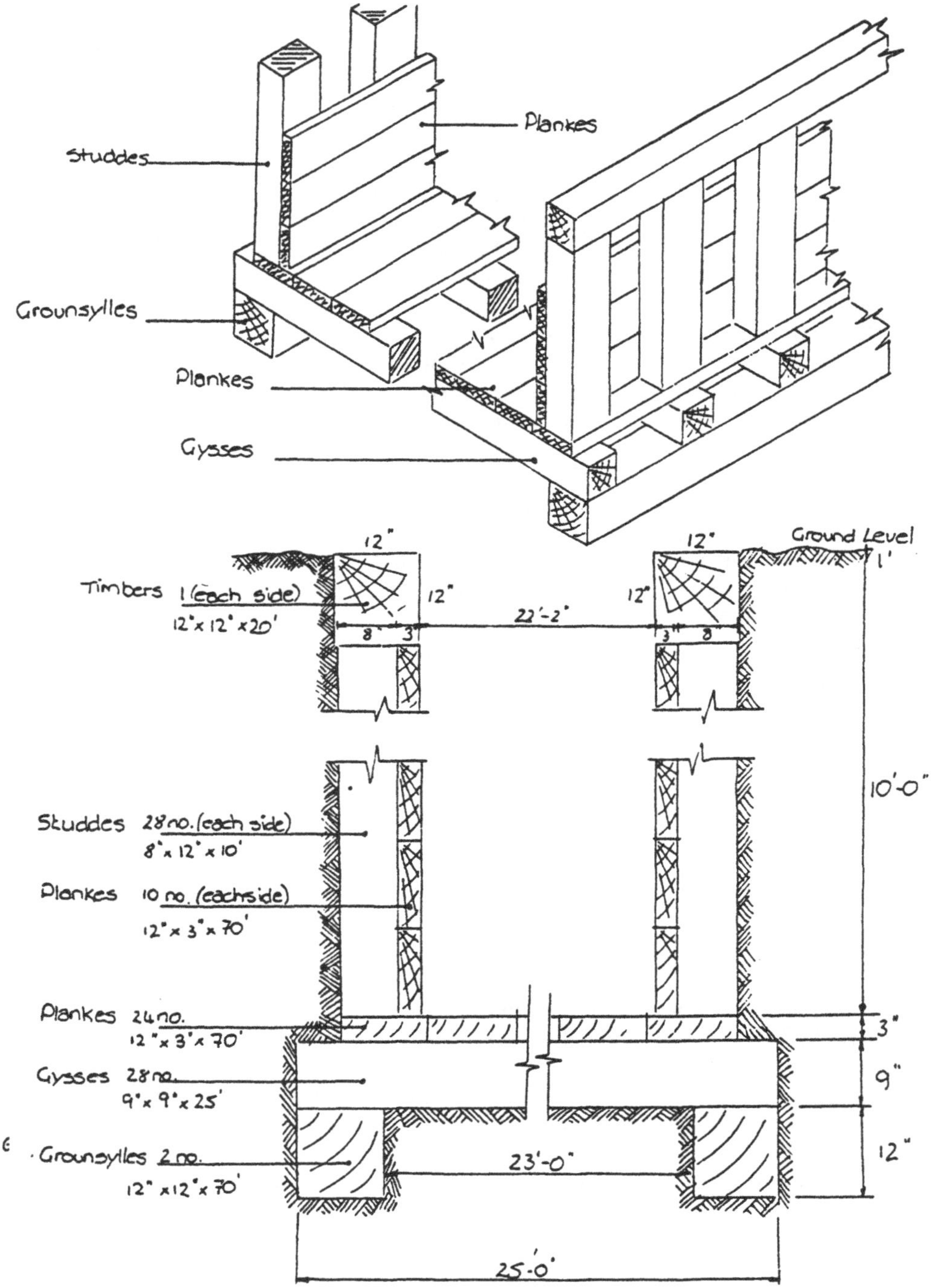

Figure 1. Pound Lock, Waltham Abbey, 1576.
(a) Isometric view. (b) Central cross-section.

lock. The floor and walls of the enclosure were made up of planks, each measuring 70 feet by 1 foot by 3 inches, laid side by side. The floor planks were nailed to the 'gysses', whilst the wall planks were nailed to vertical 'studdes' which were fixed at regular intervals along the sides of the lock. In this manner a frame 70 feet long, 24 feet broad, and 10 feet high was built up, and at the ends of this frame posts were fixed from which the four gates were hung.

Vallans mentions that the gates were opened by a 'sleight and strange devise', but unfortunately the estimate gives no idea what this might have been.* Conceivably chains were used because in 1581 it was alleged that 'Aron Yong of Waltham Abbey Tailor solde a chayne that belonged to the said lock to one Davie of Waltham Cross smyth for sixe pence'.[15] On the other hand such a chain might have been used to lock the gates so that they could not be opened except with the cooperation of the miller.

It seems likely that the Commissioners opened this new route after representations from the owners and occupiers of Waltham Mill for it was the owners who were made to bear the cost of the new cut and lock. The mill was owned by the Denny family as part of the manor of Waltham. At the time of the alterations Henry Denny had just died and his heir, Edward Denny, was still a minor and as such a Ward of Court whose interests were looked after by Queen Elizabeth. In view of this the Commissioners at their meeting in the Star Chamber decided that the cost of the work should be split equally between the executors of Henry Denny's will and the Queen in her role as Edward's guardian. All further costs of maintenance were to be borne by the miller, Thomas Perrott, during the remaining term of his lease after which they became the responsibility of Edward Denny and his heirs.

Since it was also reported that, despite their previous orders, the old flash lock was still standing, the Commissioners further decreed that the miller was to pull it down and build a loweshare by 14 February 1578 or else face a fine of £10.

Within a couple of years, however, major problems had to be faced. In 1579 full-scale repairs had to be carried out and the fact that the cost of this work was borne by the Treasury rather than the miller does suggest faults in the original construction. From hints in an account book[16] detailing these repairs it seems that one side of the wooden foundations had settled so far into the bed of the new cut that the lock had become lopsided. To rectify this the Commissioners were forced to close down the new cut so that the lock could be dismantled and masonry foundations constructed. The wooden frame was then replaced on top of these new foundations.

On the back of the account book a total cost of £96 13s. 3d. is recorded of which £40 6s. 9d. is attributed to 'redy money' and the remaining of £56 6s. 6d. to 'bills of Woork'. No further details are given about the 'redy money' but the account lays out in great detail the various charges which

*** Information from 1579 notes the purchase of soap 'for the trundles Cogges & gates', which means that the method of opening the gates is now known, see footnote 16.**

constituted the 'bills of Woork'. These include the wages paid, the amounts spent on transportation and sums for many small items such as nails and planks, the need for which must have arisen during the actual course of the reconstruction.

Work commenced on 18 May 1579 and was completed by 13 June. During the first few days temporary dams were built and preparations made so that five carpenters could be employed at 'pulling upp the plankes and laiing them agayne'. In between two masons and their assistants were employed to lay the masonry foundations. These carpenters and masons, being skilled workers, were paid 14d. per day.

Most of those employed, however, were unskilled men taken on either as general labourers or 'skavellmen'. The latter were so named because they were hired to bale out any water which seeped through the temporary dams using a tool known as a skavell. The two types of unskilled labour were interchangeable, men working at whatever job was available on any one day. Labourers worked during the day only while the skavellmen were needed right round the clock. Since the men were paid either 10d. or 12d. per shift it seems likely that they worked for either 10 or 12 hours and were paid at a rate of 1d. per hour, the variation in pay per shift not depending on the type of work done. Some men only worked the odd shift whilst others took advantage of the availability of employment to put in extremely long hours. One man, John Foster, earnt 9s. one week and 11s. the following, which means that in the second week he worked the equivalent of 7 day and 4 night shifts.

The account book also mentions that a Mr Trewe was paid £6 13s. 4d. for directing and overseeing the work. It is conceivable that this was the John Trew who was responsible for improving the River Exe. Such a high salary does suggest that he was being rewarded for professional skills over and above the mere ability to oversee a work-force. Whether he introduced any basic changes into the design or operation of the lock cannot be said.

The problems experienced with the Waltham lock arose no doubt because of the technical novelty of the work. The Commissioners, however, enjoyed far more success with their other improvements. Their success, however, only aroused opposition. Not only did the millers, the fishermen and the riparian landowners feel that their traditional rights had been interfered with but the local badgers felt that their livelihood was directly threatened. Traditionally grain and malt had been brought to markets in the lower Lee valley, particularly Hoddesdon, where local dealers and badgers bought it in order to resell in the London markets. This traditional pattern was threatened by a growing barge traffic, not only because barges could carry at cheaper rates, but also because markets further up the valley, particularly those at Ware and Hertford, could intercept much of the trade at the expense of Hoddesdon. The badgers, especially those living in Enfield, Cheshunt and Waltham, were to become the most persistent and vociferous opponents of the navigation.

In 1580 the badgers petitioned Lord Burghley requesting that the navigation be closed down. He advised them that he did not have the necessary authority and suggested that they approach Parliament instead. This they did but before their petition could be heard Parliament was prorogued.[17] Having failed by legal methods the badgers turned to violence and during the summer and autumn of 1581 destroyed many of the works along the river. A full-scale enquiry was held at which the badgers were given full opportunity to express their grievances. Nevertheless the Commissioners obtained the authority to carry on with their work without any of the badgers' objections being met. It was felt that the national interest warranted the expansion of barge traffic along the Lee even though it was admitted that the badgers did suffer as a result. For the rest of the decade the badgers continued to petition the authorities but to no avail and so they turned once more to violence. The outbreak in 1592 was better organized and barges were forced to stop using the river while appeals to the authorities to restore the navigation were made. Once more the verdict was in the bargemen's favour.

During both outbreaks of rioting Waltham Lock was a prime target. In May 1581 an approach was made to an employee at Cheshunt Mill for the loan of a handsaw with which to damage the lock but come July the less energetic course of setting the lock on fire was preferred. William Shanbrooke journeyed to London to purchase 'Rosseyn and brymston' for a groat. The lock was eventually fired in August but the damage was not as severe as hoped. One local inhabitant, Christopher Pennyfather, told bargemen who were using the lock several days later that he wished there had been a barrel of gunpowder in the lock when it had been set on fire.[18]

In 1592, however, the lock was completely destroyed.[19] In June of that year Edward Denny ordered his servants to dismantle the lock, to block up the new cut and to lower Waltham High Bridge to the height it had been prior to 1576. Denny argued that the Commissioners had made their alterations whilst he had still been a minor and therefore it could not be assumed that he had given his permission to changes affecting his freehold property and traditional rights. There were, however, veiled accusations that he had received money from the badgers to encourage him to take the steps he did. Whatever the case, however, his orders were quickly carried out.

Deprived of access to the new route the bargemen tried to take advantage of the provisions made by the Commissioners for just such a contingency. When, however, they tried to pull up the loweshare and proceed down the old river channel they met with violent opposition from large gangs of badgers and other local inhabitants who had gathered on the banks to thwart the bargemen's efforts. Throughout the summer, autumn and early winter, the violence continued. Boats were damaged, one was even sunk, and men on both sides suffered injuries. Eventually the bargemen found it impossible to continue, so once more they had to appeal to the central authorities for help.

In their case before the Star Chamber, however, the bargemen made no complaint whatsoever about the destruction of the pound lock; they concentrated solely on establishing their rights to use the original river channel. Indeed the destruction of the lock is only mentioned in passing by one of Edward Denny's servants under cross-examination. This suggests either that the bargemen felt their case to be stronger if they accepted Denny's arguments about his freehold property and concentrated on their own traditional rights, or that they were dissatisfied with the workings of the lock. Unfortunately there is no evidence to establish conclusively which was the case.

Though the bargemen succeeded in reopening the navigation the pound lock was never rebuilt. This was because the navigation reverted to the traditional route and the new cut in which the pound lock had stood was no longer used. The remaining evidence is incomplete[20] and on some points contradictory, but what does emerge is that soon after the end of the Star Chamber case some form of compromise was effected between Denny and the bargemen.

The evidence definitely establishes that Edward Denny built a turnpike, in fact a flash lock, at which he collected a toll of five shillings from each passing barge. This turnpike was still in existence in the eighteenth century despite many complaints from the bargemen that the lords of the manor of Waltham had no right to collect such a toll. The position of the eighteenth century turnpike on the map (Figure 2) is taken from a contemporary survey[21] but the turnpike which Denny built must have been roughly the same position, as must also the flash lock which was replaced by a loweshare in 1576–7. In evidence before a 1682 Commission of Sewers[22] certain local inhabitants claimed that Denny had built a new cut of over a mile in length in which the turnpike was sited but this seems very unlikely. It is much more probable that the old river channel was scoured and cleaned and that local memory had confused the measures taken by the Commissioners of Sewers in 1576–7 with those taken by Denny after 1595.

The confused evidence given to the 1682 Commission also mentions a 'Longe Poole very neere the said Corne Mills' through which barges had passed prior to the making of Denny's turnpike. P.J. Huggins[23] has already suggested that this 'Longe Poole' must have been the same as the cut built by the Tudor Commissioners and this is confirmed by measurements taken in the area. The new cut was said to have been 36 poles in length,[24] which would have made it just under 200 yards if it is assumed that 1 pole equals 16½ feet. This approximates to the distance along the 'Longe Poole' as it is shown on the map in Figure 2. Any cut between the mill stream and the river beginning at a point any further up the mill stream would have been some 40–50 yards longer. It thus seems indisputable that the pound lock was situated at the place marked on the map.

There is no trace of this 'Longe Poole' today and it seems unlikely that

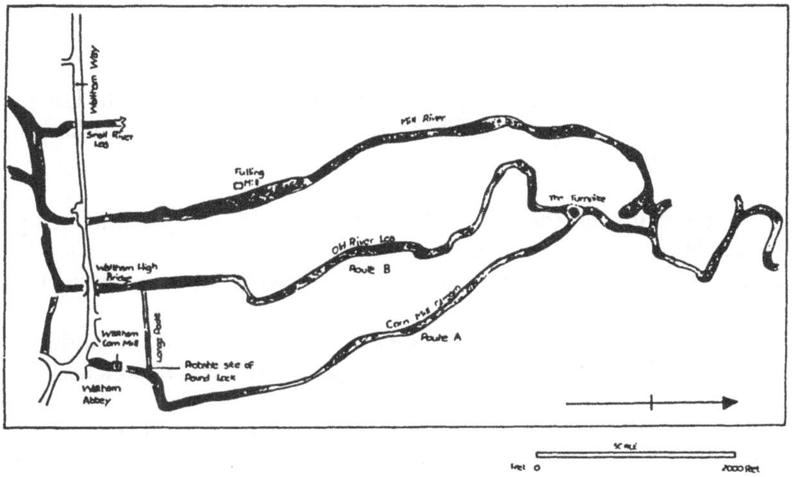

Figure 2. Branches of the River Lee near Waltham Abbey showing the site of the pound lock and of the turnpike. From a contemporary survey.

any remains of the lock would be found even if the site were excavated. However there is evidence that barges could still use the 'Longe Poole' until well into the eighteenth century. Notes in the margin of the 1682 Commission of Sewers' records mention that barges used the route within living memory though they had to carry much lighter loads than if they had used the turnpike.

Appendix

The estimate of costs to construct the pound lock at Waltham is taken from Public Record Office, State Papers Elizabeth, Domestic Series, Vol. 109 no. 133. A modern English version follows the original.

26 November 1576

An estimate of the Charges of the newe Locke
that shalbe made at Waltham yf yt be all of
tymber as followeth, and beinge in leyngthe
lxx foote and xxiiij foote in breathe

Imprimus xxviij gysses to lye in the bottom, under the plankes, every gysse being ix enches square and xxv foote longe so every gysse will Conteyne xvj foote, one quarter of a foote whiche will amownte to ix Lodes of square tymber.

Itm for the bottom xxiiij[ti] plankes, every planke Conteyninge lxx foote Longe, and one foote brode + iij enches thicke, so every planke will

Conteyne xvij foote &, whiche will amownte to viij Lodes xx foote of square tymber.

Itm ther, moste be ij grounsylles, every grounsyll of lxx foote Longe and one foote square, which will Conteyne vij^{xx} foote of square tymber, that is ij Lode, forty foote.

Itm for every syde ther moste be xxviij studdes, every studd beinge xij enches brode and viij enches thicke and x foote heygh, so every studd will Conteyne vii foote in square tymber, which will amownte to viij Lodes.

Item ther moste be ij peces of tymber to Lay upon the sayd studdes, every pece of lxx foote longe, and one foote square, which is ii Lodes xl foote.

Itm every syde will aske x plankes every planke beinge lxx foote Longe, one foote brode, and iij enches thicke, so every planke will Conteyne xvij foote &, whiche will amownte to, for bothe sydes, vij Lodes of tymber.

Itm the iiij gates wth the postes to hange them upon, and Certeyne Lande keyes for the same, will aske vi Lodes of tymber.

Suma of the Lodes of tymber xliiij } xxij^{li}
every Lode Rated at x^{s} the lode }

Itm the sawinge worke of all the foresayd tymber will amownte to lvj hundred at xx^{d} the hundred iiij^{li} xviij^{s} iiij^{d}

Itm the Carpenters worke for the same Locke will Coste xx markes so that the Carpenter be put to no other Charge but onlye the framynge, and settynge upe of the same.

Itm yf the sydes and endes of the sayd Locke be made of stone or brycke, then the tymber y^{t} shale go to the same will amownte to xxiiij Lodes or ther abowtes.

Itm the workmanshipe of the same walle, beinge x foote heyghe, and iij foote thicke will Coste xvj^{s} a pole, so that the mason be putt to no other Charge, but only the Reysinge of the same walle, and the same worke will Conteyne by estimacon x pole, every pole being xvj foote.

Itm ther will go to the same walle viij Lodes of Lyme whiche will Coste xvj^{s} viij^{d} the Lode, whiche is vi^{li} xiij^{s} iij^{d}.

Itm ther moste be to the same worke xvj Lodes of sande at ii^{s} the Lode, xxxij^{s}

Itm the heyghtinge of the hyghe brydge at Waltham to Reyse yt ii foote hygher will Cost by estimacon iiij^{li}.

Item the Charges of the Cuttynge of the newe Cutte from the myll ryver to the ryver of Lee, which Conteynes xxxvi pole, every pole rated at xxvj^{s} viij^{d} the pole, whiche will amownte to xlviij^{li}.

if it be don with timber	Tymb xliiij lodes at x^{s}	xxii^{li}		
	Sawinge + Carpenters woorke	xviij^{li}	v^{s}	
		xl^{li}	v^{s}	
the same with stone	Tymber xxiiij lodes	xii^{li}		
	Lyme viij lodes at 16.8	vi^{li}	xiii^{s}	iiij^{d}
	Sand xvj lodes at ii^{s}		xxxii^{s}	
	The walling	viij^{li}		
	Carpenters woork			
	Stone			
		xxviij^{li} [*sic*]		
	The heightening of the bridge	iiij^{li}		
	The chardge of the Cut	xlviij^{li}		
		lij^{li}		

MODERN VERSION

26 November 1576

An estimate of the cost of the new lock that shall be made at Waltham if it be all of timber as follows, and being in length 70 feet and 24 feet in breadth

First of all 28 joists to lie in the bottom, under the planks, every joint being 9 inches square and 25 feet long so every joist will contain 16 feet and one quarter of a foot which will amount to 9 loads of square timber.

Item, for the bottom 24 planks, every plank being 70 feet long, and 1 foot broad and 3 inches thick, so every plank will contain 17+ feet, which will amount to 8 loads and 20 feet of square timber.

Item, there must be 2 ground-sills, every ground-sill of 70-foot length and 1 foot square, which will contain 140 feet of square timber, that is 2 loads and 40 feet.

Item, for every side there must be 28 studs, every stud being 12 inches broad and 8 inches thick and 10 feet high, so every stud will contain 7 feet in square timber, which will amount to 8 loads.

42 *The Waltham Pound Lock*

Item, there must be 2 pieces of timber to lay upon the said studs, every piece of 70-foot length, and 1 foot square, which is 2 loads and 40 feet.

Item, every side will require 10 planks, every plank being 70 feet long, 1 foot broad and 3 inches thick, so every plank will contain 17+ feet, which will amount to, for both sides, 7 loads of timber.

Item, the 4 gates with the posts to hang them upon, and certain 'Lande Keyes'† for the same, will require 6 loads of timber.

Total loads of timber = 44
every load rated at 10s. per load } £22

Item, the sawing work of all the aforesaid timber will amount to 56 hundred at 20d. the hundred £4 18s. 4d.‡

Item, the carpenters work for the same lock will cost 20 marks§ so that the carpenter be put to no other charge but only the framing, and setting up of the same.

Item, if the sides and ends of the said lock be made of stone or brick, then the timber that shall go to the same will amount to 24 loads or thereabouts.

Item, the workmanship of the same wall, being 10 feet high, and 3 feet thick will cost 16s. a pole, so that the mason be put to no other charge, but only the raising of the same wall, and the same work will contain by estimation 10 poles, every pole being 16 feet.

Item, there will go to the same wall 8 loads of lime which will cost 16s. 8d. the load, which is £6 13s. 4d.

Item, there must be to the same work 16 loads of sand at 2s. the load, 32s.

Item, the heightening of the high bridge at Waltham to raise it 2 feet higher will cost by estimation £4.

Item, the charges of the cutting of the new cut from the mill river to the river of Lee, which contains 36 poles, every pole rated at 26s. 8d. the pole, which will amount to £48.

		£	s.	d.
if it be done with timber	Timber, 44 loads at 10s.	22	0	0
	Sawing & carpenters' work	18	5	0
		40	5	0

the same	Timber 24 loads	12	0	0
with stone	Lime 8 loads at 16s. 8d.	6	13	4
	Sand 16 loads at 2s.		32	0
	The walling	8	0	0
	Carpenters' work			
	Stone			
		28	0	0 [*sic*]
	The heightening of the bridge	4	0	0
	The charge of the cut	48	0	0
		52	0	0

Explanation of symbols (provided by Dr Norman Smith)
* One 'lode' of timber equalled 50 cubic feet.
† 'Lande keyes' probably means timber-work anchored into the ground to brace the posts; or, conceivably, it refers to some wooden device to open and close (hence key) the gates.
‡ The amount given is in fact incorrect, it should be £4 13s. 4d.
§ One mark was equal to two-thirds of a pound sterling, i.e. 13s. 4d.

Notes

1. 13 Eliz. 1 c. 18.
2. Bodleian Library, Rawlinson MSS, Essex 11.
3. John Norden's 'Description of Middlesex', British Library, Harleian MS 570.
4. For a fuller account see K.R. Fairclough, 'A Successful Elizabethan Project: The River Lea Improvement Scheme', *Journal of Transport History*, third series, Vol. 11 no. 2 , 1990, pp. 54–65.
5. See A.W. Skempton, 'Canals and River Navigation before 1750' in Vol. 3, *A History of Technology* (ed. C. Singer *et al.*, London 1957, p. 456); Philip Chilwell De la Garde, 'Memoir of the Canal of Exeter 1563–1724' in the *Proceedings of the Institution of Civil Engineers*, Vol. IV, 1845, pp. 90–102; George Oliver, *The History of the City of Exeter*, Exeter 1861–84.
6. William Vallans, 'A Tale of Two Swannes. Wherein is comprehended the original and increase of the River Lee' [1590]; in J. Leland, *The Itinerary*, Vol. V, Oxford 1710.
7. In 1594–5 a manuscript book entitled 'Proceedings in the Star Chamber' was compiled by William Harte. It is now kept in the Guildhall Record Office. Besides giving the 1594 Star Chamber case in great detail, it also provides copies of many other ancient documents relating to the Lee. Folios 174–7 provide a record of a Commission of Sewers' meeting held in the Star Chamber on 19 November 1577. This is the only remaining evidence to the Commissioners' decisions relating to Waltham.
8. *ibid*
9. *ibid*
10. British Library, Landsdowne MS 22 no. 48. Thomas Fanshawe wrote to Lord Burghley that he had just visited Waltham 'to see whether passage might not be made from the mylne Dame thorough the towne by some of the channelles there'.
11. Public Record Office, State Papers Elizabeth, Domestic Series, Vol. 109 no. 33.
12. *The Oxford English Dictionary* mentions a statute in 35 Eliz. 1 which defined a pole as being 16½ feet. If this measurement can be applied to this particular estimate then the length of the new cut would have been 176 yards.

13. Harte, *op. cit.*

14. I would like to thank Dr Norman Smith for his helpful discussion and Dr Denis Smith for drawing figures 1 and 2.

15. British Library, Landsdowne MS 32 no. 41.

16. Public Record Office, State Papers Domestic: Addenda 1580–1625, Vol. 27 no. 6.

17. British Library, Landsdowne MS 32 no. 40, paginated 109.

18. British Library, Landsdowne MS 32 nos. 35 and 41.

19. Harte, *op. cit.* folios 1–38 deal with the Star Chamber case.

20. British Library, Landsdowne MS 77 no. 16; British Library, Add MSS 33576 fo. 63; Public Record Office, Exchequer 178/4965; and Northamptonshire Record Office, WC 244.

21. Northamptonshire Record Office, YZ 6097.

22. Northamptonshire Record Office, WC 244.

23. P.J. Huggins, 'Excavations of a Medieval Bridge at Waltham Abbey, Essex in 1968', *Mediaeval Archaeology*, XIV 1970, pp. 126–47.

24. Public Record Office, State Papers Elizabeth, Domestic Series, Vol. 109 no. 33.

3
Rivers and canals
Charles Hadfield

Inland navigations engaged Smeaton's attention during practically the whole of his career and were a major preoccupation of the years 1760–73. The Calder & Hebble Navigation was his second big civil engineering job, following on from the Eddystone Lighthouse; three other navigations were constructed or improved in accordance with his designs and for 6 years he devoted much time to planning and building the Forth & Clyde Canal, the largest of all his works. In addition he reported on five other canal schemes and at least as many river navigations.

Calder & Hebble Navigation
When in the autumn of 1756 a group of men met at the "Talbot", Halifax, to revive the idea of making navigable the Upper Calder River[1] from Wakefield (where it would connect with the existing Aire & Calder Navigation) upwards to the neighbourhood of Halifax, their minds turned to Smeaton, the local man. Probably some present knew him, for his home at Austhorpe Lodge was not far away. He was then too busy upon the Eddystone, but when in June 1757 they asked again, he agreed to come, and made his survey in October and November.

By that time many English rivers had been improved by building locks, usually at situations where short side-cuts could carry the waterway past water-driven mills, and by cutting off the bigger meanders: among them the Wey, Warwickshire Avon and Welland in the seventeenth century, and in the eighteenth the Mersey & Irwell, Aire & Calder and Don. Thomas Steers had been concerned with the Mersey & Irwell, and when Smeaton arrived in the Calder Valley,

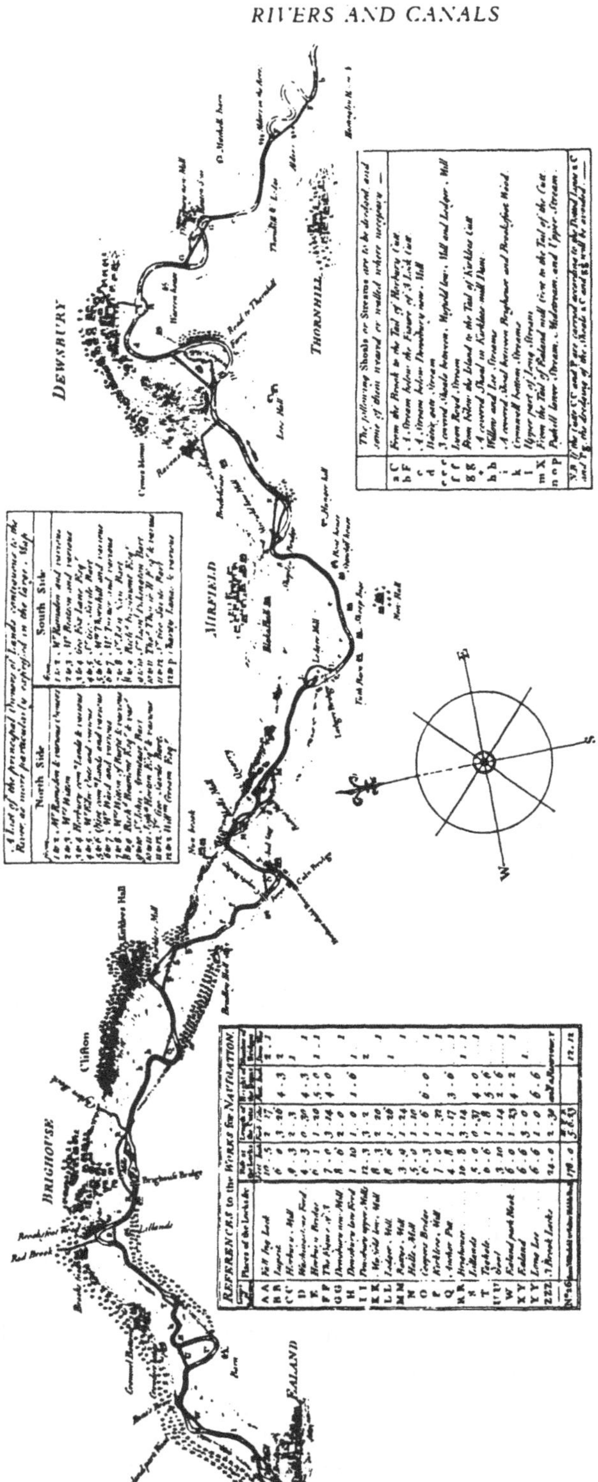

Fig. 20. Part of Smeaton's plan for the Calder & Hebble Navigation, 1757 (engraving by R. W. Seale)

surveys by John Eyes and Steers in 1740 and 1741 were made available to him.

Smeaton faced a more difficult problem than had his predecessors. They had all, like him, to avoid or placate millowners whose livelihood depended upon the maintenance of a head of water while that of boatmen consisted in depleting it, but none before—or after—faced such a steep declivity (178 ft fall in 24 miles) or the threat of floodwater off the Pennine slopes.

His plan was to dredge shoals, construct 26 locks and make 5¾ miles of artificial cut between Wakefield and Brooksmouth (Fig. 20) and then for half a mile up the Hebble valley to Salterhebble near Halifax, where a reservoir would be built. Such a navigation would take 20–25 ton boats drawing up to 3 ft 6 in. (such as could navigate the Aire & Calder also in most seasons) and not harm the thirteen mills concerned "either by Back Water, or Loss of Water, or the Meadows adjoining by occasioning or continuing Floods thereon sensibly longer than usual . . .". The engineering cost would not be less than £30,000; construction time 7 years.[2]

An authorising Act, amended to extend the navigation beyond Brooksmouth to Sowerby Bridge, was obtained in June 1758. Smeaton gave evidence upon it,[3] saying he had seen locks in Flanders and Holland and that "he proposed to make the Locks better than most of the Locks he had seen on any River in England".

The Act passed, and £36,000 subscribed, Smeaton was appointed engineer at a salary of £250 a year from 25th November, 1759; this to be a part-time job but one demanding (and receiving) frequent attendance on site.[4] As an assistant or resident engineer he had Joseph Nickalls, paid 20 guineas to cover the cost of moving from London to Wakefield and an annual salary of £100 raised, in August 1760, to £120. Nickalls arrived at the end of December and work on the Wakefield–Dewsbury stretch probably began early in 1760. By the summer Smeaton had done many of the working drawings of locks, weirs (or "dams" as they were called) and bridges, of which examples still survive in the Royal Society collection,[5] see Figs 21 and 22.

Smeaton thought highly of Nickalls but for reasons that are not clear, and possibly without real justification, Nickalls was dismissed in November 1761. To take his place John Gwyn, already on site as a foreman, was appointed superintendent of carpentry, and Matthias Scott of masonry and digging; their appointments dating from March 1762, at annual salaries of £50 and £60 respectively, and they to work directly under Smeaton's direction.

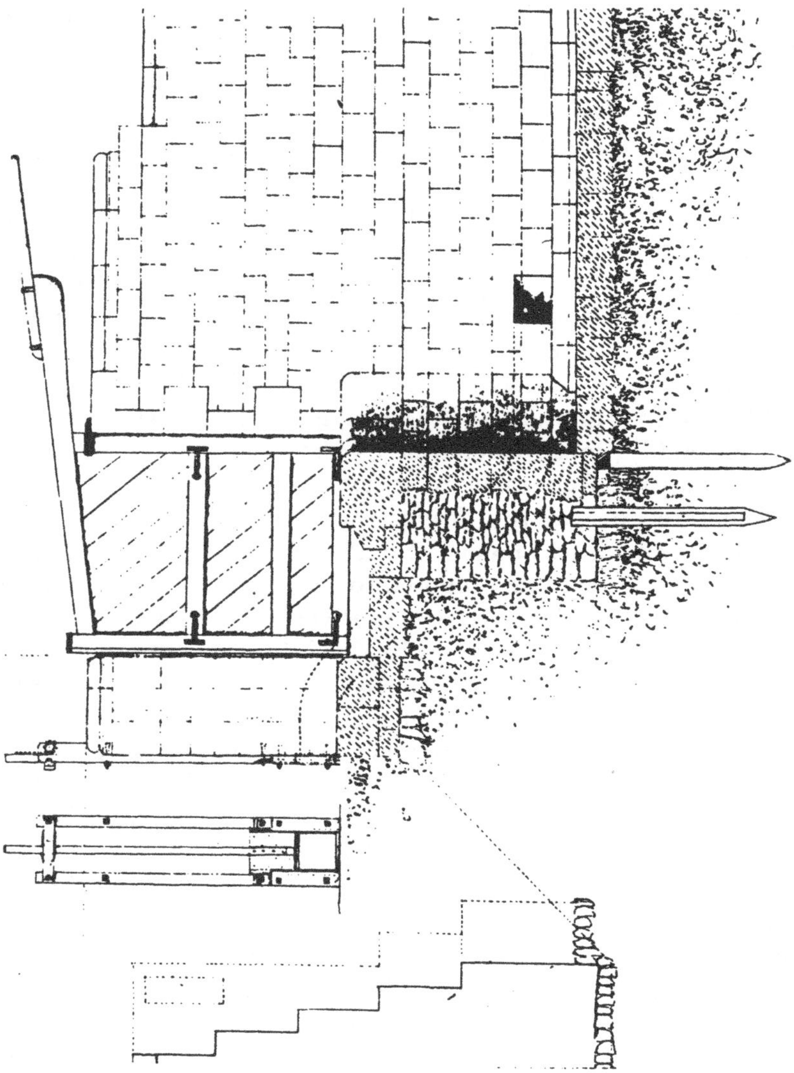

Fig. 21. Section of a lock gate on the River Calder, 1760 (Designs, **6**, *f.* 21)

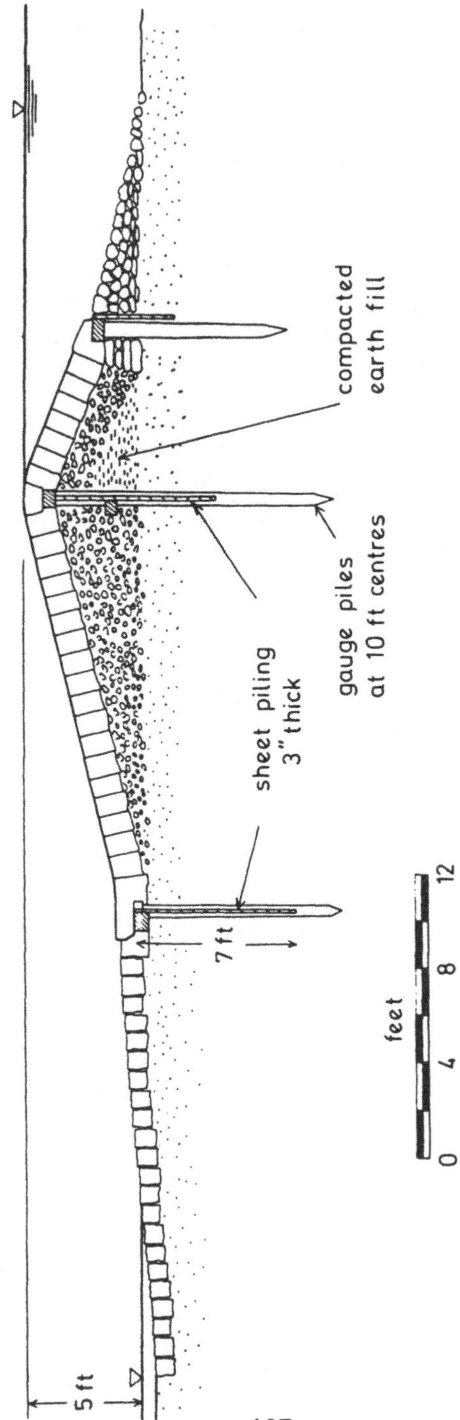

Fig. 22. *Section of dam at Coopers Bridge (Calder & Hebble Navigation) (redrawn from* Designs, **6**, *f.* 40)

By July 1764, after a total expenditure of £56,000, the navigation was open from Wakefield to Brighouse: a distance of 18 miles with seventeen locks, the locks having a length of 64 ft between gates, a width of 15 ft and an average fall of 8 ft. Not a great deal remained to be done and Smeaton hoped everything would be finished within the next year.

In September, writing to John Holmes, he remarked that:

> "the works are going on very well but we are still subject to the same rubbs, that have ever tortured the undertaking from the beginning, and will continue to the End; that is, difference of opinion, partys and disputes amongst our Rulers."[6]

A month or so later a new committee, Rochdale- rather than Halifax-based was elected. In December they told their clerk to write to Brindley and offer him the job of finishing the navigation. He was appointed at a meeting on 31st January, 1765, to which he produced a plan, and Smeaton was discharged on the same day, along with Gwyn and Scott.

Smeaton had worked hard on the Calder; but he took his dismissal well, saying:

> "I shall never envy any Man the praise for doing better than myself while I am conscious of having done as well as those that have trod the same (or perhaps less difficult) Steps before me."[7]

He was to come back.

Consulting jobs, 1758–63

Before we turn to the Forth & Clyde Canal, let us glance at other waterway work of Smeaton's in these earlier years.

John Grundy had in 1756 surveyed for a canal from Louth in Lincolnshire to Tetney Haven whereby keels from the Humber could reach the town.[8] Four years later enough support had shown itself for the town clerk to write to Smeaton, who agreed to go over Grundy's proposals. His own report of 14th July, 1761[9] broadly confirmed them, landowners' backing was forthcoming, and in 1763 Smeaton gave the Bill Parliamentary support.[10] It was passed and the canal, taking keels 72 × 15 ft, was opened in 1770. It was 11¾ miles long with eight locks, one a sea-lock having gates pointing in both directions.

In 1759 Smeaton was in Parliament to support another difficult proposed river improvement, that of the Upper Wear.[11] His scheme provided for twelve locks and several cuts between Biddick Ford near Washington and the city of Durham.[12] The Act passed, but without result. Also in about 1759 he was called to southern England to get the extension of the Wey Navigation upwards from Guildford to Godalming, surveyed by Richard Stedman, through Parliament.[13] It was authorised in 1760, but thereafter he excused himself on account of "a large affair in the North".

In Lincolnshire the Fossdyke left the Trent at Torksey and ran to Lincoln. There it joined the River Witham which, passing under the shallow-channel medieval High Bridge, ran to Boston and the Wash. Action by landowners between Lincoln and Boston produced a report in November 1761 by Grundy, Langley Edwards and Smeaton.[14] They were to cure a situation they described as follows:

> "This once so flourishing river and country have ... been falling into decay, by the banks ... being suffered to become ruinous ... so that those flood waters which were necessary and used heretofore, by their velocity and weight, to cleanse out the sand and sediment brought up by the tides, have been, and now are, suffered to ... expand over the adjoining fens and low grounds, whereby those sands ... have now so much choaked up the haven from Boston to the sea, that for several years last past the navigation thereof has been lost to shipping, and it is now become even difficult for barges of about 30 tons burthen to get up to the town on neap tides."

Their solution was a sea-lock and sluice at Boston, a new cut thence to Anthony's (now Anton's) Gowt, another to rejoin the river at Chapel Hill, and thence to dredge the river up to Lincoln and build four locks. Their estimates were £37,850 on account of drainage and £7,370 of navigation.

Smeaton and Grundy appeared on the subsequent heavily contested Bill.[15] It was passed. Over the following years the lock at Boston (the Grand Sluice) and three of the four others were built, and the long new cuts were made, with Edwards as engineer in charge. As soon as the Act had passed, Smeaton and Grundy reported on the Fossdyke, proposing to improve both drainage and navigation, mainly by dredging and embanking, for £3,817.[16] The Fossdyke lessees seem to have done little about it. In 1782 he returned to make another report, whose main purport was to improve land drainage.[17]

Smeaton reported upon another river in 1762: the Chelmer.[18] He concluded that the river could be made navigable from Chelmsford to Maldon, "yet there is so small a proportion thereof that is at present adapted to navigation, that it will prove the least expence, to perform the same chiefly by canals . . .". The navigation was eventually authorised in 1793.[19] So much for rivers; in these years Smeaton was also concerned with two schemes that were to grow into major canals.

In 1758 James Brindley, not yet employed by the Duke of Bridgewater, had been commissioned to survey for a canal upwards from Wilden Ferry (Cavendish Bridge near Shardlow) on the Trent to Stoke. In the same year Henry Bradford had proposed to make the Trent and Tame navigable upwards from the present head of the Trent Navigation at Burton to Tamworth, and Smeaton had worked over it, estimating for eight locks and a staunch.[20]

Brindley's reply was a scheme for a canal from Wilden Ferry to King's Bromley Common near Lichfield (roughly, the present Fradley), whence he proposed branches to Longbridge near Burslem (an arm would run to Newcastle-under-Lyme), Lichfield, and Fazeley near Tamworth, in all over 71 miles of canal. Smeaton, called in on this scheme also, went over the lines in November 1760 with Hugh Henshall and Brindley, the latter now in the Duke's employment, and approved them with emendations:[21] all became the bases of later canal building. Brindley had also envisaged an extension past Longbridge to a junction with the Mersey. Of interest to us is Smeaton's comment, that the Longbridge branch could indeed be continued over the watershed by lockage and "a deep cut through the summit, which will be of moderate length". Water could be provided, he thought, by colliery drainage, a reservoir, and if necessary "the help of a fire engine, to return the waters from a low level . . .". He therefore did not consider a tunnel, such as Brindley afterwards built at Harecastle, necessary.

Early in January 1762 Smeaton met Brindley again, this time to measure water that could be made available from the colliery soughs at Worsley to supply the Duke of Bridgewater's proposed extension to Runcorn. Given the heavy opposition that was expected from the Mersey & Irwell Company, it was essential to prove that they would in no way be deprived. Later that month Smeaton appeared before the Commons Committee on the Bill, to prove adequacy of supply. Brindley's evidence followed, the Bill passed, and the canal extension was built.[22]

It may have been on one of these jobs that Smeaton and Brindley made their well-known calibration of a water gauge for measuring the flow of water in a stream. This took the form of a rectangular notch cut in a long board 1 in. thick placed on top of a temporary dam.

RIVERS AND CANALS

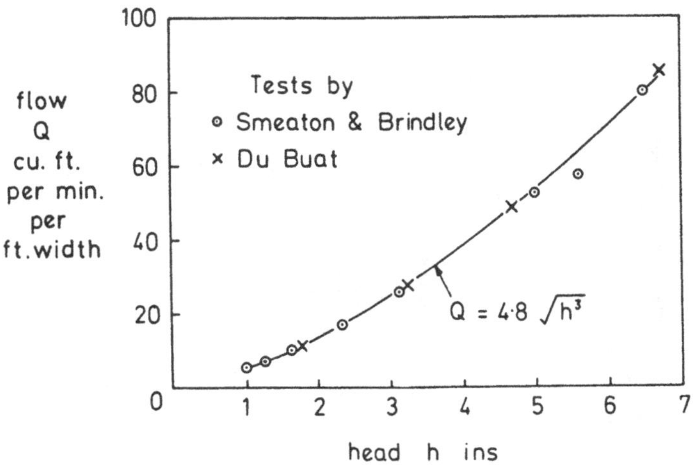

Fig. 23. Calibration of rectangular notch gauge

They measured the time required for 20 cu. ft of water to flow through the notch at various heads from 1 to 6½ in. The results (Fig. 23), together with some later tests by Du Buat, were still being quoted as providing the most reliable data available in the early nineteenth century.[23] It is also worth noting that in his 1762 evidence Smeaton gives a sensible figure for the loss by evaporation: 15 in. of water in the 5 months May–September.

Forth & Clyde Canal

And so to the "Great Canal" in Scotland. Earlier, Smeaton had been concerned with the navigation of the Clyde itself, reporting in 1755 and 1758; in the second report proposing dredging, and a river lock and dam below Whiteinch, downriver from Glasgow.[24] What he proposed resulted in the 1759 Clyde Act, to improve the navigation up to Glasgow, and build a bridge, for he appeared in Parliament upon the subject.[25] Eventually, deepening alone was decided upon, without a lock; the work was carried out successfully under John Golborne's direction.

When, in the early 1760s, interest revived in building a canal across the waist of Scotland, Smeaton was the obvious man to consult. Such a canal had been suggested earlier. Alexander Gordon did a survey in 1726 and later, in 1762, Robert Mackell and James Murray were commissioned to survey a line from the Carron River at Abbotshaugh to the Clyde at Yoker Burn below Glasgow.[26] Their report was favour-

RIVERS AND CANALS

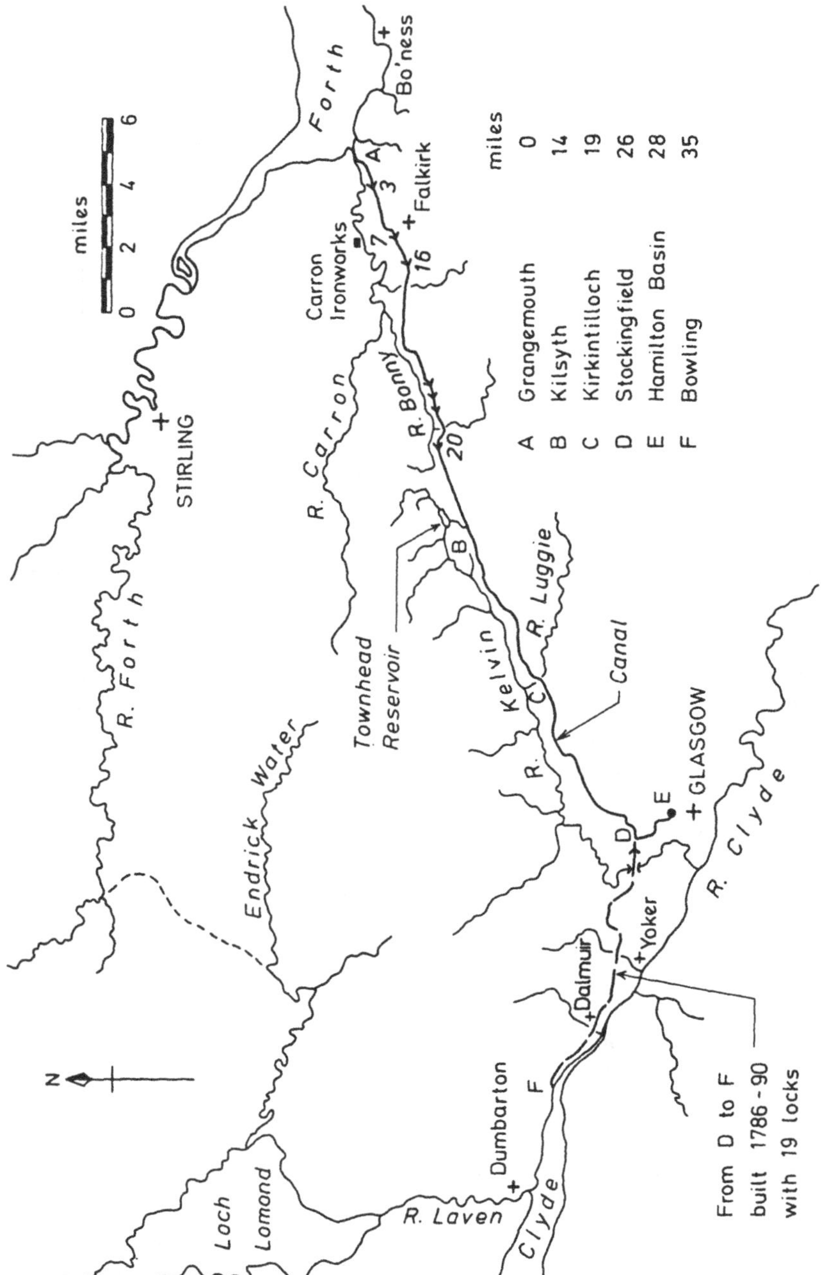

Fig. 24. Forth & Clyde Canal, 1777

able, whereupon in 1763 the Board of Trustees for the Encouragement of Fisheries, Manufacturers and Improvements in Scotland asked Smeaton to do another study. He said later[27] that the request was verbal, "to give them a design for the sort of canal, that in my judgement would best suit the country, when taken in every point of view". His long and careful report was dated 1st March, 1764, with a postscript of 22nd December.[28]

There were then many canalised rivers in the British Isles, some incorporating considerable lengths of artificial cut, but few canals. In Devon, the 4-mile long Exeter Canal with one pound-lock, which Smeaton had seen[29]; in Lancashire, the 10 miles (with branches) of the Sankey Brook from the Mersey to St Helens, with ten locks, and the lockless 7¾ miles of the Bridgewater Canal from Worsley to Longford Bridge, which the Duke was beginning to extend for another 25¾ miles to the Mersey; its ten locks at Runcorn would not be begun for some years yet. In Ireland the Newry Canal with fourteen locks had been completed in 1742 from Newry to the Upper Bann, and the complementary Lagan Navigation had been begun from Belfast towards Lough Neagh. None of these was a watershed canal, i.e., one climbing up a river valley to pass a summit and run down another, with the attendant problem of supplying the summit with water. One such canal had, however, been begun in 1756: the Grand from Dublin to the Shannon; by 1764 none of it had been opened, though much money had been spent on scattered work by the engineer Thomas Omer. Later, Smeaton was to be called to advise upon its future.

There was no precedent, therefore, within the British Isles for what Smeaton was to propose. However, in the south of France ran Riquet's great seventeenth-century achievement, the Grand Canal of Languedoc, now the Midi Canal. This was a true and very impressive watershed canal, complete with an elaborate reservoir and feeder system to supply the summit level, and at Malpas the only canal tunnel in Europe.[30] It had not been a financial success, however, so as an appendix to his report Smeaton set out the differences, ending his comparison with: ". . . the inference is very plain, that the same tolls which will hardly keep the French canal in repair, will make this a very beneficial undertaking to the British adventurers".

What, then, did Smeaton propose? There were two possible routes. The first ran up to the Forth to beyond Stirling, cut overland past the Bog of Bollat to the Endrick River, and thence by Loch Lomond and the Leven River to the Clyde at Dumbarton: a circuitous line involving difficult land cutting and rivers awkward to make navigable. The second, which he chose, used the Carron River and then the valley of the Bonny to Dullatur Bog, his summit level. Thence it ran down the Kelvin Valley to the Clyde at Yoker Burn below Glasgow (Fig. 24).

Such a canal would be 28¼ miles long, 40 ft 8 in. at surface and wider where possible to allow passing places, and 5 ft deep, with 77 (amended to 72) locks to surmount a rise and fall of about 147 ft, a short tunnel, two considerable and one smaller aqueducts, some sizeable embanking, and a towpath throughout. It would take existing Clyde river craft, 17½ ft wide, 56 ft long, drawing 4ft and carrying 40 tons or more. The cost he estimated at some £74,000. If, however, it were to be of reduced dimensions to take boats 9½ ft wide, the cost would fall to about £51,000.

Smeaton paid particular attention to the problem of supplying the summit with water, for his supplies had to provide lockage in both directions. He satisfied himself that there was ample water, by diverting streams near Kilsyth into the canal, for a much greater trade than was envisaged, without affecting local mills, but safeguarded supplies further by providing that locks should not exceed 4 ft fall, and that the summit level pound should be cut 1 ft deeper than the rest to provide a water reserve. He noted that should water supplies turn out to be more abundant, or traffic less than he had estimated, the falls of the locks could be increased and their number reduced, or even halved.

Mine adits, and indeed, at Worsley, small-bore canals tunnelled into mines, had been built, but Smeaton's was to be the first large-bore tunnel after Malpas. His paragraph, proposing to make it by cut and cover, is worth quoting:

> "Where the depth exceeds fifteen feet, which will not be above two hundred yards, I would propose the passage to be by a vault under ground; and though the matter of the hill seems to be a loose gravel, yet I apprehend the means are not very difficult by which it might be perforated. I would begin at an end and open the ground from the top, the same width I intended it at bottom, that is, eighteen feet, like the locks, with an allowance for the thickness of the walls; and cutting down the sides perpendicular, would shore the same from side to side with boards, beams, and braces, of ordinary Scotch fir, these keeping up the matter till the arch is built, the shores that are above the arch may be taken away, and replaced as the work advances, and the cross braces that interfere with the walls of the arch may be walled in, and afterwards cut out when the whole is completed."

Here, then, was a comprehensive plan for Britain's first watershed canal.

Smeaton's report caused years of argument as competing interests supported or criticised it. Chief among these were the Glasgow merchants, who wanted the canal nearer their city, and the Carron Company, who wanted the other end near their ironworks.

After a new survey by Mackell and James Watt on behalf of the Glaswegians and Carron Company, which brought the canal out by the Broomielaw, close to Glasgow, and near the ironworks at the other end, a Bill was introduced in March 1767 for a small boat canal 24 ft wide and 4 ft deep, a branch from Carronshore to Bo'ness being then incorporated. At that point interests based on the Forth area and Midlothian began a powerful opposition which was supported up and down Scotland's east coast. A rival subscription for "the great Canal, planned by Mr Smeaton",[31] that would take seagoing ships, was begun in London. Petitions in its favour poured in to Parliament. On 29th April the small canal's supporters fought off an attempt to kill the Bill by 201 votes to 141, but the great canal's supporters then stated they had raised £100,000 and promised that "the Canal shall be at least as deep and as wide as that proposed by Mr Smeaton" with cuts to Glasgow and Bo'ness. The Bill was thereupon defeated on third reading and withdrawn. The subscribers, with the Duke of Queensberry in the chair, then commissioned Smeaton in May 1767 to re-plan for a canal able to take craft 60 ft long and 20 ft wide, with enough water to carry 100,000 tons of traffic, two-thirds of which would pass from west to east. The Clyde junction was now to be further down river, near Dalmuir Burn-foot, the eastern end near the mouth of the Carron and below the works, and there would be a branch to Glasgow.

Smeaton now plans, in his second report,[32] a canal some 37 miles long having 42 locks, 70 ft long and 20½ ft wide. He reckons upon needing nearly three times as much water for the enlarged canal as before; therefore he lists the available streams and their annual supply, but says also that two reservoirs could be created: one at Townhead and the other by widening the canal where it crosses Dullatur Bog (west of lock No. 20). He concludes that there will be ample water to protect the supplies of mills and works.

Careful also to be informed of what difficulties lay ahead, he says:

> "the ground has been carefully examined by boring, by Mr Mackell . . . who has delivered to me an account of the qualities of the soil for twelve feet deep, in no less than fifty-eight different divisions; together with many other measures of the hills and hollows through which it has been designed to pass . . .".

The result is largely to confirm his 1764 route, except at the two ends.

RIVERS AND CANALS

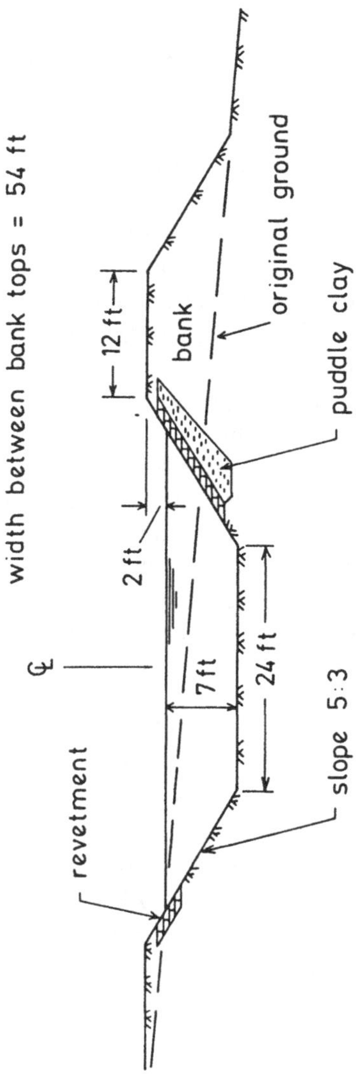

Fig. 25. Section of Forth & Clyde Canal for inclined ground, 1768

There he accepts the terminations suggested to him, but adds a second at the eastern end, at Grange Burn-foot (later Grangemouth). He now needs an extra aqueduct, but the tunnel has gone. His engineering estimate is £147,340 for a canal 7 ft deep, exclusive of parliamentary, legal and other charges.

So, as he proposed, it was enacted on 8th March, 1768, Smeaton giving evidence on the Bill.[33] It was the world's first sea-to-sea canal. On 14th March he submitted to the General Meeting of the Proprietors, at which Queensberry was elected chairman,[34] a scheme of management of the engineering work (see Chapter X). This was adopted so far as circumstances permitted and, it is interesting to note, introduced the term "resident engineer". At the same meeting Smeaton was appointed "Head Engineer" or "Engineer in Chief" at £500 a year, a most generous salary though it had to include his not inconsiderable travelling expenses. Mackell became "Sub-Engineer" or resident engineer (Smeaton's terminology rapidly came into use) at £315 a year, and on the administrative side Alexander Hart was made agent in Scotland at £200 a year. Shortly afterwards Alexander Stephen joined the team as "Clerk & Overseer", under Mackell, at £1 a week (raised in 1771 to £70 a year). In addition the land surveyor John Laurie was employed on an *ad hoc* basis, and in November 1770, with work well under way, Archibald Hope came in as "Comptroller & Auditor" at £60 a year. These men, with the active and intelligent support of a small management committee, were to direct a major part of one of the largest civil engineering projects in the eighteenth century.

Mackell and Laurie began detailed surveying in April. By the end of May Smeaton had staked out the eastern end of the canal line. He also submitted nine typical cross-sections of the canal for level and sloping ground (Fig. 25). In June the first excavation contract was let to John Clegg; the second, to Edward Smith, followed in July. William Charnley, a mason from the Calder sent to Scotland by Smeaton, began opening and working quarries; materials, including timber and pozzolana, were arriving. On site again in September and early October, Smeaton staked out a further length and found "every thing going on as well as can possibly be expected, in the infant state of the works". Four contractors were now engaged on spadework.

On the day before leaving Edinburgh, at a committee meeting, Smeaton was handed copies of reports by Brindley, Yeoman and Golborne.[35] Seemingly at the instance of the Carron Company, who considered themselves bypassed, those three engineers were commissioned to re-survey for a small canal; Brindley was probably there because Samuel Garbett and Charles Gascoigne of Carron Company and the Carron Shipping Co were also committeemen of the Birmingham Canal, whose Act had been passed a few days before that for the Forth & Clyde, and whose engineer Brindley was. Their reports, which *inter alia* proposed a narrow-boat, 4 ft deep canal and a re-routing nearer Glasgow, annoyed the canal company, who regarded the matter as settled by their Act, and were answered by Smeaton in a review on 28th October.[36] We need merely note one or two points. First, his own position, independent of any special interest:

> "whatever character my brethren may be ambitious of acting, I leave that to them; for my own part, I shall take this opportunity . . . to declare, that I do not act in consequence of any public authority or commission, to enquire, determine, and declare what is best for the Public, or what in its consequences will assuredly promote its good . . . I consider myself in no other light than as a private artist who works for hire, for those who are pleased to employ me, and those whom I can conveniently and consistently serve."

Therefore, he says, he could not have objected if the triumvirate had been sent for to examine his 1767 report; but: "to send for engineers to determine, whether the law itself be rightly founded . . . is . . . a sort of proceeding which, I think, every well-wisher to the success of the scheme should set his face against."

Smeaton respects Yeoman and Golborne, but by now he has had enough of Brindley, maybe for having said that he could build his proposed canal in four years:

> "As no difficulty is too great for Mr Brindley, I should be glad to see how he would stow a fire engine cylinder cast at Carron, of 6½ ft diameter, in one of his seven feet boats, so as to prevent its breaking the back of the boat, or oversetting . . . Mr Brindley recommends to begin at the point of partition (the watershed) because, he says, 'it is his constant prectice to do so . . .' but pray, Mr Brindley, is there no way to do a thing right but the way you do? . . . do you usually begin at the most difficult part of a work first, with raw hands, before they are trained to business? I have sometimes done so, and repented it."

Though we may agree with Smeaton that a narrow-boat canal was nonsense, and that 4 years was a foolish boast, Brindley was in fact right on two points: first, in suggesting that the River Carron should be straightened between the mouth of the Grange Burn and the Forth (this was done in 1783–85); and second, in recommending a change of route to take the line nearer Glasgow. Smeaton had rejected this, for it involved difficult ground, a substantial cutting, and a very large aqueduct over the Kelvin. Smeaton suspects that Brindley has only suggested the route in order to build the aqueduct:

> "That my brother Brindley should prefer the Printfield passage, I can readily comprehend . . . every man, how great soever his genius, has a certain hobby-horse he likes to ride; a large aqueduct bridge over a large river does not happen to be mine."

This was probably a reference to Brindley's proposed aqueduct over the Mersey. However, in 1770 Mackell was to suggest a similar diversion which, after re-appraisal, Smeaton approved.

For the time being this affair caused only a minor interruption. By the end of 1768 Smeaton had prepared drawings for swing and drawbridges, inverted syphons, etc. and for three typical locks. Work started in March 1769 on the first "land" lock (No. 3) using direct labour, not contractors. John Gwyn came down from Perth to superintend the carpentry and Charnley supervised the masonry. On site again in July, Smeaton reported in detail on water supply to the summit. By November work was in hand on Lock No. 5 and the first of the small aqueducts had been built; also, Smeaton decided on the layout of the flight of ten locks (Nos. 7–16) at Camelon just to the west of Falkirk. The contract for the first two of these went in February 1770 to William Gibb, the founder of the famous civil engineering family. In the previous month Charnley contracted for locks 17–19. A little later that year John Laurie surveyed for the reservoir at Townhead; by May rather more than 1,000 men were employed, including those in the quarries, and the first 9 miles of the canal had been cut.

It was in June 1770 that Mackell submitted his report on a new line for the canal west of the summit at Kirkintilloch, Laurie having taken the levels. This, as realised by Brindley, had the advantage of passing a good deal nearer to Glasgow than in Smeaton's plan and, moreover, would go on a constant level from the east summit lock No. 20 to Stockingfield. On the other hand an aqueduct of unprecedented dimensions would be necessary to take the canal over the Kelvin on its way to the Clyde. Partially to offset this, however, the 15-mile summit level would itself act as a reservoir and thereby simplify the water supply problems; Townhead reservoir alone probably being a sufficient supplement. Mackell considered that, on balance, there might be a saving in cost.

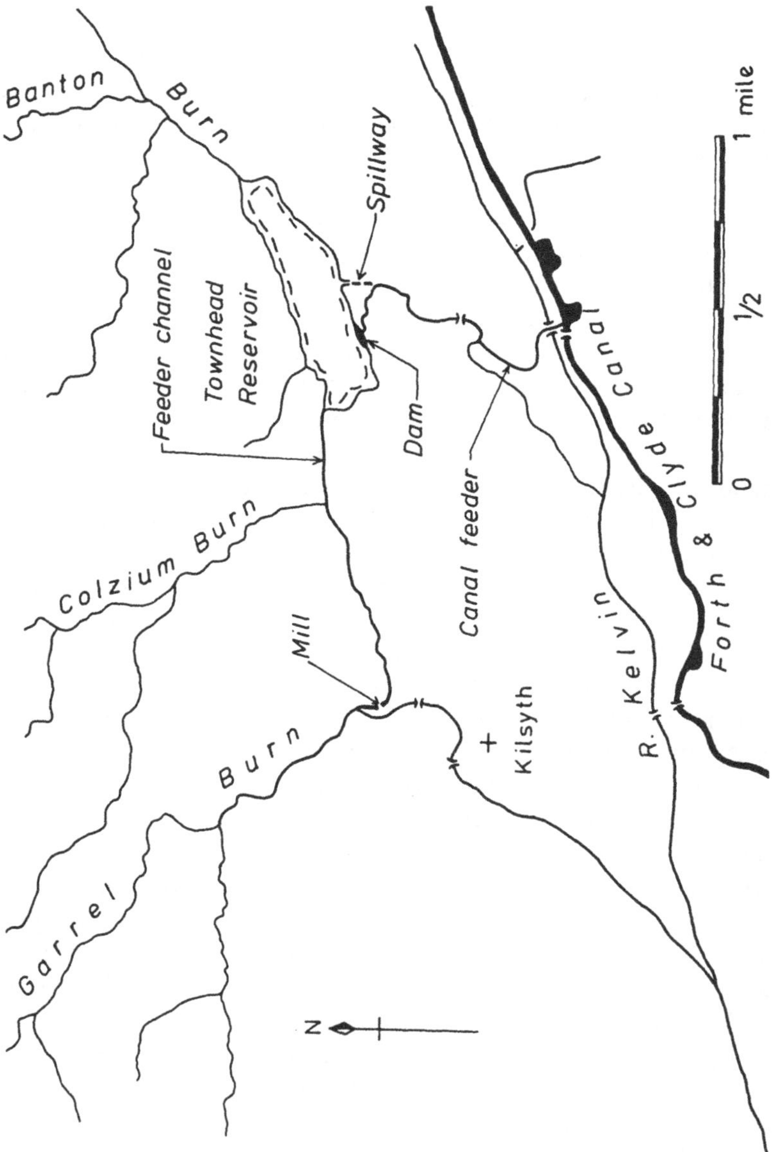

Fig. 26. Townhead reservoir, 1773

RIVERS AND CANALS

During his visit in July and August Smeaton gave careful consideration to this proposal. As to the cost, he showed that Mackell had been too optimistic, the new line would in fact be appreciably more expensive than the original plan; but the undoubted advantages outweighed the extra cost, and he gave the scheme his support, both in committee and, next year, in Parliament[37] where the necessary bill for re-alignment was passed in March 1771. Concluding his long report, dated 6th August, 1770, Smeaton says:

> "In the whole of this Business Mr Mackell has shown a great deal of Judgement & attention to the Interest of the Company. And I have the further pleasure to acquaint the Company that having carefully inspected the Works now going on, that they are in general, and in things the most material, exceedingly well executed and much to my satisfaction."[38]

Clearly, in Robert Mackell, Smeaton had a treasure of a resident engineer, and Mackell could scarcely have had a better or more understanding chief.

We may now consider in some detail the Townhead reservoir, that vital point of the scheme later described as the "sheet anchor of the navigation". An exceptionally favourable topographic situation had been found (see Smeaton's 1767 report) where the construction of an earth-fill dam, with a maximum height of about 25 ft and a crest length of less than 100 yd, would impound a reservoir covering 54 acres and containing 26 million cu. ft of water equivalent to 2,245 lockfulls (Fig. 26). From the sluice in the dam, water would be taken to the canal partly in an existing stream bed and partly in a new channel passing over the Kelvin in a small aqueduct. There was also a convenient place for the spillway. Banton Burn fed directly into the reservoir, but was not by itself sufficient. Colzium Burn, in its natural state, turned west and fell into Garrel Burn just below the tailrace of Kilsyth mill. To increase the supply of water to the reservoir, therefore, the gradient of this part of Colzium Burn was reversed, the tailrace was diverted into it, and a new feeder channel cut to the reservoir. By this ingenious system use was made of the entire flow of two streams and a good part of a third, without in any way reducing the power of the mill.

Cutting the channel to the reservoir began in May 1770; by August 1772 the dam was finished and the canal feeder in "great forwardness", and in March 1773 the reservoir had nearly been filled. A slip in the feeder embankment caused some delay, but all was ready by the summer of that year.

Meanwhile, in 1770 and 1772, Smeaton designed the aqueducts over the road at Camelon and over Luggie Water; and, after borings some 27 or 28 ft deep had been made, submitted plans for the two sea locks (Nos. 1 and 2) at Grangemouth. The contract for these went to John Easton in February 1771. William Gibb finished the Camelon aqueduct—one of Smeaton's more "architectural" designs—later that year. By Christmas 1771 the engineering expenditure amounted to £72,000 and, reporting this fact in July 1772, Smeaton could record that very good progress had been made along the whole line to Kilsyth. Finally, on 3rd September, 1773, the canal was opened from the Forth right through to Kirkintilloch, Gibb having completed the Luggie aqueduct there.

In this distance of 19 miles there were twenty locks accommodating a rise of 155 ft to the summit, the locks having a length of 74 ft between gates and a clear width of 20 ft; there were two sizeable aqueducts, numerous small ones, and draw and swing-bridges, as well as Townhead reservoir.

On several occasions Smeaton had suggested to the Company that his salary might be reduced, or even that he might resign, to reduce expenditure and as he had full confidence in Mackell. Each time his offer was very politely refused. However, in February 1773 he wrote again, saying that now the canal had been virtually completed to the summit level, and Townhead reservoir would shortly come into operation:

> "and being very desirous of relieving the Company from all unnecessary Expenses, I beg leave to aquaint the Committee that the 14th of March coming will compleat 5 Years since I had the Honour of entering this Service and that I am ready and desirous of laying down my Salary at that time, and shall be ready to give my attendance so far as it shall be wanted . . . upon my usual terms."[38]

The Committee decided, however, to continue his salary to September, and so in that month Smeaton finally left his great canal, with the warmest thanks from the Company for the very masterly manner in which he had planned and conducted his works. He kept in touch by letter for several years. His letter books of letters sent and received (which probably include his site and progress reports), running from March 1768 to April 1779, covered 770 pages (see Chapter XI). He had during the course of the work submitted at least 50 plans and working drawings,[39] and attended more than 20 committee meetings in London and Edinburgh, as well as giving evidence in Parliament on two Bills. It only remains to add that Mackell finished the canal to Stockingfield in November 1775 and to Hamilton Basin at Glasgow exactly 2 years later.[40] By that time the total expenditure had reached £164,000 and the Company could do no more for the present.

Mackell died in 1779. Asked in 1785 to resume operations, Smeaton replied: "it will be better to make a fresh beginning with a fresh engineer".[41] Robert Whitworth was accordingly appointed, and it was he who completed the line from Stockingfield to the Firth of Clyde, with the Kelvin aqueduct as his masterpiece, during the years 1786 to 1790. The canal then had a length of 37 miles, including the Glasgow branch, and the total expenditure came to £305,000.

River navigations 1766–74

Let us now return to the Upper Calder, where Brindley had taken over early in 1765, by the end of which year the navigation was probably open to Brooksmouth near Halifax. By June 1766 he seems to have left—his departure is unrecorded—for then he was engineer to the newly authorised Trent & Mersey and Staffs & Worcs Canals. The company's clerk, Thomas Simpson, became superintendent, and took the navigation to Salterhebble, probably during 1767.

Then disaster came. Early in October 1767 a flood damaged the works. The subscribers asked Smeaton to return and, generously, he did, reporting on 30th November.[42] The flood, he assured subscribers, was "higher than any flood in man's memory", higher than an extraordinary one of 40 years before. "These facts", he said, being considered, "it seems more marvelous that the works have not been wholly swept away, rather than that they have been broken in particular places." However, he allows himself to observe that, had he not been dismissed, he had intended (and had said so) to build flood-gates at the head of three cuts, and also to reinforce certain banks: "had this been put in execution, I cannot help saying, that the damages on the present occasion had laid in much less compass, though probably not entirely prevented". However, the damage has been from flooding, not from burst banks or collapsed locks: indeed only one lock had fallen, and that by external erosion. Therefore he proposes to build flood-gates (at more places than previously intended) and to raise and strengthen banks. Preventive work should be begun forthwith, the rest as soon as possible.

Emergency work did begin, but before much had been accomplished, another flood followed in February 1768, doing more damage, and yet a third in June. The concern now found itself with a badly damaged navigation, a debt of £63,000, and unable to raise more money. In 1769, therefore, an Act incorporated the Calder & Hebble Company to take over the works, convert loan stock to shares, raise more funds, and increase tolls. With Simpson back as superintendent, and Luke Holt and Robert Carr as resident engineers, work began again along the lines Smeaton had recommended, and with his flood-gates; it was completed to Sowerby Bridge, in September 1770. He does another survey in December 1770,[43] telling shareholders in January that he "finds the river now put into as good a state of security as could possibly be expected at this time, and is, indeed, in a very defensible condition", and goes on to make a number of recommendations for improvement and bank strengthening.

Thereafter, the company more than once consulted Smeaton upon further improvements to meet their quickly growing trade. One such was Smeaton's suggestion to a meeting in July 1781, when he put forward a plan[44] to replace Brindley's staircase pair of locks near Salterhebble by two single locks to save "half the lockage water". They were so replaced in 1783 with Jessop as engineer in charge.

In early 1776 N. I. Korsakov, a Russian engineer officer, was sent abroad to study canal works. The Russian Consul General, Alexander Baxter, wrote to Smeaton to ask what Korsakov and his companion should see. Smeaton replied interestingly:

> "I would advise their beginning with the Forth & Clyde Canal, for though it is a Canal of the largest dimensions, yet as everything is done in the strongest and simplest manner, without any Pedantry or affectation, my belief is that they will get more real knowledge by considering the Construction of this Canal than all the rest putt together . . . it might not be amiss to cast their Eye upon Upper Calder Navigation, which is a specimen of the making a River Navigable according to its own Course, and this is upon one of the most rapid Rivers in the Kingdom, of its Size, and quantity of water that comes down its Valley in time of Floods. . . .[45]

In the 1760s Smeaton was also concerned with another important river navigation, the Lee.[46] The commissioners of this thoroughly old-fashioned navigation that ran from the Thames tideway for some 31 miles to Hertford (with a fall of 111 ft) had, in August 1765, asked him to "settle the navigation on a new plan as will be most conducive to the good of the public". They tried again in July 1766, this time offering Thomas Yeoman to help him. He agreed, and reported in September 1766.[47] The Lee was indeed old-fashioned:

> "The present navigation . . . above the tides way, being in all probability

formed upon such expedients as occurred before the invention of locks . . . its present principle . . . is that of pens and flashes from eighteen weirs and turnpikes."

He proposed, as he told the Commons committee, to convert the river to a 28-mile long navigation "which is extremely practicable, will be of much greater Certainty, not liable to Obstruction . . .". It would give 3 ft navigable depth, with 2 ft 6 in. in the driest seasons. This could be done without affecting adjacent lands by building cuts to shorten the river, a towpath, and 21 pound locks at an engineering cost of £23,023, excluding land and other costs,[48] and also the Limehouse Cut, from the Lower Lee to the Thames at Limehouse, which Yeoman had separately estimated at £5,310, including land.

"The navigation will be not only improved by being rendered certain at all times, except in frosts and extraordinary floods, and by the shortening the distance, but by navigating in more still water, and in the straighter courses much labour in towing will be saved, and will be rendered so expeditious, that a vessel at moderate work will make her way upwards at the rate of two miles per hour, passing the locks included; that is . . . in about thirteen hours. Hence the ruin of horses, straining of barges, and great wear and tear of the tackle will be saved, as well as greater expedition be secured."[47]

The report formed the basis of the 1767 Act, upon which he appeared in Parliament.[49] Yeoman became engineer in charge. The Limehouse Cut was opened in 1770, and by 1771 about 11 miles of cuts and at least twelve locks had been built: a major modernisation. Smeaton returned to the Lee to advise on special points of difficulty in 1771, 1779 and 1782.[50]

Back in Yorkshire Smeaton was involved at about the same time in schemes for making navigable the upper Ouse and the River Ure with a canal branch to Ripon. He surveyed in 1766[51] and gave evidence to Parliament in March 1767.[52] The Acts passed, work began with John Smith as resident engineer and Joshua Wilson, another of the Calder-trained masons, as contractor for the locks.[53] By the end of 1769 the two big river locks at Linton and Milby (near Boroughbridge), both having a fall of about 10 ft and correspondingly large weirs, were completed. The whole 18-mile navigation was opened early in 1773 at a total cost of £16,400, with six locks, three weirs and a basin and warehouse (designed by Smeaton in 1771[54]) at Ripon. Jessop, as Smeaton's assistant, certainly did some of the levelling, and soundings in the rivers, and may have been deputised to take charge of the work; but Smeaton himself made several site visits and five of his reports are available.[55]

His last river job involved improvements of the Aire & Calder. This navigation, from the lower Aire above its junction with the Yorkshire Ouse to Castleford, and then along the Aire to Leeds and the Calder to Wakefield (where it joined Smeaton's Calder & Hebble), had been made navigable under an Act of 1699. It became most successful, and by the late 1760s was leased to the efficient but unpopular Peter Birt. From 1766 the idea of a trans-Pennine canal had been canvassed, to run from the Aire at Leeds to the Mersey at Liverpool, and in 1770 was authorised. Opposition to Birt, and enthusiasm for canals, combined in a plan to continue the Leeds & Liverpool Canal by another, the Leeds & Selby, to the Yorkshire Ouse, so bypassing the Aire & Calder for Leeds and canal traffic.[56]

Birt first, and then the company, in 1771, sought Smeaton's help upon improving the navigation in order to fight off the canal threat. His report of 28th December[57] was crucial to the future of this, one of Britain's most long-lived and important navigations. He begins:

> "it appears to me, that the original projectors of the navigation, not having had any notion of the extensive trade that was likely to be carried on by means thereof, have formed their plan upon too diminutive a scale, and particularly with respect to depth or draft of water [some locks having only 2 ft 6 in. over their sills] and also not being aware that it would become the practice of millers to draw and keep down their ponds, in order to levy contributions on the navigation."

In other words, to extract dues from boatmen for providing water that should have been available to them anyway. To these impediments shoals were added, and water shortages caused by giving the flashes necessary to get boats over them. Constant delays to craft resulted. Therefore he says:

> "I shall not take upon me to consider this matter in so extensive a light, as if the necessary navigation had never been made; but keeping to the present tract and works wherever it can be properly done, I shall content myself with such alterations as appear necessary to procure the essentials of a navigation, viz. the means of keeping vessels always afloat, so as to move freely in either direction when in the proper channel of navigation, and when not loaded beyond a proper draft of water."

He therefore plans for a minimum depth in dry seasons of 3 ft 6 in. This would involve extensive dredging, and also new locks and cuts, together with a long bypass cut to the lower Aire below Haddlesey. The company then petitioned for leave to bring in a Bill for improvements along Smeaton's lines; but by now opposition was widespread, and other canal schemes were being talked of that would completely avoid the rivers. So they withdrew from Parliament, and went back to Smeaton on the bypass canal. He, being too busy, put his assistant William Jessop on the job; the latter proposed a substitute cut nearly twice as long, from above Haddlesey to a point about 3 miles from the Aire's confluence with the Ouse. Smeaton approved it in December 1772,[58] and back to Parliament the company went, simultaneously with the Leeds & Selby Canal Bill, and battled well enough to postpone a decision to the following session.

By then the Aire & Calder had taken over part of its opponents' plan, itself proposing a canal from Haddlesey on the Aire to the Ouse at Selby. Smeaton and Jessop supported the new Bill in 1774,[59] making clear that the Aire cuts had been surveyed under Smeaton's direction, but that the Selby Canal was Jessop's own. There the Leeds & Selby Canal Bill was lost, though the promoters had tried to disqualify Smeaton as a witness because he held shares in the Calder & Hebble. The Aire & Calder's was carried, authorising the main Aire cuts Smeaton had recommended,[60] and the Selby Canal, at a cost of some £40,000. Jessop then built these, working with the navigation's resident engineer, John Gott.[61]

Canals, 1773–91

After the Aire & Calder, Smeaton's remaining waterway concerns were to be with canals, not rivers.

In 1756 the Grand Canal of Ireland had been begun by the Commissioners of Inland Navigation. Fifteen years later much had been spent, but no part was open. A company being planned to take over the work, one of its promoters, Redmond Morres, in June 1771 wrote to ask Smeaton to come over. Too busy then, Morres tried again in September 1772. This time Smeaton agreed, taking young Jessop with him. Both crossed to Ireland in June 1773, where Smeaton did a report;[62] another followed in 1775.[63] Thereafter the Grand Canal became Jessop's responsibility.[64]

Five canal surveys followed, for the Greasbrough in 1775, and in 1778 the Bude, Kingston & Ewell, Tyne bypass and Cann quarry.[65] None need detain us.

Later, before the passing of the Thames & Severn Canal's 1783 Act, Smeaton had been asked by a local landowner, Lord Eliot, to advise on the effect of the canal on his estate's water supply. Smeaton discussed this with the canal's engineer, Whitworth, but his advice had only partially been taken. When, therefore, cutting reached Lord Eliot's estate, within which lay three of the canal's locks (Cerney Wick, Latton and Eisey) Eliot stopped the company digging a feeder from the Ampney Brook and again asked Smeaton's advice. The latter did not mince his words, criticising the unequal lock falls of Whitworth's layout and suggesting an alternative supply.[66]

Smeaton's remaining canal work was in the Birmingham area. In 1771 he had reported briefly on a problem concerning the Birmingham Canal,[67] but his main contributions were subsequent to 1776. By that year water was becoming increasingly scarce; Samuel Bull, the company's engineer, therefore ordered from Boulton & Watt a small (20 in. dia. cylinder) steam engine to pump into the canal's Smethwick summit level. This began work early in 1778 and in April Smeaton tested the engine, finding it very satisfactory (see Chapter VIII); whereupon, in June, another engine was ordered for summit pumping. In 1782 he was consulted again, on possible improvements and extensions. He wrote:

> "I can sett forward from hence for Birmingham on Monday the 7th October and stay there one week if necessary; but that my stay may not be protracted beyond what is necessary shall be glad that every thing preparatory be done in the way of common surveying and Levelling; and in case there is anything critical in the latter, will carry my own level along with me."[68]

The company's request arose partly from their own wish to expand their branch network, partly to prevent the building of a rival promotion, the Bilston Canal from near Wolverhampton past Wednesbury to near Birmingham and then to Fazeley on the Coventry Canal. Smeaton saw at once that availability of water supplies governed all expansion of an already big canal network, much of it lying upon a plateau. Natural streams were few; most supplies had to come from mine drainage pumping and spring-fed reservoirs.

His report is a fascinating analysis of an industrial canal's sources of water.[69] Its conclusion was that expansion must be limited, even after all additional sources of mine pumping had been utilised and as much leakage (heavy in a mining area) as possible stopped. In 1783, by intense lobbying, the Birmingham Company managed to get the Bilston Bill shelved and, with Smeaton's support,[70] authority for a new company with virtually the same shareholders as the old (and soon to be amalgamated with it) to build their extensions from Wednesbury and also a Birmingham & Fazeley Canal. Upon the latter, the shareholders' meeting of 26th September, 1783 resolved that "it might be right to consult Mr Smeaton again relative to some material points in the execution of the intended Canal...". Smeaton, however, refused on grounds of ill-health. He was consulted later, always with water in mind, in 1786 when the company decided to lower their 491 ft summit by 18 ft between Smethwick and Spon Lane, so eliminating three locks at each end, and in 1789–91 when he advised the Birmingham Company in connection with the proposed Worcester & Birmingham Canal.[71]

On this affair he seems to have been employed by them to survey and estimate the Worcester & Birmingham's likely cost, which he put at £285,000 against £186,000 from John Snape, the Worcester & Birmingham's surveyor.[72] The latter company went ahead on Snape's figure, but indeed Smeaton's itself turned out to be an underestimate. Then in June 1791, having devoted much time and trouble to the scheme,[73] he appeared for the Birmingham Company upon the Bill;[74] his last attendance in Parliament and the end of a long career on inland navigations. His great pupil and assistant, William Jessop, was the leading river and canal engineer of the next generation.

NOTES AND REFERENCES

Notes

1. For the Upper Calder (later Calder & Hebble) Navigation, in its early days, see Hadfield, C. *The Canals of Yorkshire & North East England*, Vol. 1 (Newton Abbot, 1972) pp. 44–61.
2. Smeaton's printed report and map of 1757 are in *Designs*, **6**, f. 14 and 14 v.
3. *Journal of the House of Commons (JHC)*, 14 Mar., 1758. The locks in Flanders and Holland he had seen on his journey to the Low Countries in 1755; he had also studied Belidor's *Architecture Hydraulique*. The Act for the Upper Calder Navigation is 31 Geo. 2, c. 72.
4. Unless otherwise stated, details are taken from the Calder & Hebble Navigation Minute Books and Letters (PRO). See also ref. 1.
5. Smeaton's *Designs*, **5**, contains about 40 drawings for the Calder & Hebble.
6. Letter Book of 1764 (Inst. Civ. Eng. library). Smeaton to Grundy, 4 Sept., 1764.
7. Calder & Hebble Minute Book, 11 Apr., 1765.
8. For the Louth Canal, see Boyes, J. and Russell, R. *The Canals of Eastern England* (Newton Abbot, 1977) pp. 304–315.
9. *Reports*, **1**, pp. 22–25; 14 July, 1761.
10. House of Lords Committee, 14 Mar., 1763.
11. *JHC*, 16 Mar.; Lords Committee, 1 May, 1759.
12. *Designs*, **6**, f. 72.
13. *JHC*, 6 Feb., 1760.
14. *Reports*, **1**, pp. 26–35; 23 Nov., 1761.
15. Lords Committee, 14 May, 1762.
16. *Reports*, **1**, pp. 55–69; 16 Oct., 1762.
17. *Reports*, **1**, pp. 72–87; 31 Dec., 1782.
18. *Reports*, **1**, pp. 36–38; 21 June, 1762.
19. Boyes, J. and Russell, R. *The Canals of Eastern England* (Newton Abbot, 1977) pp. 62–75.
20. *Designs*, **6**, f. 77.

NOTES AND REFERENCES

21. *Reports*, **1,** pp. 13–17; 11 July, 1761. *Designs*, **6,** f. 75.
22. Malet, H. *Bridgewater: the Canal Duke* (Manchester University Press, 1977) pp. 71–73. Dr Malet has kindly made available a copy of the Parliamentary evidence, now in the Huntingdon Library, California.
23. Farey, J. Article on "Water" in Rees' *Cyclopaedia*, Vol. 38 (London, 1818). See also Brewster, D. on "Hydrodynamics" in *Edinburgh Encyclopaedia*, Vol. 11 (Edinburgh, 1817).
24. Smeaton's reports of 1755 and 1758 are in the Boulton & Watt Collection, Birmingham Reference Library. For the 1758 plans, see *Designs*, **6,** f. 54–58.
25. *JHC*, 19 Feb., 1759; Lords Committee, 12 Apr., 1759.
26. For a general account of the Forth & Clyde Canal see Lindsay, J. *The Canals of Scotland* (Newton Abbot, 1968) pp. 15–51.
27. *Reports*, **2,** p. 73.
28. *Reports*, **2,** pp. 31–65.
29. He refers to it in his evidence on the Upper Calder Bill.
30. For the Languedoc Canal, see Skempton, A. W. "Canals and river navigations before 1750". *History of Technology*, Vol. 3 (Oxford, 1957) pp. 464–468.
31. *JHC*, 28 Apr., 1767.
32. *Reports*, **2,** pp. 73–97; 8 Oct., 1767.
33. Lords Committee, 25 Feb., 1768. The Act is 8 Geo. 3, c. 63.
34. All details of the Forth & Clyde Canal proceedings are from the Minute Books (West Register House, Edinburgh) and are given here by courtesy of Dr Jean Lindsay and Professor Skempton.
35. These reports were published in Edinburgh on 29 Oct., 1768. Smeaton's reply was printed in the following month.
36. *Reports*, **2,** pp. 98–120; 28 Oct., 1768.
37. Lords Committee, 21 Feb., 1771. The Act for the new line is 11 Geo. 3, c. 62.
38. Most of Smeaton's reports and some of his letters, written during the progress of the works, are copied into the Canal Minute Books, see ref. 34.
39. *Designs*, **5,** f. 2–37, and **6,** f. 59–68.
40. These and other key dates and statistics of the canal are given in appendices to the Superintendent's Report, 31 Dec., 1791 (West Register House).
41. Machine Letter Book (Inst. Civ. Eng. library). Smeaton to the Lord Provost of Edinburgh, 7 Mar., 1785.
42. *Reports*, **2,** pp. 125–127; 30 Nov., 1767.
43. *Reports*, **2,** pp. 128–130. For some details of the later work on the Calder, see Hadfield, C. and Skempton, A. W. *William Jessop, Engineer* (Newton Abbot, 1979) pp. 20–22.
44. *Designs*, **6,** f. 37.

NOTES AND REFERENCES

45. Cross, A. "A Russian engineer in eighteenth century Britain: the Journal of N. J. Korsakov, 1776–7". *Slav. East Eur. Rev.*, **55** (1977).
46. The river is spelled "Lea", the navigation "Lee". For this navigation see Boyes, J. and Russell, R. *The Canals of Eastern England* (Newton Abbot, 1977) pp. 13–39.
47. *Reports*, **2**, pp. 155–163; 24 Sept., 1766. *Designs*, **6**, f. 85.
48. This estimate, dated 26 Feb., 1767, is revised upwards from that in the original printed report of 1766.
49. *JHC*, 15 Apr., 1767, Lords Committee, 3 June. The Act is 7 Geo. 3, c. 51.
50. *Reports*, **1**, pp. 282–294: three reports dated 27 Mar., 1771, 17 Apr., 1779, 12 Nov., 1782.
51. *Designs*, **6**, f. 79.
52. *JHC*, 2 Mar., 1767. The Acts are 7 Geo. 3, c. 93 and 7 Geo. 3, c. 96.
53. Hadfield, C. *The Canals of Yorkshire & North East England*, Vol. 1 (Newton Abbot, 1972) pp. 111–112.
54. *Designs*, **4**, f. 48.
55. *Reports*, **1**, pp. 310–312 on Linton Dam, undated but *c.* 1768; **2**, pp. 172–184 four reports on the works, May, 1770 to Jan., 1772.
56. For a general history on the Aire & Calder in the eighteenth century, see Hadfield, C. *The Canals of Yorkshire & North East England*, Vol. 1 (Newton Abbot, 1972), pp. 17–43.
57. *Reports*, **2**, pp. 131–140; 28 Dec., 1771.
58. *Reports*, **2**, pp. 151–153; 5 Dec., 1772.
59. For Smeaton's evidence, see *JHC*, 22 Feb., 1774 and 23 Mar., 1774.
60. Smeaton's plans of the new cuts and locks are in *Designs*, **6**, f. 16, 17, 18 and 19(ii).
61. More details of Smeaton's and Jessop's work on the Aire & Calder are given by Hadfield, C. and Skempton, A. W. *William Jessop, Engineer* (Newton Abbot, 1979) pp. 13–20.
62. *Reports*, **2**, pp 252–260; 6 Oct., 1773.
63. *Reports*, **2**, pp. 262–277; 3 Apr., 1775.
64. For Smeaton on the Grand Canal, see *Letters between Redmond Morres and John Smeaton* (Dublin, 1773) and Delany, R. *The Grand Canal of Ireland* (Newton Abbot, 1973).
65. The Greasbrough report is in Sheffield Public Library; the four reports of 1778 are in *Reports*, **2**, pp. 213–241.
66. Household, H. G. W. *The Thames & Severn Canal* (Newton Abbot, 1969) pp. 84–85. Smeaton's report of 14 Aug., 1789 is in Gloucestershire Record Office.
67. *Reports*, **2**, pp. 189–193; undated but 1771. Lords Committee, 15 Feb., 1771.
68. *Machine Letters*, 16 Sept., 1782. Smeaton to Meredith.
69. *Reports*, **2**, pp. 194–212; 16 Dec., 1782. For the complicated background

NOTES AND REFERENCES

to Smeaton's proposals see Broadbridge, S. R. *The Birmingham Canal Navigations 1768–1846* (Newton Abbot, 1974) pp. 22–24.

70. *JHC*, 14 Feb., 1783.
71. Birmingham Canal Proprietors' Minute Book, 26 Mar., 1790.
72. *Aris's Birmingham Gazette*, 15 Mar., 1790. For the Worcestershire & Birmingham Canal, see Hadfield, C. *Canals of the West Midlands* (Newton Abbot, 1966).
73. Hutton, C. "Some account of the life of Mr John Smeaton". *Reports*, **1**, p. xx.
74. Lords Committee, 7 June, 1791.

4
The construction of the Huddersfield narrow canal, 1794–1811: with particular reference to Standedge tunnel

R.B. Schofield

INTRODUCTION

This vile canal as I'm a sinner
It steals alike my cash and dinner.[1]

The Huddersfield Narrow Canal was one of three canals which linked Lancashire across the Pennines with the West Riding of Yorkshire, the others being the Leeds and Liverpool, and Rochdale Canals, both broad navigations which followed longer but easier, more northerly routes.

It was formally proposed in May 1793 when there was no other completed waterway between Manchester, the Yorkshire navigations and ports of the east coast. Its promotion was mainly due to the success and high premiums attached to shares of the Ashton Canal Company, the proprietors of which were seeking extension of their trading link from Manchester. The two canals were to share management structures and the future was promising except from the danger of a possible Rochdale Canal[2] which, had it been built earlier, would have extinguished enthusiasm for the Huddersfield route, which had the disadvantages of a narrow canal connecting broad navigations on each side of the Pennines.

Irrespective of these factors, the Huddersfield Narrow Canal gained enthusiastic support at a meeting in Huddersfield on 30 May 1793, when £150,000 was promised of a total of £200,000 required for the project. Temporary officers appointed included Worthington and Battye, Clerks, Nicholas Brown, Surveyor, and "Mr. Jessop if at liberty, Engineer".[3] Later it was found that the latter was not available and presumably he recommended his friend Benjamin Outram, who had been trained under Jessop's patronage on the construction of the Cromford Canal[4] before embarking on a civil engineering career on canal and tramroad construction.

CONTRACT DOCUMENTS

Documentation prepared by Outram and Brown for the construction of the canal was restricted to the minimum required by law[5] and comprised a survey and plan of the canal with a book of references to land occupations, also a rudimentary bill of quantities and estimate. Outram's report[6] was accepted by a public meeting on 22 October 1793 with little dissension, although mill owners in both the Tame and Colne valleys insisted on augmentation of reservoir capacities and other measures to safeguard their rights. Their resolutions[7] were accepted and feeder reservoir capacity was increased, thus lifting the canal estimate by £4,000 to £182,784, a figure which Outram stated excluded expenses of obtaining the Act, but "fully considered the high price of labour and materials".

THE CONSTRUCTION OF THE HUDDERSFIELD NARROW CANAL 1794 – 1811

Fig. 1. The Trans-Pennine Canals.

The Surveyor, Nicholas Brown, a native of Saddleworth, was to become an accomplished land surveyor and draftsman in later years[8] but the Huddersfield Canal survey, an early example of his work, was crude and of limited value for engineering purposes. This was probably a compass and chain survey because its inaccurate alignment suggests that a theodolite was not used for horizontal control. It is no more than a good diagram, its use only suitable for perusal by interested investors and Members of Parliament who gave the scheme no opposition when the Bill was reported without amendment on 27 March 1794, becoming law on 4 April. Outram recommended a route from Sir John Ramsden's Canal (the Huddersfield Broad Canal accommodating vessels of 58 ft × 14 ft 2 in) in Huddersfield, to follow the Colne valley to Marsden, climbing 436 ft (438 ft 10 in as built) to the summit level at 648.6 ft A.O.D., then proceeding by tunnel over three miles in length (his estimate gives 5380 yds but on completion it measured 5477 yds[9], its approximate length today is 5698 yds[10] due to modifications during the last century) then "to excavate in Brun Clough" from which place it pursued the Tame valley descending via Stalybridge to the Ashton Canal 334 ft 8 in (338 ft 1½ in as built) with a tunnel of 200 yds. at Scout Mill (220 yds as built). The Engineer did not refer to a further tunnel near Ashton, which, though completed, was opened up in later years. The total length of canal was given as 19¾ miles (19⅞ miles in fact). The number of locks was not stated although the "lockages" of 436 and 334½ feet were costed at £65 and £70 per foot respectively. Outram evidently intended to leave such important matters to the setting-out phase, and yet he could include such trivia as "26 pairs of fence gates and posts at £39". In the event 42 locks were built in Yorkshire and 32 on the Lancashire side with average fall of about 10 ft 6 in per lock.

He stressed that the route was "the shortest communication yet between Manchester and the eastern navigations and that it will pass through a country full of manufacturers.... and by the vicinity of the proposed canal to the rivers, the mills upon them would obtain their articles free from land carriage". This was true; the numerous mills shown on the survey on each side of the mountain could hardly have been located closer to the canal because of the narrow valley floors. Outram was aware of the mill owners' fears for the future of their water supplies, and he sought to reassure them regarding his plans for conservation and distribution. Reservoirs would be situated only in large and deep valleys "where collected waters frequently produced torrents of flood". He was to take no water from the brooks and would allow them to fall freely between succeeding mills and works. This was fair and reasonable, and there was little opposition to the scheme although some foresaw difficulties in passing to and from the broad canal

THE CONSTRUCTION OF THE HUDDERSFIELD NARROW CANAL 1794 – 1811

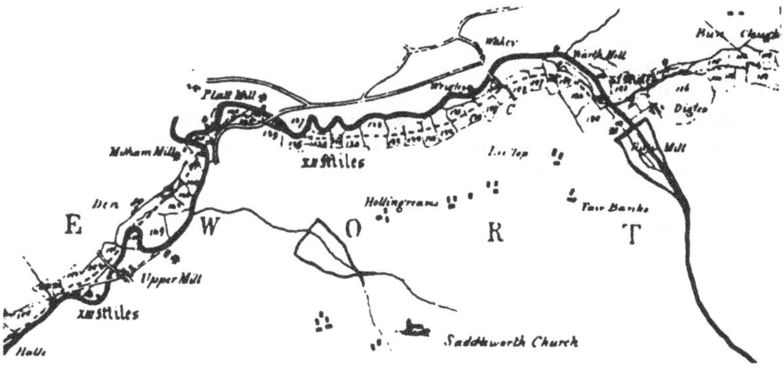

Fig. 2. Part of Nicholas Brown's 1793 plan from Upper Mill to Standedge Tunnel. (Canal is shown as a dotted line).

system,[11] a factor never mentioned by Outram. He continued, stating that the great tunnel at Standedge "appears favourable; the strata consists of gritstone and strong shale and the low ground in the centre, near Red Brook, will afford the opportunity of opening the works by means of steam engines, so as to greatly facilitate completion of the tunnel in five years". His estimate included 5380 yards of heading at £7 per yard (£37,660); sinking 18 shafts (£2767.50), although 32 were actually commenced on the main line, and adits; "two thirds loss on three steam engines to raise the water" (£3,000) and "for coals, day labourers and engine tenter at £2 per yard" (£10,760). Unwisely he did not include for any lining or 'arching' assuming that even the blue shale would require no supports. He was proved incorrect in that assumption.

Of the total estimate of £182,748, £55,187 was allowed for the construction of Standedge Tunnel. Writing in later years, the engineer John Sutcliffe[12] stated that Outram had offered to execute and find security to drive this in five years, but that some of the Committee were of the opinion it could be completed for a smaller sum and declined. Sutcliffe considered (in view of the fact that the tunnel took 16½ years to finish at a cost of £123,804)[13] that this was fortunate for Outram, though he had "heard his friend Jessop say he might complete it in the time proposed and get money by the contract".

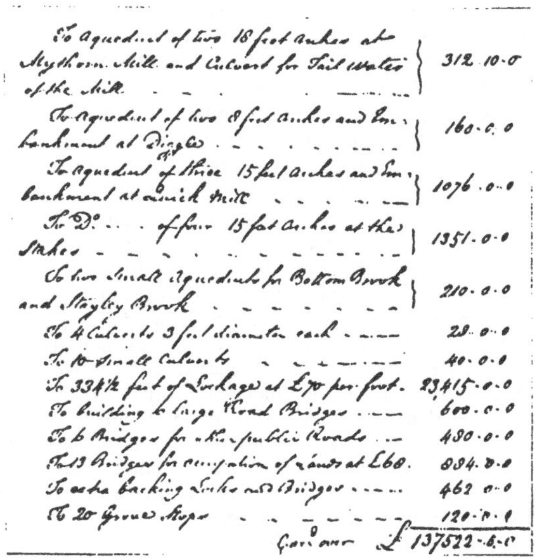

To Aqueduct of two 18 feet Arches at Mytham Mill and Culvert for Tail Water of the Mill	312.10.0
To Aqueduct of two 8 feet Arches and Embankment at Diggle	160.0.0
To Aqueduct of three 15 feet Arches and Embankment at [illegible] Mill	1076.0.0
To D° of four 15 feet Arches at the Stakes	1351.0.0
To two small Aqueducts for Bottom Brook and Stayley Brook	210.0.0
To 4 Culverts 3 feet diameter each	28.0.0
To 10 small Culverts	40.0.0
To 334½ feet of Lockage at £70 per foot	23415.0.0
To building 6 large Road Bridges	600.0.0
To 6 Bridges for other public Roads	480.0.0
To 13 Bridges for occupation of Lands at £68	884.0.0
To extra backing Locks and Bridges	462.0.0
To 20 Grooved Stops	120.0.0
Carried over	£137522.6.0

Fig. 3. Form of Outram's Bill of Quantities and Estimate.

THE CONSTRUCTION OF THE HUDDERSFIELD NARROW CANAL 1794 – 1811

Although many engineering reports of the period included rudimentary economic surveys with speculations as to likely developments in local agriculture and industry, Outram wisely avoided them. He did "not think himself competent to form an accurate estimate of the trade to be expected" and left local industrialists to decide on these issues.

The documentation was brief, in several respects inadequate, and left many engineering problems to be examined and resolved as works progressed. This was unwise but there was nothing unusual about such methods at that time and even the best known engineers did no differently. It is not surprising that estimates were often optimistic and frequently exceeded, and when this happened with the Huddersfield Canal, that ardent critic, William Pontey, wrote that "its ultimate effect has served only to show the truth of the observation—engineers are not infallible".

COMPANY MANAGEMENT STRUCTURE

After the Act of 1794 [14] became law, the Huddersfield Canal Company was formally established, and George Worthington, Solicitor, of Altrincham was appointed Clerk to the Company (simultaneously he acted in that capacity for the Peak Forest and Ashton Canal Companies) assisted by John Battye to manage daily affairs. Bankers, Treasurers and a Valuer were also engaged. Benjamin Outram of Butterley Hall, Ripley, was appointed Engineer, and was also nominated, as required under the Act,[15] to ascertain with John Holt, Surveyor, the discharge of streams and their usage by the Company and riparian owners. His salary was fixed at £3.3.0 per day plus £10.10.0 expenses for each journey to the canal,[16] and since his fees varied from £150 to £175 every six months[17] it suggests he would spend 4 or 5 days a month on site and also attend the monthly Management Committee meetings. In assessing recompense, one must take into account his many unpaid hours off the job spent in planning, preparing designs, estimates and reports, writing letters and visiting companies to examine machinery and materials. Hence it does not seem that engineers were highly paid by today's standards.

Outram's was an exacting task and he had to rely heavily on the abilities of the Management Committee and full-time staff on site. As a young man of 29 years, he was comparatively new to the profession and unlike Jessop and other senior engineers, did not have an entourage of well-tried resident engineers, overlookers and contractors who could be called upon to execute works under his overall control. His resident engineer, Nicholas Brown, was a youthful and inexperienced surveyor selected for the job mainly through the patronage of local mill-owners and merchants who were proprietors of the Company. Brown was appointed "Surveyor, Book-keeper and Superintendent" of the canal works and required to devote all his time to the task for a salary of £315. He was permitted to employ a book-keeper to help in his formidable appointment but wages would be paid from his own salary, as well as all expenses, except when on business 15 miles from the canal.

Brown was expected to supervise 20 miles of canal construction in that remote country. The project was understaffed from the onset; it called for at least three resident engineers to assist the superintendent and although Brown was allotted overlookers as work progressed, initially he had all the work to do himself. And of course even the conditions of his appointment discouraged him from spending freely on a capable assistant and on extensive travel about the line. All this, coupled with his inexperience, led to ineffective management. The Company Minutes are uncritical of Brown, but the signs of inadequate supervision are all too plain. There were setting-out errors: one of these was the result of misunderstanding Outram's written instructions, for which Brown was required to pay compensation; another involved the incorrect location of weirs; yet another the building of culverts too small for their intended purpose. Each month there was a list of claims for damages to lands and property, and for trespass. Two major claims went to arbitration. It seems that some land-owners and industrialists along the line, proprietors among them, made claims for both real and imagined injuries to their properties. Considerable sums were paid out on the flimsiest of evidence, the details of which were not recorded in the Minute Books. It was too easy to win claims against the Company, but this was not surprising when judgements were made by friends or fellow proprietors. A competent resident engineer could have soon put a stop to such claims, and many of the disputes could have been avoided by prior consultations with landowners, by accurate setting-out and temporary fencing, and by written instructions to contractors and direct labour gangs. Brown seemed unable to cope with such problems. Outram must be blamed in part for tolerating this situation, but he appears to have persevered and helped Brown in the first 18 months of the project until a serious illness incapacitated him during 1795 – 6. Towards the end of that period, Brown was also absent through illness which for a time left the Management Committee in charge of the works.

The Committee had always seen fit to meddle in the canal's management, and they even had a sub-committee for setting out accommodation and other engineering works. Perhaps it was indicative of the confused state of their management policies when, in 1797, they called in Robert Whitworth to report on the work and "the propriety of the line as laid down by Mr. Outram".[18] That report was published[19] and Whitworth stated "the masonry and earthworks were the worst executed of any he had ever seen". Privately he said to fellow engineer John Sutcliffe that "the work will be nearly all to do over again, particularly the locks".[20] Outram was criticised on his design of lock abutments, built with foundations "very short towards the hill" such that on completion the top of the abutment was four feet beyond the foundation. Also the waste water channels built around locks were not watertight so there was seepage into the foundations and through the walls such that during the severe frosts the masonry collapsed "into little better than a heap of rubbish". Yet a further fault was the very short length of canal pounds set-out on the steep inclines because they contained too little water for a busy navigation.

Just before Whitworth's visit, the Committee lost patience with Brown after his building of a culvert which was found to be unnecessary![21] He was given 6 months notice to quit, and although the period was later extended, he was replaced by William Bailiffe of Marsden who, though fairly successful, resigned in 1801. [22] His place was taken by John Rooth of Manchester, an entrepreneur with a dominating personality who was superintendent, with fair but not outstanding success until completion in April 1811. Afterwards he was retained as 'Agent' until his dismissal in favour of a younger man in 1817. [23]

Outram withdrew as Engineer during 1801 (he also resigned from the Peak Forest Canal Company) when the Huddersfield Canal Company had all but ceased construction due to lack of finance, although he was occasionally to assist with specialist advice on contract specifications, until 1802. His place was never taken by another engineer, although the Committee considered asking Jessop to help them in 1802. [24] Thomas Telford surveyed the canal works and planned their completion [25] after visits in late 1806. [26] His report, estimates and programme were presented to the Company and followed most carefully by Rooth in the years ahead. Rooth saw fit to forget the engineer's involvement in this important phase of the scheme, because when he retired he described his part in the canal's construction[27] and claimed the works had been done "without the help of any engineer".

SETTING OUT AND CONSTRUCTION

The present day procedure would be to set out the alignment of the canal with theodolite or E.D.M. equipment using co-ordinates computed from a drawing based on a land survey. Brown's elementary chain survey, on which the canal line appears to have been sketched in freehand, would not permit this, and Outram would set out the works using the plans only as a guide. Centre line pegs, the tops of which were levelled to the proposed water surface, were driven in probably at 2 to 3 chains[28] intervals or closer where the canal was built on embankment or in cutting. In the former case, stakes of considerable length may have been used and in the latter, holes were first dug before pegs were set in them at the correct level.[29] Alignment between principal setting-out pegs was left to the overlookers to determine, a practice which in places led to short meanders which are common enough on many canals, although unnecessary and avoidable. Generally the route selection was economical and imaginative and Outram took care to reduce earthmoving to a minimum. It is a simple matter to set out batter profiles to define the slopes and toe positions of earth banks and such additions to basic setting-out are always desirable because land take can be agreed and fenced in before work commences. Outram, in common with other engineers of the period, preferred not to do this and left land assessment until work was finished. Not surprisingly this practice led to disagreements and claims for trespass and damage to property.

A difficult matter was the establishment of vertical and horizontal control of the great tunnel through Standedge and the superintendents seemed to find it so judging by the mistakes they made. The problems comprised:

(i) an accurate surface alignment between the tunnel mouthings at Diggle and Marsden in order to locate the position of the construction shafts;

THE CONSTRUCTION OF THE HUDDERSFIELD NARROW CANAL 1794 – 1811

(ii) the transfer of bench marks across the mountain to ensure a common basis for levelling;

(iii) the transfer of the surface lines and levels down the shafts as work progressed.

Surviving records fail to describe how any of these problems was resolved although it seems that instrumentation was of the simplest kind. Surface alignment could have been achieved by an iterative process of open traversing with magnetic compass or even by reciprocal ranging, using very long poles. Accuracy was poor and the surface position of shafts, so far as can be judged,[30] is within 3 and 9 feet laterally of the correct points. One of the original canal shafts (the downcast or 'engine' pit at Red Brook) is offset by a distance of about 26 feet which was perhaps deliberate so that survey lines were not obscured by the engine housing. Transfer of a parallel line to the base of a shaft was achieved by the so-called 'Simms Method'[31] (which is still used today for shallow shafts) whereby plummets hang freely at pit bottom on twin cords suspended from hooks located on the surface survey line. Thus a parallel line could be projected into the tunnel mouthings. Following the transfer of bench marks across the mountain, the depth of each shaft was calculated and vertical level controls set by direct measurement using Gunter chains hung down the shafts.

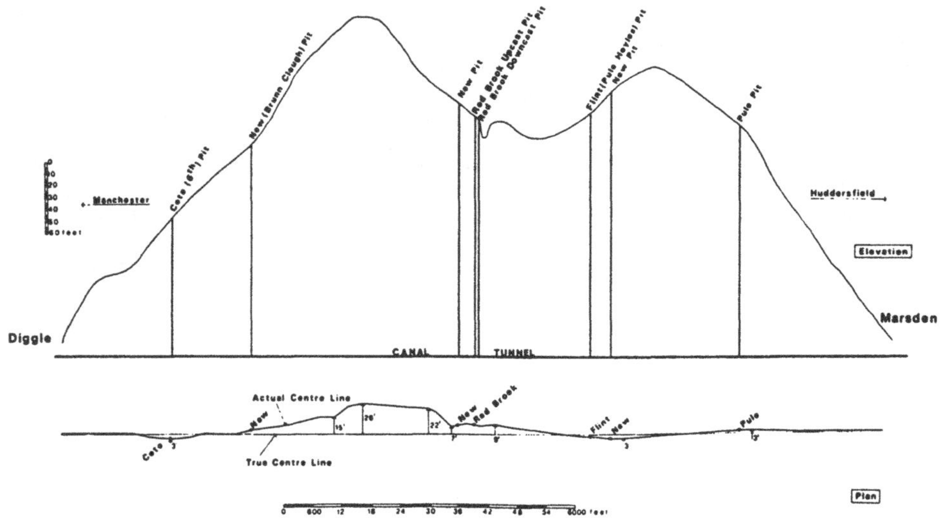

Fig. 4. Longitudinal section and plan of Standedge Tunnel showing misalignment of tunnel centre line.

In later years it became evident that there had been serious mistakes in surveying Standedge Tunnel. After visiting during the winter of 1806 Telford wrote[32] that:

> The season was totally unfit for proving the general line of direction over the mountain and dialling and levelling below, but each end appears to be very direct in itself.

He was confident that all was satisfactory because he understood that "they have been tried and found accurate by several properly qualified persons." Telford was correct at the time because it was not until after his visit that the section from Red Brook to Brun Clough was driven with a maximum error of 26 feet to the west.[33] And so it remains to this day, although the matter was never reported in the Company minutes! That was not all; there were errors of 40 and 14 yards in the positions of two shafts, undiscovered until the hill was pierced in 1809 [34] when the engineer stated that the corrected length was 5477 yards.

THE CONSTRUCTION OF THE HUDDERSFIELD NARROW CANAL 1794–1811

As regards levelling, it was as late as 1802 when David Whitehead,[35] a surveyor associated with Outram on the Peak Forest Canal and the South Wales Tramroads, was asked to check the levels at each end of the tunnel. Serious discrepancies were found of which details were recounted by Rooth[36] in later years. At the Marsden end the pool by the first lock was two feet shallower than the tunnel invert, but far more serious was the difference between the surface levels at the tunnel ends. Taking the Marsden surface level as correct, that at Diggle was much higher and, on excavating there to the proper depths, the foundations of the sidewalls were undermined. These, as well as the completed arching, had to be taken down and re-built. There was a long headed section beyond and that too had to be cut down to the true level. Even at the supposedly correct eastern end, the headed work dipped 3 ft deeper than it should have done!

PROGRESS IN STANDEDGE TUNNEL

Outram's report of June 1796 [37] referred briefly to the tunnel but much can be gleaned from this and the longitudinal section prepared by Nicholas Brown at the request of the Committee.[38] Shafts were sunk at 100 to 180 yards intervals from each end initially; six pits were completed at the Diggle end (depths varying from 33 ft to 80 yards); and two at the Marsden end (depths 45 and 85 yards). Pumping costs were very high (the cost of working the engine at the sixth pit was 11 guineas per week at one stage) due to the unexpected ingress of water from the millstone grit. This led to a reduction in the number of main line shafts, and the only others sunk were at Brun Clough (140 yards); two at Red Brook each of 160 yards; Heathby Lee (146 yards) which was subsequently ordered abandoned by Telford;[40] Pule Hoyles (160 yards) and Pule Hill (156 yards).

Nine adits were driven into the main shafts close to ground level in order to draw off surface water[41] and to provide a discharge route for 'water engines'. The latter operated probably as illustrated. Spoil loaded into a bucket at tunnel level was lifted by a water-filled kibble moving down a shallow balance pit. The system was controlled by head gears, ratios of which were related to the depth of shafts. This simple system was installed in a few shafts initially. Elsewhere water-wheels provided motive power, but these were unsuccessful and were abandoned in favour of the water engine. Water for the latter was ducted to cisterns adjacent to the balance pits, from nearby streams or feeder reservoirs. These supply systems appear on the early O.S. plans[42] and their remains can still be found today at Flint and Pule Hill pits. An interesting ventilating system is used at Red Brook pit where a duct, still in use, discharges 27 gallons of water per minute[43] as a fine spray down the engine pit to induce a strong draught of air into the workings.

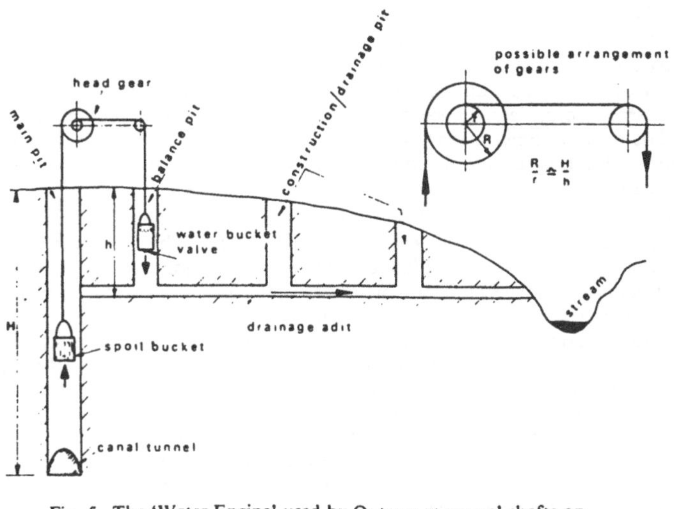

Fig. 5. The 'Water Engine' used by Outram at several shafts on Standedge. Diagram prepared from descriptions in various Company reports.

THE CONSTRUCTION OF THE HUDDERSFIELD NARROW CANAL 1794 – 1811

In addition to these economical hoists (preferred by Outram to horses which were "a great and unavoidable expense on most works of this nature"), four steam-engines were used for pumping out the wettest pits, the biggest being located at Red Brook.

Fig. 6. Engine House at Red Brook which encloses the 'Engine' and 'Bye' pits. Pule Hoyles pits can be seen in the far left.

It is difficult to unravel the sequence of construction in the Standedge Tunnel during those sixteen years. Work should have been carefully preplanned but instead it proceeded in a haphazard manner. Minds were changed, there were frequent stoppages; sections and machinery were abandoned for long periods and labour dismissed or switched between working faces. The reason for this chaotic working was the constant shortage of capital mainly through the inability, or unwillingness, of shareholders to pay their calls. For example by June 1796 [45] when working output should have been rising to a peak £92,000 of share capital had been called, of which £22,650 was in arrears and there was only £889 in the bank! In spite of this the Management Committee persisted in maintaining several working faces with all the associated equipment and labour force this entailed and with only empty promises of capital to come. A better plan would have been to drive the tunnel from each end with a small labour force, relying on gravity to draw off the water and to open up Red Brook as funds increased, as originally proposed by Outram. But the proprietors were impatient to start business in minimum time, and were misguided enough to believe their foolish practices were appropriate. It was not surprising there was little progress until Telford came to survey in 1806. His recommended plans[46] were based upon what he recognised as limited resources, and when his time and cash estimates were acted upon the works were completed on schedule. Would that the Company had used the same methods in 1794!

In the first working season of July 1794, the proprietors' plans were determined by the availability of specialist contractors able to take on such extensive, difficult work. The age of the general contractor had not arrived, so that contracts were for labour only and the Company provided temporary working materials, powder and other facilities. Fortunately they were able to agree with one Thomas Lee on a rate (undisclosed in the Minutes) for the first 1,000 yards of main-line tunnel at the western end. Ostensibly this was a straightforward section through alternating strata of shales and millstone grit. Drainage by gravity eliminated pumping and the task should have required only a steady determined effort. Lee's progress was very slow and by September 1797 Outram reported to the Committee that the contractor had met unforeseen difficulties, his losses were considerable and he was facing ruin. The Committee sent a member to investigate the reasons and at the November Meeting they offered Lee £300 compensation for extras over contract, and an additional 15 s per yard for his remaining work. He was allowed an extra year to complete and penalty clauses were waived. This was generous but Lee did not accept and the Company had no option but to terminate his contract.[47]

Meanwhile the eastern end was opened up late in 1794, by direct labour, the heading driving through relatively easy blue shale until a fault was crossed into millstone grit between Pule Hoyles and Pule Hill pits. Although the strata are apparently of the Namurian series[48] a few thin coal measures were encountered and indeed those found at Red Brook were used to fuel the steam engines.[49] The eastern end excavations contained some geological curiosities described by Outram in a paper submitted to the Royal Society in 1796. [50] The miners apparently encountered greyish shale for 240 yards where they discovered, on the north side of the workings, a fault full of shale mixed with softer earth and in some places with coal. After a further 40 yards the fault occupied the whole tunnel, 4 yards in width, and five feet from the southern edge it contained a rib of limestone 4 feet thick at the bottom, thinner at the roof. On each side it contained balls of limestone of various sizes from one ounce to 100 lb weight.

> The workmen have now pursued the tunnel to near 350 yards from the entrance and the rib of limestone and balls continue the same. [The balls are] not perfectly globular, being flattened a little on two opposite sides, which appear to have been the poles when in a revolving state.

The paper noted the absence of limestone nearby, and none nearer than Clitheroe or Buxton.

Fig. 7. Unlined section of tunnel. View from under Pule Hill towards Marsden.

The progress report made to the Company shareholders in 1796[51] was optimistic. Although there was evidence of achievement on the canal sections, it reported only limited progress in the 'Great Tunnel'. After two years, merely 445 yards had been opened up at the west end and 350 yards at the east, and even these sections required finishing and arching. Much labour had gone into adit construction (1485 yards) and the sinking of pits. Four of the latter were complete and another ten in "tolerable forwardness" and it was explained that "the progress of the main tunnel must necessarily be slow till the deep pits are sunk down to the level as in the mean-time but few miners can be employed at the two ends". Much was made of the excavation of new quarries, the getting of stone and the building of haul roads, cottages and workshops on the mountain which facilities it was claimed, would prevent a repetition of those delays and loss of time which have been already experienced in sinking the deep pits". These were, of course, excuses. The Company was in difficulties and were not prepared to admit it. There was even an attempt to turn their major misfortune (the "very copious supply of water from the tunnel") into a useful asset because this was "probably sufficient for the summit of the canal, without the aid of reservoirs". In conclusion, this disturbing report noted that costs for the tunnel stood at £20,049. In two years they had expended more than one third of the original estimate on less than one seventh of the work in the tunnel, and yet no warning by way of revised estimates was given. It is regrettable that Outram lent his name to this misleading document because as a professional engineer he was obliged to write the truth and not allow his judgement to be influenced by the financial circumstances of the proprietors.

Meanwhile, by September 1796, a contract had been signed with one George Evans 'for cutting and making the part of the tunnel under Pule Moss and Brun Top'. This was roughly the central section based on Red Brook which included several pits and the notoriously wet 6th pit. Further disappointment was to follow as Evans broke his agreement in May 1798.[52] The details were not explained, but the Committee complimented Outram on the manner in which he cancelled the agreement to the Company's advantage.

THE CONSTRUCTION OF THE HUDDERSFIELD NARROW CANAL 1794 – 1811

Outram then made the offer, unusual by today's professional practice, to contract for completion of the tunnel. Next followed an invitation by the Committee[53] to prepare his quotation "to finish and complete the tunnel at Brun Clough and Pule Hill in a given time, taking engines etc. provided by the Company", and also to complete the remaining short length of cut from the Diggle entrance down to Woolroad near Saddleworth. The estimate was presented to the Committee next month,[54] and they set it aside for consideration by a special general meeting. The details of that estimate have not survived, but no doubt it is the measure referred to by Sutcliffe[55] which he assumed occurred before the canal works commenced in July 1794. The Committee should have accepted the offer, but they hesitated for a while and before long they were prevented from doing so by unprecedented rains and floods which destroyed the earthworks of the canal and reservoirs. That calamity cost them several thousand pounds of money they did not possess, and a special appeal had to be made[56] to shareholders. In retrospect there is little doubt that Outram could have undertaken the contract because he had experience of similar work at the Butterley Tunnel on the Cromford Canal. But the opportunity was lost, the matter was never raised again, and Standedge Tunnel languished, almost untouched, for several years. There were attempts to let the tunnel works two years later, and in January 1801 several contractors visited the workings, three of their number presenting quotations one month later.[57] The quotation of John Varley was accepted in preference to that of Pritchard[58] and Williams who were paid £21 for their time and trouble in preparation. Unhappily Varley deceived the Committee regarding his sureties and the offer was then rejected[59] although six months later he reappeared with fresh guarantees which were proved false only after the Committee accepted yet again.[60] That deceit closed the matter and there were no further dealings with this unreliable fellow.

By that time John Rooth had been appointed Superintendent for the entire canal, and John Booth, a local man from Greenacres had accepted the job of tunnel inspector in anticipation of Varley's contract. Booth was retained and probably undertook minor maintenance tasks about the tunnel whilst the hunt continued for another contractor. In late 1802 Samuel Marsland, a Committee Member, informed his colleagues he had succeeded in locating a suitable contractor, but their expectations were short-lived on learning of his terms. The Committee had estimated that it would be worth while to pay up to £8000 extra over the cost of just working from both ends, provided they could expedite completion by 2 years. A bid for such a time schedule was invited from a Mr. Jonathan Woodhouse of Ashby-de-la-Zouch. His figures were outrageous and clearly to his advantage. He offered to complete in three years for a price of £13 per yard and the extra sum of £8000 quoted. All company equipment would be made available for his use and he would retain everything at the end of the project. Not only that, he insisted on a downpayment of £5000 on commencement of the work, £2000 at the end of 4 months and £2000 at the end of every 3 months after, plus bonuses. This received a courteous but guarded welcome;[61] it would be referred to a Special Assembly for consideration and this took place two months later,[62] by which time Woodhouse had changed his mind "and had other proposals to offer". A motion requesting Woodhouse to bring these forward was defeated.

Another two years passed with little progress except that Rooth was instructed to dismantle and sell off the steam engine at Red Brook. This seems a surprising decision but the pumping costs were prohibitive, indeed Sutcliffe claimed these were as high as £50 per week![63] In earlier years Outram had tried to improve the performance of this engine, and his efforts led him to contact the Boulton & Watt northern agent, James Lawson,[64] for advice on its conversion to the Watt patent. Instead of offering to modify the 70 inch cylinder engine, James Watt Jun. suggested its replacement with a 52 inch cylinder to work at 10 strokes per minute, with the existing 14 inch pump of 6½ feet stroke. The cost would be £1170[65] but this was more than the Company could afford and Outram did not trouble the Committee with the idea. After a further cautious enquiry from Outram, Watt terminated correspondence abruptly when he realised that the Canal Company were in no position to buy anything except on the cheap! It is likely that the water engine was used at Red Brook for drainage purposes after 1803.

In 1805[66] the Company decided to apply for their third Act[67] "to raise a further sum of money for the discharge of their debts and to finish and complete the Huddersfield Canal...." and Rooth was required[68] to furnish estimates for the application. That matter concluded, the eminent engineer Thomas Telford agreed to visit the canal to assess the remaining tasks and to prepare a programme to completion. Work had started on the tunnel again and was proceeding at 11 yards per week. On arrival Telford found the following had been achieved:

THE CONSTRUCTION OF THE HUDDERSFIELD NARROW CANAL 1794 – 1811

	Finished	Headed	Unbroke	
	yds	yds	yds	yds
East	331	1531	452½	
Red Brook	107	50		
West	510	1535	934½	
	948	3116	1387	5451

In his report[70] Telford claimed a "thorough knowledge of the state of the works because he had examined everything twice and had even checked the filling and emptying of every lock". His efforts were concentrated on a construction programme and estimate for the remaining work in the tunnel, but he also planned completion of the canal from Diggle to Woolroad, and recommended more reservoirs. He confirmed that the drainage problem in the tunnel was not as acute as formerly; "the water has either drained off, or now falls in chiefly near the bottom, by which means the shale is in general becoming dry and firm and the whole promises to form a perfect tunnel".

Thus Telford could confidently predict progress in the coming years as, for example;

West End

....

XI. 280 yards, commencing at the termination of the last length and ending at the fourth pit, at the rate of 8 yards per week, which will occupy up to the 1st of November 1807.

XII. 564 yards from the fourth pit to Brun Clough pit, at the rate of 8 yards per week, from the 1st November 1807 to the 9th March 1809, being 70½ weeks".

This programme was accompanied by a longitudinal section, based on the earlier drawing of Brown, a bill of quantities and an estimate of £55,290.7.4 for the tunnel and a total of £98,945.12.5 for the remainder of the entire works. The Company followed these directions carefully in the ensuing years until their canal opened in April 1811, just 5 months later than anticipated by the engineer.

Through all these troubled years, the unfortunates who suffered most were the work-force of miners, tradesmen and labourers who toiled in brutal conditions for small recompense. By the turn of the century the Company was often in their debt;[71] indeed at one stage the wage was only paid to those owed less than £30, greater sums being paid at five shillings in the pound! The Company Minutes only occasionally refer to welfare matters; the surveyor on one occasion was instructed to set up a sick fund (the Company subscribing merely 5 shillings per week) and on another the Committee granted but "£1.1.0 towards the expense of burying a workman who died today on the line of the canal".[72] Of the workmen's origins little is known, but something can be gleaned from local parish records[73] of births and deaths. The day of the Irish itinerant had not yet arrived, and names mostly suggest persons of northern stock. A random selection is

Robert Whitehead of Stonefolds, mason, killed in tunnel 10.3.1795; John Kell who was killed in the tunnel, a native of Stanhope-in-Weardale, 19.3.1810.

It is likely that upwards of 50 men lost their lives along the line of the canal during the construction years and, of course, one should add to this number the mothers and children who died, the latter often within weeks of birth in those squalid cottages and shanty towns set up adjacent to the pit heads on Puleside, at Gilberts Intack and elsewhere, the foundations of which can be found to this day.

BRIDGES AND AQUEDUCTS

Outram[74] initially estimated for 49 occupation and highway bridges, 30 culverts and 9 aqueducts. In accordance with the practice of the day there had been no formal negotiations with riparian landowners prior to the passing of the Act, and eventually more bridge crossings were agreed than originally intended, probably totalling more than 60. These were mostly stone arches (costing from £64 to £100 depending on type), built in squared, regular-coursed masonry. Their design is a standard pattern which can be attributed to Outram,[75] and occupation bridges rebuilt in later years compare closely with this style.

THE CONSTRUCTION OF THE HUDDERSFIELD NARROW CANAL 1794 – 1811

The standard design approximates to a semi-elliptical arch opening (rounded at the shoulders to permit freer access along the towpath), with a narrow string course above to define the highway surface level. The wing walls curve in plan to terminate in square, capped pilasters; the spandrel walls also curve in sectional elevation. Built with indigenous millstone grit and lime mortar, these simple inexpensive bridges are immensely strong and have had to withstand long periods of neglect. Several timber footbridges and swivel bridges were built before 1800 for the sake of economy, but were mostly replaced by durable masonry structures even before the canal opened in 1811. Farey[76] noted, following a visit in 1808, that:

> bridges over this canal have very injudiciously from motives of economy been made without towing paths under them, even at crossings of some of the turnpikes.

This situation was later changed by John Rooth[77] who claimed to have rebuilt "nineteen bridges over the canal with towing paths under them".

Some surviving occupation bridges were built skew to the canal (maximum angle 19°). None was built on correct principles because the longitudinal arch jointing in all cases is at right angles to the springings rather than parallel to the direction of the skew. This mode of construction gives rise to zones of weakness at the haunches of the arch which may crack and fail under load but none of the skew bridges today show any such defects, possibly because they sustain only light live loads. A partially correct solution to the skew arch problem was demonstrated by Chapman[78, 79] in 1787 and Outram used the method in 1798 on the Store Street Aqueduct (Manchester and Ashton Canal);[80] presumably the Huddersfield Canal skew bridges had been completed by this time, because the new technique was not applied to any structure so far as is known.

Outram was to set out seven aqueducts, the details of which are given below (the Engineer's named location noted in parenthesis):

Aqueduct	As planned	Estimated Cost	As built	Arch Pitch-Ratio (rise/span)	Remarks
Paddock (Longroyd)	2/18 ft arches	£298.50	Single skene 36.5 ft arch	0.25	Slight skew
Scarbottom (Skyerbottom)	not specified	£260.00	Single skene 38.5 ft arch	0.26	Slight skew
Marsden (Waterside)	2/8 ft arches	£152.00	Not known	—	Demolished or covered during railway era.
Diggle	2/8 ft arches	£160.00 (included embankment)	Single 16 ft semi-elliptical arch	0.47	Existing but disused
Uppermill (Mythom Mill)	2/18 ft arches	£312.50 (included culvert for mill tailwater)	Single skene 35 ft arch	0.22	Slight skew
Royal George (Quick Mill)	3/15 ft arches	£1076.00 (included embankment)	2/25 ft skewe arches	0.2	—
Stalybridge (Stakes)	4/15 ft arches	£1351.00	Single 55 ft span iron trough	—	Original bridge destroyed in 1799 floods

None of the bridges as built compare with the original proposals, and it is unlikely that costs bore any resemblance either. There is nothing remarkable about their designs; they are simple and functional carrying few embellishments and, with the exception of Diggle, all incorporate a skene arch. This type allows some economies in earthworks and lockage besides giving reasonable clearance below soffit level, although the low pitch ratios coupled with incorrect skew construction and crumbling mortar joints, has resulted in some arch movements at Uppermill and Scarbottom.

THE CONSTRUCTION OF THE HUDDERSFIELD NARROW CANAL 1794–1811

Fig. 8. Typical skew over-bridge built square at the springing.

Fig. 9. Stakes cast iron troughs in Stalybridge. The masonry arch carries the tow-path over the river Tame.

The floods of 1799 had a devastating effect on the canal along its entire length, destroying earthworks of the reservoirs and waterway, and washing away the two aqueducts at Marsden and Stalybridge. That at Marsden, close by the tunnel entrance, was rebuilt by John Rooth only in 1806 when funds allowed, but the aqueduct at Stalybridge was located on a section already in navigation and immediate action had to be taken. At that point the river crossing is narrow and headroom below the canal invert was limited to about 5 feet. There was too little height for a single arch and so Outram had designed a 4-span structure instead. In floods, the bridge impeded free flow and this contributed to its collapse, hence it would have been unwise to replace it with a similar bridge. Outram already had experience of a new structural type which admirably lent itself to such a situation. This was the single span cast-iron trough, one of which he had already constructed in 1794 to carry the Derby Canal across a narrow stream at the Holmes in Derby. The location at the Stakes was similar, which explains how he came to recommend "one arch of cast-iron" for the 55 ft. clear span. The new bridge was probably designed and cast at the Butterley Iron Works, of which Outram was the Managing Partner, and the high quality of the manufactured units is indicative of the advances made at that foundry in the few years of its existence. The components were assembled in situ and the whole probably commissioned by May 1800,[82] although Farey[83] states that it was finished in 1801.

THE CONSTRUCTION OF THE HUDDERSFIELD NARROW CANAL 1794 – 1811

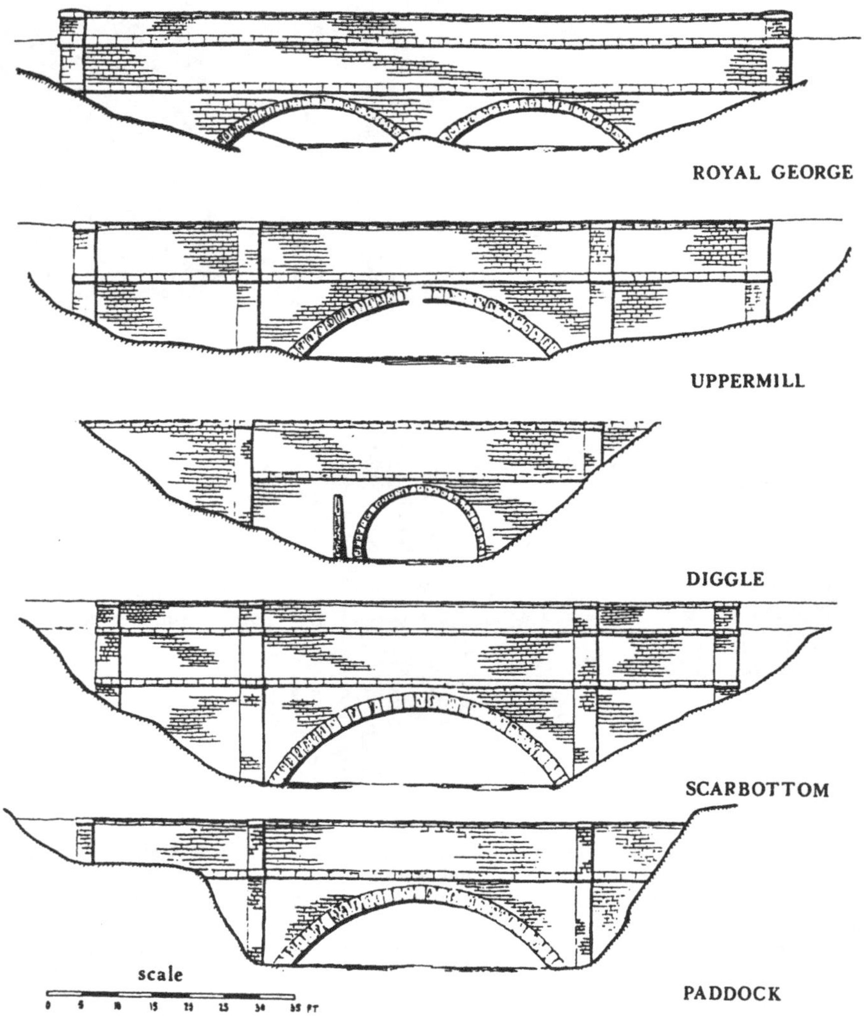

Fig. 10. The masonry aqueducts.

THE CONSTRUCTION OF THE HUDDERSFIELD NARROW CANAL 1794–1811

The Holmes aqueduct was, unfortunately, a structural failure by 1802, although its excessive deflections and inherent defects were no doubt observed before this time. Examination of the remains of the Holmes aqueduct[84] by the author showed that it was made up of light (1½" thick) plates of various sizes bolted together along scarfed joints to make up a rectangular cross-section for the canal, measuring 15 ft 6 in wide by 5 ft 6 in deep. The designer probably had little understanding of its likely action under load, and the evidence suggests that the wall plates failed in compression along the upper flanges. In turn this led to a transmission of the load such that a longitudinal tensile crack opened along the floor plates.

Outram probably realised that his first iron aqueduct was grossly underdesigned, but he still did not appear to understand its structural behaviour judging by the design of the Stalybridge aqueduct. This was stiffened in the floor and at the base of the wall plates, yet the vulnerable compression flange was not thickened as one would expect. This no doubt led to some movement in subsequent years, because a clumsy cross-bracing of 2 in diameter wrought iron bars was bolted on to the structure in 1875,[85] and the whole was partially supported by a vertical iron rod from the centre of the adjacent stone arch. Farey[86] described the aqueduct after his inspection, sufficient to confirm that the existing structure remains in its original form:

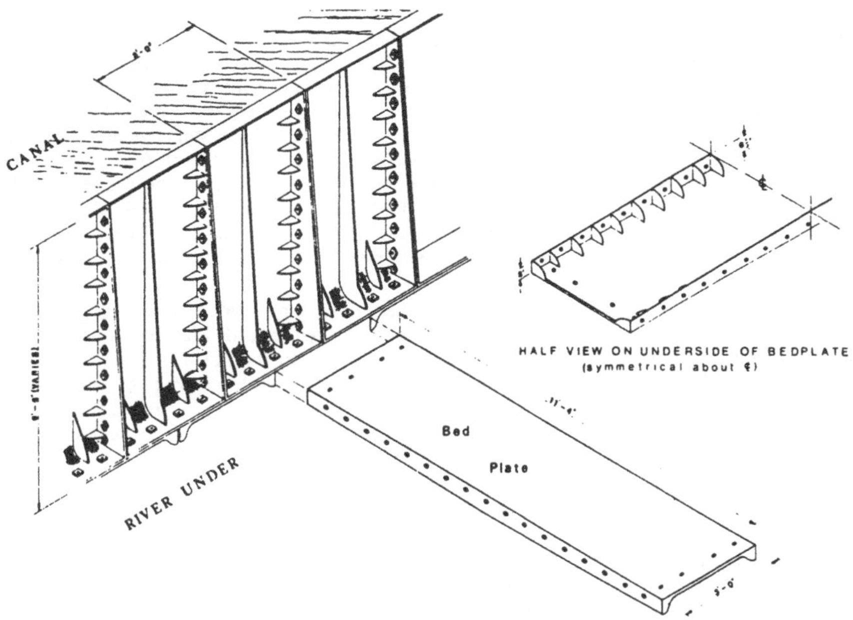

Fig. 11. Details of the cast iron trough aqueduct at the Stakes, Stalybridge.

> ...a square trough put together with flanches (sic), which are strengthened by brackets, cast at proper intervals for the rivets between them: the vertical flanches are widest at bottom and have their brackets cast between them.

He also describes the nature of the 'rust joints' between flanges which were manufactured from "iron turnings and borax, made into a paste with boiling water". Farey, not unreasonably, stated that the towpath should have been fixed to the trough instead of which "a ridiculous stone arch" was built at one side. The inaccessible and not well-known Stalybridge aqueduct is a sturdy, handsome bridge which is probably the oldest surviving structure of its type, next to Telford's Longdon-on-Tern[87] aqueduct on the Shrewsbury Canal.

THE CONSTRUCTION OF THE HUDDERSFIELD NARROW CANAL 1794–1811

WATER SUPPLY AND THE RESERVOIRS

Heavily locked routes such as the Huddersfield Narrow Canal were vulnerable in times of drought and this waterway became notorious in that respect. Outram, when presenting his report on October 22, 1793,[88] proposed reservoirs "to contain 14,900 locks full of water to supply 100 locks per day for four months together, exclusive of ample allowances for evaporation and absorption". Moreover he claimed that "this supply I conceive will be more than adequate to answer the largest trade that can be expected to be carried on the canal" and a capacious system was needed to discourage the tapping of mill waters in dry seasons. To his credit he recognised the need to protect mill-owners' interests because their continuing prosperity could benefit the canal company. The mill-owners disagreed[89] with his calculations and insisted on reservoirs for "not less than 20,000 locks of water, each lock to contain 180 cubic yards". Thus Outram, 24 hours after making his confident statement, changed his mind and accepted these proposals. It was fortunate (according to his report) that the countryside thereabouts provided ample situations in large deep valleys for up to five times the extent of the reservoirs he needed, but one wonders how much his assessment of this issue relied on guesswork rather than hydrological considerations. The industrialists insisted on other provisions to ensure that streams would flow uninterrupted and that only flood waters would be taken for the canal supply. These resolutions seem reasonable, bearing in mind that their sole source of power came from the 36 water-wheels (according to the deposited plans by Nicholas Brown) located on the rivers Tame and Colne.

The Act of 1794[90] permitted at least 607 million gallons to be stored in reservoirs within 3 miles of the line and withdrawals from watercourses a distance of 2000 yards, subject to safeguards for other users. Outram's estimate[91] submitted with the Parliamentary depositions quoted a price of £17,800 for the reservoirs without noting their positions, although the accompanying plan showed one location at Stayley Mill, another west of Uppermill village, three in the Wessenden valley above Marsden, and others at March Haigh and Slaithwaite. All but the first and last of these were on the summit level. In the event, only those at March Haigh (71 mg) and Slaithwaite (68 mg) were built by the Company. The Wessenden chain was constructed by a consortium of mill-owners before 1800[92] and re-developed by Huddersfield Corporation many years later. The amending Act of 1800[93] gave additional powers to take floodwater from the Colne and Tame, and seven years later Telford proposed even more reservoirs, thereby emphasising the inadequacy of the earliest proposals.

Slaithwaite was nearing completion by 1796 because it was needed to supply the canal from Huddersfield to Slaithwaite then building. As the canal continued towards Marsden, a new reservoir was commenced in early 1798[94] at Tunnelend (22¾ mg) immediately to the north of the Standedge Tunnel entrance. The Committee[95] was urging William Bailiffe to complete both reservoirs by November 1798 so that navigation could begin, but his attention was particularly directed to Slaithwaite dam which was leaking. Reservoirs were later built at March Haigh and Red Brook (68 mg) to augment supplies to Tunnelend via Red Brook and its tributaries.

There was a modification of the original plan when Outram recommended that the reservoir planned for Staley Mill be re-located on Swinshore Common[96] (capacity unknown) and the Committee readily agreed. Construction was well advanced by June 1796 as it was essential for the lower section of the canal to Ashton then nearing completion. Bailiffe[97] was directed "to execute the reservoir at Diggle as speedily as may be, so as not to exceed in the purchase of lands and execution the sum of £700". This was to provide for the next navigable section of canal then building from Stalybridge up to Woolroad. Obviously some business was being attracted to the waterway on both sides of the mountain because the Minutes report the purchase of trade boats and the appointment of wharfingers. The Surveyors were regularly asked for reports on business and tolls.

All reservoirs were contained by conventional earth dams with clay cores, although actual design and construction methods are unknown. Most seemed to give trouble by leakage for several years after completion, and this can only be attributed to ignorance of the basic engineering principles involved; also the lack of suitable plant and as much as anything, to inexperience and inadequacy of supervision. The state of the reservoirs in 1800 was said by John Rooth[98] to be so bad that

> "the whole of the water that could have been collected from the summit would not have been sufficient to support the then little trade upon the canal...there was not one reservoir out of five that would retain any water".

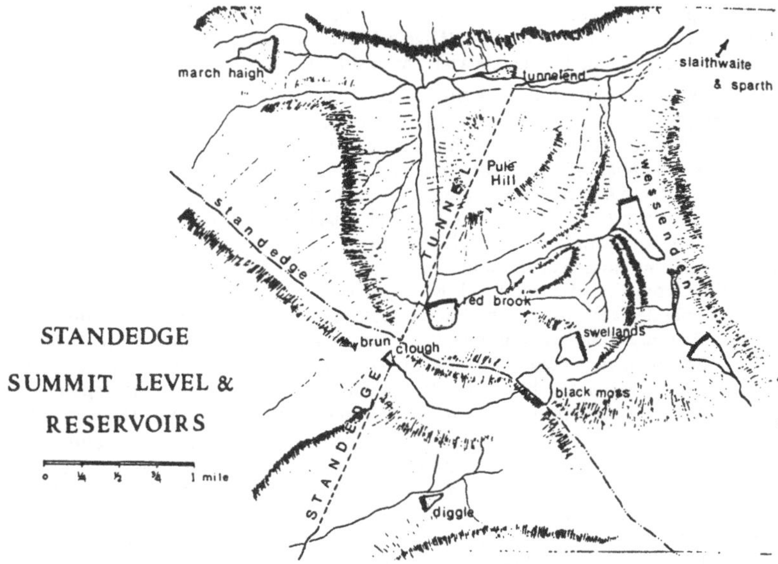

Fig. 12. Standedge summit level and reservoirs.

He stated also that there was a "fairly well-grounded apprehension of great destruction to the property of the country below the Slaithwaite Dam". In fact the Agent to the Earl of Dartmouth's estates in the Colne Valley had earlier written[99] to his Grace that

> "....there is to be a very large reservoir a little above the town which if the head *should break* would sweep away the greater part of the town, but the Engineer is perfectly confident there will be no danger......"

In fact Outram assured him that the slope on the downstream face would be so flat that "it could be mowed" and that "the inside would slope in the same manner".

Trouble persisted at Slaithwaite and Bailiffe was again instructed to secure the banks and to divert the waters of the Merry Dale Brook entering the reservoir until earthworks were sealed. Unfortunately the great flood of Summer 1799, occurred shortly after that instruction. The Slaithwaite Dam suffered badly and held after it had been hurriedly cut open in two places but the spillway at Tunnelend was too small, the dam was overtopped and the shoulders partially collapsed. The onrush of waters caused havòc in Marsden and beyond. Outram came up to view the wreckage and costed repairs at £3500 plus a further £3000 "to prevent future injury by the largest floods".[100] Committee members were demoralised by the turn of events and agreed that[101] "there are no possible means of raising money for repairing the said damages and reinstating the works of the said canal....". But money was borrowed somehow and the amending Act[102] of 1800 authorised the raising of further capital. Essential repairs were put in hand immediately and a youthful Edward Banks[103] commenced his successful contracting career on the reparations when his bids were accepted. Swinshore Common reservoir was so badly damaged in 1799 that it was abandoned for good. In 1844 J. F. Bateman,[104] in reporting on the possibilities of the Swinshore Valley for supplying the Manchester and Salford Waterworks Company came across the remains of this reservoir which still existed at that time and he stated that the failure was due to the neglect to provide an overflow.

For some time afterwards, the Company was preoccupied with maintenance and essential repairs to the battered remains of their waterway, but by June 1803 Slaithwaite was again causing trouble and a new tunnel and control cock were installed after the existing draw-off system broke under the settling embankment. In subsequent years frequent repairs have been effected there and the bank has gradually been flattened off near the toe to a slope of 1 in 4.

THE CONSTRUCTION OF THE HUDDERSFIELD NARROW CANAL 1794–1811

A spirit of optimism was to prevail after the passing of the Act of 1806.[105] At the time Rooth was engaged on[106] another summit reservoir at "Diggle Moss" (now called Black Moss, 18¾ mg) begun in 1806, as well as on rebuilding the Diggle reservoir "against floods as the one in January last injured it".[107, 108]

Telford visited and reported his new proposals in January 1807, and although much of the report concerns the tunnel, he also advised on water supplies. In considering the existing system Telford wrote,[109] "....it is now too late to express a regret that such large sums of money have been expended on such narrow dingles of small capacity", as these were liable to rapid filling with detritus washed down by the mountain streams. After computing the capacity of each reservoir he recommended construction of a further lake "to the eastward of that now forming upon Diggle Moss" and which would be connected with it. This came to be known as Swillers, or Swellands (54½ mg), which he claimed would be of some 50 acres with a head of 45 feet. Situated near the watershed, this reservoir may be discharged into the Colne Valley only, although flow from the adjacent Black Moss reservoir can be directed to either side of the divide. Telford also recommended reservoirs to be built at Bakestone Pits near Delph and noted that this, along with its network of supply channels would be expensive, but that it was a possibility if trade on the canal increased to such an extent that further water supplies were needed. Moreover the unhappy catalogue of events at Slaithwaite was extended by his quoting a sum of £2000 for necessary repairs and finally, he recommended a tiny lake at Sparth (8 mg) which Rooth was to put in hand just three months later.[110]

Although the sequence and timing of all works on the canal was to follow Telford's plan over the next 3 years, a final disaster afflicted the countryside on 29 November, 1810. Swellands earthworks failed, probably by underseepage, when the surface area was about 28 acres and the water depth was 7 feet. This took place at 1.0 am and the water rushed eastwards into the Colne Valley, inundating the valley at Marsden and beyond to Paddock. Factories and houses were wrecked and 5 people perished. The force of the waters was such that a 15-ton boulder was washed down 2 miles from the summit and the mountainside was denuded of soil.[111]

And so after 16½ years, the Huddersfield Narrow Canal was finished, to open on 4 April 1811 in the wake of that terrible tragedy on the night of the "Black Flood".

CONCLUSION

The final costs of the canal project were quoted as £396.267 (June 1811) but in view of modifications and repairs carried out during the following year it is more correct to include for this maintenance period; thus total costs were £402.653 (June 1812).[112] The cost of the Standedge Tunnel was £123,804 (June 1811) or £22.12s per yard compared with the original estimate of £9 per yard.

Several contemporary writers commented on these figures and the quality of the completed works. John Sutcliffe for example, an engineer of small reputation, gave some forthright views in his book[113] on the manner in which canal companies were established, their capital raised, and indeed how much the public had been misled on "the supposed profits that arise from the making of them". He criticised speculators who were "little better than swindlers", engineers for their "fictitious estimates" and yet had sympathy for managing committees who embarked on these schemes with little knowledge or guidance regarding their commitments. Sutcliffe singled out the Huddersfield Canal for denigration in order to support his views. The design, engineering direction and workmanship came in for severe criticism, as did the financing of the canal over those 16½ years. "The great loss of interest on the capital and time in executing this canal should be a lesson to all speculators" and "no confidence ought to be placed in the estimates of their engineers".

Similarly William Pontey,[114] a speculator himself and sometime employee of the Huddersfield Canal Company, criticised interminably and neither proprietors, engineers, nor anyone connected with the undertaking escaped his vindictive pen. John Phillips,[115] an enthusiast for inland navigations, was more circumspect preferring to note when describing the Huddersfield Canal "the wonderful sums of money raised for carrying on navigable canals is astonishing, but shews their utility and also the richness of the country".

The truth of the matter is contained in all these views.

THE CONSTRUCTION OF THE HUDDERSFIELD NARROW CANAL 1794 – 1811

As regards engineering aspects, Outram must be criticised for his sketchy proposals and unsatisfactory estimates, but his methods were not unusual by the rudimentary engineering procedures of the day. Even so, too many important details of the scheme were changed after work commenced, thus suggesting that preliminary planning owed more to guesswork than to lengthy, careful investigations. A competent contractor and resident engineer himself, it must have been obvious that Brown and others were unsuited by age and experience to their tasks and he should have insisted on changes and additions to the supervisory staff within months of commencing works. Possibly because of its remoteness and inaccessibility, it seems clear that the Huddersfield Canal did not command the same degree of Outram's attention as did his other commitments on the Peak Forest Canal and elsewhere, and by the late 1790's he became little more than advisor rather than consulting engineer with all the responsibilities that entailed. But then, serious illness over 18 months of the critical period of 1795 – 96 did little to help matters, nor did the continuing shortage of capital and the calamitous effect that had on progress, the quality of materials and workmanship. Outram was a spirited, outspoken man and it is likely that he was exasperated at the incompetent set-up of the Company and their self-inflicted problems. During the long spell after 1800, when little was accomplished through lack of finance, he does not appear to have spent much time on the project, and he finally withdrew after 1802. Until then, he had enjoyed success on major works elsewhere; but he was wasting his energies at Huddersfield and he probably felt it was time to go.

The source of many difficulties was unquestionably the Company's management structure. From the onset the Management Committee enthusiastically set about the task of engineering, supervising workmen, making agreements for damages and land sales—in fact, meddling in every task and petty detail for which they had already appointed staff. Outram did not seem to be accorded his place as director of engineering works, and the unfortunate Brown could have had little notion of the true nature of his duties. Only a weak and inexperienced man such as he would have tolerated the extraordinary decision of the Committee to appoint the ubiquitous William Pontey

> to attend to and keep memorandum of the payments to the workmen, the letting of small jobs and to superintend the conduct of the surveyor and overlookers on that part of the said canal which extends between Huddersfield and the east end of the tunnel.

The structure of management was thus divided and disorganised and for long it remained a defect in the enterprise. There was some improvement when John Rooth was appointed[116] in 1801 to superintend and manage the works of the canal. He was a dominating personality and this characteristic, in addition to the terms of his agreement, gave him an authority denied to his predecessors. In the ensuing years the Committee left many decisions to him, although it must be said that he was no engineering expert and he made plenty of mistakes during his term of office, no matter how competent he claimed to have been.[117]

Irrespective of the inherent defects of the organisation and the shortcomings of individuals involved, it must not be overlooked that the Huddersfield Canal was a major engineering enterprise even by today's standards. In retrospect therefore, was it not surprising that a group of local men comprising weavers, drysalters, farmers and gentlemen, with no previous knowledge of engineering or large scale business, should set up a Company, appoint a handful of professionals and work at this formidable task using such primitive equipment and methods? That they should complete the project even in 16½ years, is a tribute to their courage and faith in the future of their country. And perhaps the canal never brought prosperity to the proprietors and speculators, but it did much for the common people who inhabited those desolate moors, bringing employment and communications which cannot easily be valued in monetary terms. For all the failings, the Canal was successful; it was a splendid engineering concept and its completion was a fine achievement.

And in conclusion, later years saw the advent of the railway and the construction of three further tunnels through Standedge adjacent to the canal, in 1849, 1870, and 1894. All utilised the original tunnel for access and removal of spoil, and thus in a sense the waterway subscribed to its own demise and ultimate closure in 1944. Today there is fresh promise for the canal, largely through the efforts of voluntary bodies, which it is hoped will secure a future for water supply and recreational pursuits; fitting enough purposes for this great navigable waterway after 180 years of industrial life.

THE CONSTRUCTION OF THE HUDDERSFIELD NARROW CANAL 1794–1811

ACKNOWLEDGEMENTS

Research for this paper was made possible by a Grant from the Leverhulme Trust for the preparation of a biography of Benjamin Outram (1764–1805), Civil Engineer. Original material has been drawn from a number of sources including the House of Lords' Record Office; The Institution of Civil Engineers' Library; British Transport Historical Records, London; the City of Leeds Public Library Archives; Wakefield County Record Office; Kirklees Libraries and Museums Service; The Institute of Geological Sciences, Leeds; British Rail, Leeds.

A number of private persons have rendered assistance and advice, and are cordially thanked, including Dr R. C. Witter of Chester; Mr. D. M. Stakes, British Waterways Board, Leeds; Mr. R. Bennett and Mr. P. Walsh of Manchester.

For help in the preparation of material, Mrs. Mary Rea and Mr. S. D. Brown of Belfast are thanked.

REFERENCES

Abbreviations:	**CHCC**	Committee of the Huddersfield Canal Company
	GAHCC	General Assembly of the Huddersfield Canal Company
	HOLRO	House of Lords' Record Office
	BTHR	British Transport Historical Records, London.

1. Pontey W. *A short account of the Huddersfield Canal etc.* Hodson, Cambridge, circa. 1812. Library of the Institution of Civil Engineers.
2. Hadfield, E. C. R. and Biddle, G. *Canals of North West England (Vol. 2)* David & Charles, Newton Abbott, 1970.
3. Report of a meeting at Huddersfield, 30.5.1793. Radcliffe Mss. 319/C/1 Leeds City Archives.
4. Schofield, R. B. The promotion of the Cromford Canal Act of 1789: an engineering study. *John Rylands University of Manchester Library Bulletin,* Autumn 1981.
5. Standing Orders of the House of Commons, 14.3.1793.
6. Outram, B. Report 22.10.1793. See Pontey, W. op. cit. (1)
7. Resolution of mill owners on the line of the Huddersfield to Ashton Canal, 23.10.1793. Radcliffe Mss. 319/C/1. Leeds City Archives.
8. Brown practised as land surveyor in Wakefield in later years.
9. Rooth, J. Report 29.6.1809. See Pontey, op. cit. (1)
10. Hadfield, E. C. R. & Biddle, G. op. cit. (2)
11. Correspondence of W. Elmsall to the Earl of Dartmouth, 7.7.1793. DT3/9 Leeds City Archives.
12. Sutcliffe, J. *A treatise on Canals and Reservoirs,* Rochdale 1816. (see later, the evidence suggests that this offer was made in 1799, five years after the project commenced).
13. Rooth, J. Report on the Huddersfield Canal to the **GAHCC** 27.6.1811. HRP/6/8/X/K979. **BTHR.**
14. Huddersfield Canal Act 34, George III. c53, RA 4.4.1794.
15. Quarter Sessions Records 15.1.1795. Wakefield C.R.O.
16. Minutes **CHCC** 11.7.1794. **BTHR.**
17. Ibid. 17.5.1796.
18. Ibid. 29.6.1797.
19. Ibid. 15.9.1797.
20. Sutcliffe, J. op. cit. (12)
21. Minutes **CHCC** 3.8.1797. **BTHR.**
22. Ibid. 26.3.1801.
23. Minutes **GAHCC** 26.6.1817. **BTHR.**
24. Ibid. 23.12.1802.
25. Minutes **CHCC** 26.12.1806. **BTHR.**
26. Telford, Thos. Abstract of the report relative to the state of the works on the Huddersfield Canal. 29.1.1807. Kirklees Libraries & Museum Service.
27. Correspondence of John Rooth to Robert Firth 24.2.1817. Tomlinson Mss. Kirklees Libraries & Museums Service.
28. The Gunter chain of 22 yards.
29. Rees, A. (ed.) Art. on Canals. *Encyclopaedia or Universal Dictionary of Arts, Science and Literature,* Longmans Hurst, London 1819.
30. Standedge Canal Tunnel Plan (LNWR) 22.1.1894 No. 11231A. British Rail, Leeds.
31. Rees, A. (ed.) op. cit. (29)
32. Telford, Thos. op. cit. (26)
33. Standedge Canal Tunnel Plan. op. cit. (30)
34. Rooth, J. Report on the Huddersfield Canal. 29.6.1809. Pontey, W. op. cit. (1)
35. Minutes **CHCC** 1.11.1802. **BTHR.**
36. Correspondence, John Rooth to Robert Firth, op. cit. (27)
37. Report on the state of the finances and works of the Huddersfield Canal by the Committee and Engineer. 30.6.1796, Pontey, W. op. cit. (1)
38. Minutes **CHCC.** 7.4.1797. **BTHR.**
39. Ibid. 16.11.1796.

THE CONSTRUCTION OF THE HUDDERSFIELD NARROW CANAL 1794 – 1811

40. Telford, Thos. op. cit. (26)
41. Report by the Committee and Engineer. 30.6.1796. op. cit. (1)
42. 25" Ordnance Survey 1892 series.
43. Standedge Tunnel. Unpublished typescript. Tomlinson Mss. Kirklees Libraries & Museum Services.
44. Minutes **CHCC.** 10.2.1797. **BTHR.**
45. Report by the Committee and Engineer 30.6.1796. op. cit. (1)
46. Telford, Thos. op. cit. (26)
47. Minutes **CHCC.** 9.4.1802. **BTHR.**
48. Information from the Institute of Geological Sciences, Leeds.
49. Farey, J. *General view of the Agriculture of Derbyshire*, Vol. **III**, London 1811.
50. Particulars of the discovery of some very significant balls of stone found in the works of the Huddersfield Canal, communicated by Sir Joseph Banks, 21.4.1796. *Vol* **LXXXVI**. *Phil. Trans. Royal Society* London.
51. Report by the Committee and Engineer. 30.6.1796. op. cit. (1)
52. Minutes **CHCC.** 18.5.1798. **BTHR.**
53. Ibid. 21.2.1799.
54. Ibid. 20.3.1799.
55. Sutcliffe, J. op. cit. (12)
56. Correspondence, Geo. Worthington, Clerk, to Shareholders 2.9.1799. Pontey, W. op. cit. (1)
57. Minutes **CHCC.** 26.6.1801. **BTHR.**
58. Presumably of the firm of Pritchard & Hoof who constructed the New Harecastle Tunnel. See Rolt, L.T.C. *Navigable Waterways.* Longmans, London 1969.
59. Minutes **CHCC.** 22.7.1801. **BTHR.**
60. Ibid. 2.10.1801.
61. Ibid. 23.12.1802.
62. Minutes **GAHCC.** 10.2.1803. **BTHR.**
63. Sutcliffe, J. op. cit. (12)
64. Correspondence B. Outram to I. Perrins 25.6.1798. Boulton & Watt Mss. Birmingham P.L.
65. Correspondence B. Outram to Boulton & Watt 16.8.1798. Boulton & Watt Mss. Birmingham P.L.
66. Minutes **GAHCC.** 27.6.1805. **BTHR.**
67. Huddersfield Canal Act 40. Geo. III. c39 RA 30.5.1800.
68. Minutes **CHCC** 7.11.1805. **BTHR.**
69. Rooth, J. Report of 26.6.1806. op. cit. (1)
70. Telford, Thos. op. cit. (26)
71. Minutes **CHCC.** 18.12.1800. **BTHR.**
72. Ibid. 29.6.1797.
73. Parish Records 1776 – 1812 St. Bartholomew's Church, Marsden.
74. Outram, B. Estimate of the expense of making a navigable canal (Huddersfield Canal 1794). **HOLRO.**
75. The design is similar to those by Outram on the Peak Forest Canal.
76. Farey, J. op. cit. (49)
77. Correspondence John Rooth to Robert Firth. op. cit. (27)
78. Chapman, W. (Rees A. ed.). Art. Oblique Arches. *Encyclopaedia or Universal Dictionary of Arts, Science & Literature,* Longmans Hurst, London, 1819.
79. Pallet, J. E. "The contribution of William Froude to the development of oblique arch". *Trans. Newcomen Soc.* Vol. **XLV** (1972 – 73.)
80. Schofield, R. B. Benjamin Outram (1764 – 1805) Canal Engineer Extraordinary. *Proc. Inst. C.E.,* Part **I**, 1979, 66. Nov. 539 – 555.
81. Minutes **CHCC.** 18.9.1799. **BTHR.**
82. The adjacent towing path was rebuilding at that time.
83. Farey, J. op. cit. (49)
84. Stored in the Highways Depot, Derby.
85. Date is marked on components.
86. Farey, J. op. cit. (49)
87. Reported to have been completed one month later than the Holmes Aqueduct, see L.T.C. Rolt. op. cit. (58)
88. Outram, B. Report of 22.10.1793. op. cit. (1)
89. Resolution of mill owners on the line of the proposed Huddersfield and Ashton Canal. 23.10.1793. Radcliffe Mss. 319/C/1. Leeds City Archives.
90. Huddersfield Canal Act. 34. Geo. III. c53 RA 4.4.1794.
91. Outram, B. Estimate. op. cit. (74)
92. Minutes of a meeting of millowners situate on the River Colne, 4.4.1799. Dartmouth Mss. D77/5/2. Leeds City Archives.
93. Huddersfield Canal Act. 40. Geo. III c39 RA 30.5.1800.
94. Minutes **CHCC** 15.12.1797. **BTHR.**
95. Ibid. 12.10.1798.
96. Ibid. 29.7.1795.
97. Ibid. 21.2.1799.
98. Correspondence, John Rooth to Robert Firth. op. cit. (27)
99. Correspondence, W. Elmsall to the Earl of Dartmouth 9.2.1795. DT 305/7/2. Leeds City Archives.

THE CONSTRUCTION OF THE HUDDERSFIELD NARROW CANAL 1794 – 1811

100. Correspondence, Geo. Worthington, Clerk to Shareholders, op. cit. (1)
101. Minutes **CHCC.** 23.8.1799. **BTHR.**
102. Huddersfield Canal Act. 40. Geo. III. c39 RA 30.5.1800.
103. Dickinson, H. W. "Joliffe & Banks, Contractors". *Trans. Newcomen Soc.* Vol **XII.** (1931 – 32.)
104. Information from A. I. B. Moffatt, University of Newcastle-upon-Tyne.
105. Huddersfield Canal Act. 46. Geo. III. c12 RA 31.3.1806.
106. Minutes **CHCC** 26.12.1806. **BTHR.**
107. Rooth, J. Report 26.6.1806. op. cit. (1)
108. A. I. B. Moffatt stated that 5 deaths were caused by this failure; but the Company Minutes and Report makes no mention of it.
109. Telford, Thos. op. cit. (26)
110. Minutes **CHCC.** 23.4.1807. **BTHR.**
111. Rooth, J. Report to the **GAHCC** 27.6.1811. **HRP/6/8X/K979. BTHR.**
112. Ibid.
113. Sutcliffe, J. op. cit. (12)
114. Pontey, W. op. cit. (1)
115. Phillips, J. *A general history of inland navigation,* Crosby, London 1805.
116. Minutes **CHCC.** 23.4.1801. **BTHR.**
117. Correspondence, John Rooth to Robert Firth, op. cit. (27)

DISCUSSION

Mr. Peter Russell mentioned Nicholas Brown's career following his dismissal by the Huddersfield Canal Company. He was a water engineer in the early 19th century and the reservoirs at Galton, built about 1826, were among his works. **Professor Schofield** said that Brown had been retained by the Huddersfield Company to undertake land surveys, even as late as 1815. Subsequently he practised as a land surveyor from an address in Wakefield and either he, or his son, was still in business there as late as 1840. Mr. Russell also referred to investigations on the summit reservoirs now being undertaken by Sir William Halcrow and Partners, consulting engineers to the Waterways Board, from whom he had obtained much historical data and engineering information. Mr. Russell was presently researching for a book on J. F. La Trobe-Bateman, the Victorian engineer associated with several Pennine reservoirs, one of which had been built at the site of Outram's abandoned Swinshaw dam.

Dr. R. C. Witter referred to the four 'wides' in Standedge Tunnel, all of which were located between Diggle and the Red Brook shafts. They had been added he understood in 1848 when the first railway tunnel was completed and might have been used for the removal of spoil. **Mr. Neville New** asked why the tunnel had been closed. **Professor Schofield** said that there had been a steady deterioration since the canal had been abandoned by Act of Parliament in 1944. There had been many minor rock falls over the years and the tunnel was no place for the amateur boatman. The Waterways Board still maintained the tunnel as this was an important water supply route and had to be kept open.

Mr. Ronald Carr asked for information on the three railway tunnels at Standedge. **Professor Schofield** had explored the system, including many of the adits, but his principal interests were canal engineering. He nevertheless described the fine quality brickwork in the single track tunnels and the interesting cast-iron arching at the centre point known as 'the Cathedral'.

Mr. Ian McNeil in discussing the 'water engines' used on tunnel construction confirmed that similar devices had been in use for mine pumping as early as 1745, although he had never encountered the water balance system previously. **The President** added that the Agricola woodcuts showed similar devices.

Mr. R. J. M. Sutherland asked about the iron aqueduct at Stalybridge and the curious bracing added in 1875. In describing how this supported the trough at mid-span, **Professor Schofield** stated his belief that this was the third such bridge to be built in Britain, and was apparently the oldest still in use.

The President called on **Mr. R. W. Rennison** to propose the vote of thanks. He referred to the problems encountered by the early engineers, particularly when they had not the convenience of the Ordnance Survey maps for route selection and setting-out purposes. Mr. Rennison thanked Professor Schofield for a fascinating, interesting paper, and the vote of thanks was accepted with acclamation.

5

John Pinkerton and the Birmingham canals

S.R. Broadbridge

IN SOME WAYS John Pinkerton, canal contractor, would seem to have a claim to be called the Thomas Brassey of canals. With his brother he worked on the Driffield Navigation (1768), Bishop Soil sluices (1770), Market Weighton Navigation & Drainage (1772), the Hedon Navigation (Hedon Haven) (1774), the Selby Canal (1775), on other parts of the Aire & Calder (1775–8), and the Calder & Hebble Navigation (1776–80). Alone he contracted for, amongst others, the Billingham embankment and drainage (1777), the Erewash Canal (1777–80), the Birmingham & Fazeley Canal (1783–9), the Dudley Canal tunnel (1785), the Basingstoke Canal (1788), the Gloucester & Berkeley (1793), the Lancaster Canal (1794), and the Barnsley Canal (1793–9). Closer investigation reveals, however, more differences than similarities, differences which, at root, are those between the situations of railways in the nineteenth century and of canals in the eighteenth.

Rarely did Pinkerton contract for the whole of any navigation, unless it was small. This was largely because, unlike his railway successors, he lacked financial resources and stability. He had no permanent nucleus of labour, relying almost entirely upon local manpower, and no equipment of his own, even of the simple forms required at the time. Above all, however, he was totally dependent upon the advances of money made to him, weekly or monthly, by his employers. As a result he was never able to take the whole responsibility for construction upon himself; he was always subject to supervision, even in minor details. Not surprisingly, therefore, considerable friction resulted, not unaided by his character, which seems to have been prickly and litigious beyond the normal. One result was a multiplication of documents which cast a flood of light upon the details of eighteenth-century contracting.

Amongst them those which illuminate his labours in the Birmingham area are particularly prolific.

His earlier work had been in Yorkshire and eastern England, and his first foray into the Midlands was to build the Broadwaters extension to the Birmingham Canal in 1783.[1] Six months later he was mentioned as 'engineer' to the Fazeley Canal, for which the Act had been obtained at the same time.[2] He certainly did not step off on the right foot with the staid John Houghton, clerk to the company, since before tendering

> he informed [Houghton] that, in some or most cases or places, it was customary for contractors to allow a per centage of . . . about four per cent. upon the amount of the contract to the principal agent of the Company . . . and requested [him] to inform him what was customary to be given . . . before he made his estimates, as the rate . . . must be taken into account.

Later Pinkerton claimed that this was a joke, based on an anecdote he had heard about the Leeds and Liverpool, but the damage was done. Houghton virtuously informed him that the Birmingham company did not allow their servants to take perquisites and, presumably, did not trust him thereafter.[3]

'Frequent disputes' between Pinkerton and Bough, the company's superintendent of works, over the Broadwaters extension, 'particularly respecting the cement to be used', were allegedly settled by the committee in Pinkerton's favour, but they were only an earnest of what was to come.[4] In particular, Pinkerton contended that,

> After the foundations were set out by the Company's Agents, they had, properly speaking, done with him, and ought not to have further interfered with him.[5]

But he could never achieve this degree of independence. Bough supervised his work to ensure that standards did not fall below those in the contract, and friction was certain to result. Nevertheless his work was presumably satisfactory, for he was immediately employed on the Fazeley canal, for which the Act had been obtained at the same time as Broadwaters.

As was apparently normal with the Birmingham company, although work started contracts were not signed, in part because Pinkerton's estimates did not agree with Bough's and they were submitted to 'some neutral Person' for a decision.[6] Surprisingly, both he and James Watt were asked to give an estimate of the cost of the canal prior to the advertisement for tenders, which would seem to put him in a privileged position.[7] In the event his tender arrived late, but he was awarded half the contract: that from Minworth to Fazeley, including the Curdworth Rocks and the Dunton Tunnel to be completed by midsummer 1787. The other section into Birmingham, which included the heaviest lockage, was given to Thomas Sheasby, later engineer to some South Wales canals.[8]

It was not until January 1787 that a contract was prepared, and at that time a covenant was added

> enabling the Committee (in case Mr Pinkerton should neglect to proceed in the execution agreeable to their expectations) to take the Works into their Hands and get the same executed . . . and to change the expence thereof to Mr Pinkerton's Account.[9]

Not surprisingly Pinkerton, who was by now deeply involved in the construction, raised objections: 'when I consider the power it is giving a superintendent over me, I cannot, consistent with my own safety, but make some objection to it'; but it was insisted upon because of 'the circumstances in which Mr Pinkerton then stood with the Dudley Company'.[10] The contract, though dated 25 July 1786, was not executed until 18 May 1787. It committed Pinkerton to complete the works by 24 June 1788 for a price of £11,563 8s 11¾d, to be paid as they progressed, with £500 retained for three years after completion, during which period he would keep the canal in repair.[11]

We must now turn to Pinkerton's circumstances with the Dudley Canal. In 1785 this company obtained an Act to extend to a junction with the Birmingham Canal at Tipton by means of a tunnel through the spine of the Black Country plateau. In August they advertised for a contractor for the tunnel:

> The said Tunnel to be executed by the Yard forwards, including the sinking of the Shafts, driving the Headway, widening the same to contain the Brick-

work . . . to be made nine Feet three Inches wide, five Feet six Inches deep in Water, and seven Feet high above Top Water, and to be continued from End to End in a straight Line; The Centers to be made three Inches higher than the above Dimensions to allow for sinking of the Arch, the bottom to be an inverted Arch of Culver Bricks of four and a half Inches, on sound Foundations, and nine Inches on springy Lands . . . the Side Walls and Top Arch to be nine Inches thick, and where it shall be found necessary, the Side Walls to be fourteen Inches thick, the Top of the Arch to be clayed four Inches thick, and Brick Ends or other sound Materials laid thereon, so as to conduct the Water in small Streams down the Outside of the Brick Work, thence through small Inlets into the Tunnel.

The lower arch was to be clay-lined, the mortar made with lime and boiling water with rough grit sand and the bricks beaten down with a mallet.[12] The contractor was to enter into a penal bond for £4,000. Notwithstanding his other commitments, and his apparent lack of experience in tunnel work, Pinkerton tendered and was accepted, with William Jessop as his surety. John Wildgoose was surveyor, to superintend the works under Thomas Dadford the engineer, with the assistance of Abraham Lees.[13]

By August 1786 it was found necessary to order Wildgoose and Lees to

pay strict Attention to the Execution of the Tunnel by Mr Pinkerton . . . and particularly that they do not suffer any soft or bad bricks to be made use of and in general to observe that the several Works are completed according to the Tenor of the Contract.

upon pain of dismissal for neglect.[14]

Less than a fortnight later a special committee meeting had to be called to authorise the treasurer to pay Pinkerton £500 on account over and above his agreed weekly payments, fixed at £150 in June, and he was ordered, in future, to pay out

such Sum of Money only, as Abraham Lees & John Wildgoose shall certify to be actually expended on the Course of such Week, and that Mr Thomas Fieldhouse Do see to the Application of such Money Weekly, in discharging Workmens Wages, and Bills for materials, and other necessary Matters, And that no Money be Issued or paid by the Treasurer but for those purposes.[15]

This was not the end of the problem, however, and three months later it had to be resolved 'that all orders enpowering Mr Ab. Hawkes the Treasurer to pay any Money to Mr John Pinkerton be and are hereby rescinded'.[16]

Pinkerton and Jessop were ordered to attend the next meeting, all goods and materials in his possession were taken over by the company, and legal opinion was sought as to whether they could enter his works without prejudice to his bond.[17] A meeting of representatives of the committee with Pinkerton and his attorney took place at Lord Dudley's London house on 11 May, at which the company refused to accept a suggestion that his contract be waived and the value of his completed works left to arbitration. They also refused an offer of £1,500, claiming at least £2,000, which they would accept in instalments, with interest, provided there was good security, 'as they have not the least Confidence in his personal Security alone'. Even so, they claimed they would lose upwards of £2,000 since 'many Applications made by Contractors [show] that the Tunnel may be compleated at the Price agreed for by Pinkerton'.[18]

No agreement was reached then but, under threat of legal proceedings, Pinkerton offered the necessary £2,000 for release from his contract, to be paid in four instalments, the last on 24 June 1790, with three securities. The company were to have all tools and materials and enter into the works immediately.[19] In the event proceedings had to be threatened to obtain the third instalment on time and actually commenced for the last. It was not until June 1792 that payment of £2,072 8s 6d including interest was finally completed.[20]

Pinkerton later claimed that his over-expenditure was

> On account of the extreme hardness of the strata, and the immense quantity of water it made, a circumstance not apprehended from the information which had been collected from the most intelligent colliers in the neighbourhood.[21]

On their part the company complained that he left the care of the works to a nephew, 'who was a fine gentleman and neglected them' and added, for good measure, that they 'never wished to employ Mr Pinkerton again, having a very low opinion of him'.[22]

This was not Pinkerton's only other contract in the area at this time. When the Act for the Birmingham and Fazeley Canal had been passed in 1783, it had laid upon the company the obligation accepted by the original proposers in 1782, together with the Trent and Mersey Canal, of completing the Coventry Canal from Fazeley to Fradley. Early in 1784 a meeting of the four companies concerned agreed that, as a gesture of good faith, all should start the execution of their works, laying out £1,000 each. At the same time an Act should be sought to formalise the agreement. This Act (25 Geo III c 97) was passed in 1785 against considerable opposition from the Birmingham Canal, who did not want completion enforced until the route to Oxford was complete and who resented being made to pay half the costs of the Act. Nevertheless the Trent and Mersey had begun construction from their end in 1783 and Pinkerton had been appointed contractor in February 1784. By the time the Act had been passed he had completed four miles to Dennis Brook and the two companies agreed that he should continue as contractor for the remaining seven miles to Fazeley at a contract price of £8,716, the whole to be completed by Christmas Day 1786, and the two companies advancing £1,000 each in turn as the work progressed. The price included keeping the works in repair for three years, 'in order to prove that they are completed in an effectual and substantial manner,' £500 of the price being withheld for that period.[23]

It was not long before trouble arose. The Birmingham company had for some time been trying to ensure that the Broadwaters extension was completed properly, but even after Pinkerton had claimed to have repaired the defects it was reported 'left very incompleat' and the company refused to settle his account.[24] They also asked for a report on the Fradley/Fazeley works which, proving adverse, was sent to the Trent and Mersey, emphasising that

> The defects complained of are so various, and of such magnitude, as seem to call for the serious attention of both Committees, that the Undertakers may forthwith amend the imperfections of their past Works and avoid similar errors in their future proceedings.[25]

John Pinkerton and the Birmingham Canals 39

A deputation from the Birmingham committee met the agent and engineer of the other canal and went over the section.

> They found that the part under execution by Mr Pinkerton was very incompleat injudiciously laid out and in many places very defectively executed . . . particularly in the direction of the Canal which . . . is curved in a very improper manner the pudling benching and many other particulars . . . seem to have met with equal inattention . . . and in General the defects are so notorious as not to require the Eye of an Engineer to discover them.

They recommended that no more be set out until Bough was present, 'having found that what he has heretofore done . . . has been approved and scarce ever alter'd by Mr Smeaton Watt and others who Survey'd it'—a eulogy they were surely to regret before many months were past.[26] It must be assumed in this context that Pinkerton, as the Trent and Mersey's original contractor, would inherit some of the animosity felt towards the other canal for promoting the amending Act and because they did not send committee members to meet this deputation.

Although Pinkerton agreed to go over the line with Bough and set it out again, next year it was once more reported that the works were 'very defectively executed and that considerable damage [was] likely to ensue unless great attention [was] paid thereto by the Engineer'. However, by July 1788 final accounts were ready for inspection, though Pinkerton did not announce the section as complete until December.[27]

Meanwhile work was proceeding on the Fazeley Canal, though it was held up because of delays in purchasing land, another common failing of the Birmingham company. Some of this was required for clay to make bricks, but some of it was land through which the canal had to pass, often urgently.[28]

At the end of 1786 Pinkerton reported that the levels were apparently wrong between Minworth and Fazeley and was allowed a resurvey by Bough in the presence of his nephew, George Pinkerton—apparently the 'fine gentleman' before mentioned. Their report showed an error of 'as much as four feet and eight inches' and possibly more between

Minworth and Birmingham; and Bull, the company's engineer, was ordered to resurvey the whole line completely, 'without having the least reference to the Field Books Notes or Calculations of Mr Bough'.[29] According to Pinkerton, Bough's reaction was: 'Damn it, I knew I was three feet wrong, but I'd not think it had been as much as it is.' George Pinkerton thought that this was what 'occasioned Mr Bough to become a great enemy to Mr Pinkerton'.[30]

Bull's report showed that Bough had represented the fall from Minworth to Fazeley as 77ft instead of 71½. Sheasby had estimated for this when tendering for the length, whereas Pinkerton had used his own, correct, section. The difference saved £1,400 on the estimate and made a whole lock unnecessary at Minworth.[31] Bough had even more to answer for, since Sheasby found that, as a result,

> he had lost the part of the line from Minworth to Fazeley. . . . For, by deducting the sum the mistake had added to his estimate, he found that his estimate was under Mr Pinkerton's and that, therefore, he was justly entitled to have the execution of the whole line from Birmingham to Fazeley.

He was reported to be 'mortified', but the company were to have more cause to be so than he.[32]

They felt progress was too slow and Pinkerton was told to make a weekly return of his proceedings while Bough (not, one would have thought, the best person for this) was ordered to make weekly returns 'by way of cheque to Mr Pinkerton's report'. He said not enough men were employed.[33] For his part Pinkerton admitted that 'the line has not the appearance . . . of being pushed on with that resolution that you may think necessary', but claimed this was because he had felt it necessary first to complete the canal from Fradley to Fazeley so that he could 'take the water after us till we get amongst the locks'.[34]

There were also complaints against Sheasby at this time, in particular for using poor materials, but he seems to have put these right and we hear of no further criticisms of his works. Those against Pinkerton, however, grew as time passed, and were eventually detailed in full when the differences between him and the company reached the stage of legal arbitration. It was alleged that, instead of puddling, gutters were filled with 'soil, gravel or sand', 'that such bad bricks were used they could scarce find enough for fronts' and 'that bricks were laid without mortar and mortar without bricks'. In the inverted arch of a culvert, 'the bricks were so loose that they might be picked out with one hand'.[35]

Similar evidence of the state of the works in October 1788, when Pinkerton was removed, was given by John Houghton and by William Hodgkiss, bricklayer, who 'refused to work for Mr P. because the materials were so bad' and affirmed that, at the fifth lock, 'when a boat went in and out, the whole face and back of the walls would move'.[36]

> There was such constant failures in the embankment, bridges, locks, &c that it never could be clearly ascertained when the canal was completed.

Many locks had to be taken down and rebuilt, and in some 'the mortar was as soft as when first laid'.[37] According to Bull, the brickwork in the Dunton tunnel was not in line and was 'daubed over . . . with clay, in order, as he supposes, to prevent the joints being seen'. It fell in while building.[38]

Clearly, at this stage the company was trying to make out a case. Nevertheless, when in the 1960s Staffordshire County Council had occasion to take down one of his bridges for road widening, they found it only a half-brick thick on the arch,[39] and similar faults occurred on other contracts carried out by Pinkerton.[40]

Whatever the workmanship, in late 1788 he began to exceed his estimated price; money was cut off except by special decision and soon afterwards cut off entirely, pending 'a particular Acct of the nature of His Extra Works' because 'the Money he has already receiv'd greatly exceeds what ought to have been advanc'd, considering the quantity of Work yet remaining to be done'.[41]

Clearly the committee were tightening financial control. Shaw, the walling surveyor responsible for superintending the works, was dismissed for his 'connivance at the late transactions in the Works under [his] immediate inspection'. One must assume he had been receiving his share of the 4 per cent, for he was previously warned:

no one can doubt from the experience you have had that you must know it to be wrong yet you suffer such work to be done without the Committee hearing a word from you. . . . Your past conduct clearly shows you can but have little if any regard for the interests of your employers . . . let not the Undertaker . . . any more make you believe that Sand is Lime or that Clay unburn'd can be bricks.[42]

Notwithstanding all this, Pinkerton claimed in December that the canal would be open in six months, but in February 1789 he was told that, unless he could give 'sufficient and undeniable security' that he would complete in three months, he would be removed from the works, which would be carried on by the company at his expense. He 'declin'd looking out for any Security' and was allowed to appoint an agent, at his own expense, to keep an eye on the company's proceedings, though there is no evidence that he took advantage of this. Work continued and the first commercial boat passed on 11 August; the committee travelled over the line in their own boat six days previously.[43]

It might have been thought that all troubles were over, but in many respects they had only just begun. In December navigation was stopped by the failure of two of Pinkerton's locks at Curdworth. He was called back from the Basingstoke Canal to repair them and inspect the others 'to prevent as much as possible further damages of a similar kind'.[44] This was the beginning of a long history of collapses and repairs, which was not ended until thirty-five years later, when a report on repairs could at last state:

As respects the Locks at Curdworth & Dunton which were built by the renowned Mr Pinkerton & which have in truth proved so costly to the Company it is proper to remark that the work necessary to put them in a tollerable state of security may now be considered as done.[45]

The financial problems were settled earlier, but required probably as much effort, at all events from senior officials and the committee. What was involved was the amount done by Pinkerton over and above his contract, for which he was owed by the company, and the amount carried out by the company in order to complete the canal, for which he owed them. It was not helped by the accounting system in use, which involved a detailed double-entry system but with no attempt to check on costs or even, as will appear, to find what lay behind the receipts given in return for expenditure.

After the canal was complete Bough was ordered to make accounts for settlement with both Sheasby and Pinkerton. The former was settled by August 1792, but Pinkerton refused even to see the committee to discuss the question. Relationships were presumably not improved when, faced with a demand from the Trent and Mersey for their share of the money withheld for three years from Pinkerton, which amounted to £287 10s, including interest, the company paid by a bill drawn on him on the grounds that he was 'considerably in the Company's debt'. Not surprisingly he refused to accept it and the account was not finally closed until all others were settled.

At length, in November 1794, legal proceedings were instituted, and Pinkerton eventually met the committee and agreed to comment on the accounts after going over the line with Bough. Nevertheless, it was not until December 1795, following more legal threats, that he sent his version of the accounts. They were referred to a sub-committee, which ordered a resurvey and called him in for consultations to reconcile the considerable differences between the two accounts.[47] Because of his other canal obligations these consultations took place at long intervals.

Pinkerton alleged that the accounts, by which the company claimed £5,870 11s 0¾d, were 'complicated'.

> The bills are in general without quantity or price and some of them, in one line, saying bricks, timber, &c so much money.[48]

Certainly the examples he gave seem to bear this out; it would be difficult, if not impossible, to see from them to what extent he had completed, or overfulfilled, his contract. He was allowed to see the originals (though we do not learn whether they were any more detailed) and presented his case. It was referred to a 'select committee' of three, who reported

> That he received from the Company much more indulgence than he was entitled to in deferring to take the completion of the canal into their Hands. . . .

> That when [they did] Mr Pinkerton had then receiv'd of the Company £12263 15s 6d being £700—6¼ [sic] more than the total of his contract
>
> That from that time payments made by the Company . . . were correctly and minutely kept and clearly prove what Summs have been paid . . . in the completion of the Canal & in the repair thereof . . . [being] £5447 13 2½.

Pinkerton had claimed £3,051 2s 8½d as 'extras' which, if admitted, left him in debt to the company £3,096 10s 11¾d. According to Bull and Bough the 'extras' were only £1,337 13s 4½d. For his part Pinkerton claimed the company should have been able to complete the canal for £1,883 8s 8½d, but the sub-committee recommended that, as the company had an 'unquestionable right' to £3,096 10s 11¾d, Pinkerton ought to be 'required forthwith to make good that sum', while the account of extras went to arbitration.[49] As might be expected Pinkerton could not 'conform to the terms of the report . . . being fully convinced that there is a considerable balance due to me from the said Company'. He claimed £3,716 16s as extras, together with £1,380 caused by advance of prices due to delay in starting, making a total of £2,456 5s 3¼d due to him.[50]

When the suit came to an issue much argument took place about this matter of wage levels which served mainly to show that, whatever happened to wages and prices in general, the beginning of construction meant an increase in the prices of labour and materials in that locality. Thomas Dutton, a canal cutter working on sub-contract, said Pinkerton allowed him, at first, to draw 9s a week per man, then 10s and finally 12s as wages rose. Men on day work—which Pinkerton called 'a way no contractor can support'—received 1s 6d per day, later rising to 2s. When the company had tried to speed the work up by raising wages it

> did not produce the effect desired; for it was not given to the workmen generally but to the gangsmen. . . . These gangsmen gave the men under them no more wages than . . . before, so that the advances of prices did no more good, than that of adding to the pockets of the gangsmen, who were, in general, distinguished by the name of Bough's Darlings. . . . The foreman used, when he reckoned with them, to put down as many cube yards as would answer the wages he gave them.[52]

Whatever the true position, Pinkerton was not prepared to accept any of the figures put forward by the select committee and therefore it was not possible to agree on a reference to arbitration.

In any case Pinkerton wanted engineers as arbitrators, which the company refused, on grounds which are far from clear. Legal action was therefore commenced and, after several delays, looked like reaching an issue when the suit was withdrawn on Pinkerton's attorney agreeing to be bound by the decisions of three mutually chosen arbitrators.[51] Messrs George Humphreys, John Lawrence and Richard Poyton were agreed upon and Pinkerton entered into a £10,000 bond to abide by their award, which was to be completed by 23 January 1800. In fact, it was postponed five times and finally delivered on 10 September 1801.[53]

At the arbitration Pinkerton's main technical arguments were faults in the specifications and delay:

> It was impossible for him to begin till the Co's engineer had duly set out the line, and the line was not set out, so that he could begin, till eight months after the time in which everything should have been ready . . . the most material . . . parts of the line were not purchased . . . till eighteen months after the date of the agreement, even those very portions of land which he intended first to have begun upon.

As a result prices rose and he lost £2,750 15s 9d in increased costs of earthworks, buildings and 'agency' and in his own time and expenses.[54] However, his main fire, or that of his attorney, was reserved for the company's servants:

> nothing which, in the strictness of language, can be called blame attaches to either party . . . the Company . . . never in their collective capacity wished anything harsh, unjust or oppressive. . . .

The fault lay with

> falsehood and calumny, with which little, mean and envious individuals assaild his character, and poisoned the minds of the Company's Committee.[55]

The role of chief devil was reserved for Houghton:

> the man who could perform many of the labours of his occupation . . . unassisted and alone, and who, perhaps, never in his life, would have been laid

open to the inspection of the public, had not the Company, fortunately for themselves, contracted with the Defendant.

The rooted hatred which . . . this witness cherished against the Defendant, might be watered from many springs, but derived its chief strength from the character of the Defendant.

Such men as Bough . . . were . . . men at once vicious and needy, too ignorant to penetrate through a hypocritical disguise, and too low bred not to consider the head clerk . . . as a very great man.

Unfortunately . . . the Defendant came . . . and from his talents; and demeanour . . . was beginning to have considerable weight with the Committee . . . the head clerk . . . saw . . . that either the Defendant's character must be destroyed or his own consequence lost.[56]

Apart from the fact that the committee's minutes reveal no time at which Pinkerton was beginning to have considerable weight with them, the character of Houghton is not convincing. As revealed in his letters and memoranda, which are rather more wordy than pure business would necessitate, he seems rather likeable than otherwise, though a trifle pompous, and dedicated to the honest service of the company. Certainly there was no conspiracy by him about land purchases, as Pinkerton alleged; these were a perennial problem, caused mainly by the slackness of Meredith, at that time the company's solicitor. It is possible that Houghton was guilty of suppressing a rough account of extras submitted by Pinkerton in 1788 and found by accident by his attorney when looking, with permission, in the company's solicitor's box—probably to save himself the trouble of redrawing the accounts. However, it was of no import and the manner of its rediscovery does not suggest a deep-laid plot.[57]

Much argument raged round the question of responsibility for bad work. It has been suggested above, and this can be abundantly supported by entries in the committee's minutes at the time, that much of the blame must be placed upon Pinkerton. Some of it may have been caused by interference by the company's agents; Pinkerton makes a persuasive case that they adversely affected the mortar-mix. Nevertheless there seems no doubt that Pinkerton scamped his work, that there was little supervision of it by the company's servants on the spot, though much bickering, and that there was not adequate control of costs when he had gone. The episode is certainly revealing in the inadequacies of the men available for technical posts at a time when demand was fast outrunning supply and training.[58]

At the end of it all, the arbitrators ordered the company to pay Pinkerton £436 in full settlement, each side to pay its own costs. In the company's case these amounted to £334 5s 8d for solicitor's fees and £38 2s 4d for the arbitrators' expenses.[59]

The matter was not going to end this quietly, however. Immediately it was over Pinkerton published an account in full of the proceedings, accompanied by a preface in which he complained

> that this action was commenced against the Defendant for not cutting a Canal through land which the Prosecutors did not allow him to enter, for not using good workmanship and materials, in buildings they did not suffer him to erect, and for not coming to a settlement, when they never furnished him with an intelligible account of the expenditure they called on him to defray,

and accused the company of using forged documents from dead men. The publication was a

> vindication of his character, to explain how it came to pass that such an award was given, in a case so plain and clear, [as a result of which he] lost all his labour, and a considerable part of his property.

The company, however, had lost more, 'by employing such a miserable description of agents as those who have occasioned all this mischief'.[60]

Since this publication included all the attacks on Houghton which have been quoted, and many more, the latter took legal action, as a result of which the book was found 'a scandalous and malicious Libel', Pinkerton being fined £200 and imprisoned for a month. The general assembly of the company resolved that this action was 'highly proper', paid Houghton's costs and gave him £500.[61]

What happened to Pinkerton we do not know; certainly he does not appear to have executed any more contracts for a long time after the lawsuit.[62] In that sense it is quite clear who suffered most from the episode.

References

1 British Transport Historical Records, minutes of committee of Birmingham and Birmingham and Fazeley Canal Navigations, BCN 1/4, 15 August 1783. (Submission of tender.)
2 Ibid, 2 February 1784.
3 J. Pinkerton, *An account of the cause lately arbitrated between the Birmingham Canal Navigation and J. Pinkerton* (1801), 314–15 (evidence of J. Houghton).
4 Ibid, 259 (Pinkerton's evidence, not borne out by committee minutes).
5 Ibid, 252 (Pinkerton's evidence).
6 BTHR/BCN 1/4, 23 April 1784.
7 BTHR/BCN 1/5, 11 February 1785.
8 Ibid, 2 January 1786, 22 April 1786. For Sheasby in South Wales cf C. Hadfield, *The Canals of South Wales and the Border*, passim.
9 Ibid, 12 January 1787, 26 January 1787.
10 Pinkerton, op cit, 130–1 (letter of 16 March 1787, 281).
11 BTHR/BCN 1/5, 18 May 1787.
12 *Aris's Birmingham Gazette*, 9 September 1785.
13 BTHR, minutes of the committee of the Dudley Canal DDC 1/3, 26 September 1785, 12 December 1785 (contract signed), 25 April 1786.
14 Ibid, 29 August 1786.
15 Ibid, 8 September 1786.
16 Ibid, 8 January 1787.
17 Ibid, 23 January 1787, 6 February 1787, 26 February 1787.
18 Ibid, 17 May 1787.
19 Ibid, 3 July 1787, 20 July 1787. Jessop was one of the securities.
20 Ibid, 12 February 1789, 20 June 1790 and minutes of assembly of Dudley Canal, BTHR/DDC 1/2, 25 June 1792.
21 Pinkerton, op cit, 232.
22 Ibid, 292.
23 BTHR/BCN 1/5, 28 October 1785.
24 Ibid, 4 May 1786, 19 May 1786, 24 July 1786.
25 Ibid, 30 August 1786.
26 Ibid, 13 October 1786.
27 Ibid, 3 November 1786, 3 August 1787, 10 November 1787, 11 July 1788, 12 December 1788.
28 Pinkerton, op cit, 87 (letter of 14 July 1786 from Pinkerton), 4, 84 (letter of 26 April 1787 from Pinkerton—the quotations on the two pages cited do not exactly tally).
29 BTHR/BCN 1/5, 8 December 1786, 22 December 1786.
30 Pinkerton, op cit, 234, 230.
31 BTHR/BCN 1/5, 5 January 1787.
32 Pinkerton, op cit, 234.
33 BTHR/BCN 1/5, 20 April 1787, 27 April 1787.
34 Pinkerton, op cit, 84 (letter of 26 April 1787 from Pinkerton).
35 Ibid, 8–9 (evidence of Lee, the company's solicitor and member of their committee).
36 Ibid, 29–30, 36–8.

37 Ibid, 17 (Lee), 39–40 (evidence of Blastus Hughes and John Taylor, bricklayers).
38 Ibid, 54–5.
39 Information from Mr Granville-Edge. It must be noted that it had, notwithstanding this, survived nearly 200 years.
40 Eg, within six weeks of opening the banks of the Basingstoke Canal collapsed twice (P. A. L. Vine, *London's Lost Route to Basingstoke*, 57) and the Greywell Tunnel on that canal seems to have been as badly built as that at Dunton (R. Harris, *Canals and their Architecture*, 112).
41 BTHR/BCN 1/6, 4 October 1788, 18 October 1788.
42 BTHR, Letter-book of the Birmingham Canal, BCN 4/371A, 4 October 1788, 23 September 1788 (Houghton to Shaw). Though catalogued in BTHR this is in private possession but can be obtained through Warley Public Library, to whom I am grateful for making it available.
43 BTHR/BCN 1/6, 26 December 1788, 13 February 1789, 17 February 1789, 5, 6 August 1789; Pinkerton, op cit, 17.
44 BTHR/BCN/371A, 19 December 1789 (Houghton to Pinkerton).
45 BTHR/BCN 1/11, 26 September 1823.
46 BTHR/BCN 1/6, 12 November 1790.
47 BTHR/BCN 1/7A, 21 November 1794, 5 December 1794, 23 January 1795, 27 February 1795, 18 December 1795, 26 February 1796, 9 September 1796.
48 Pinkerton, op cit, 565 (Letter of 1 November 1796 from Pinkerton).
49 BTHR/BCN 1/7A, 8 December 1796, 20 January 1796, 7 April 1797. It is to be noted that there are often minor arithmetical errors.
50 Pinkerton, op cit, 122, 124–5 (letters of 20 April 1797 from Pinkerton).
51 BTHR/BCN 1/7A, 1 April 1797, 30 June 1797, 6 June 1798; minutes of assembly of Birmingham Canal BCN 1/42, 28 September 1798, BCN 1/7B, 8 March 1799.
52 Pinkerton, op cit, 185–6, 304, 565.
53 BTHR/BCN 1/7B, 22 March 1799, 18 October 1799, 1 November 1799, 15 November 1799, 13 December 1799, 24 January 1800, 30 May 1800, 5 September 1800, 12 December 1800, 12 June 1801.
54 Pinkerton, op cit, 157, 549–50.
55 Ibid, 57–8.
56 Ibid, 527–9.
57 Ibid, 430–3.
58 Eg, 'A mistake having been made in fixing a mark for the top water line on the Fradley Canal, previous to it being cut, it occasioned the levels for the Birmingham Canal to be carried on, from the junction, upwards of nine inches too low' with a consequent need to raise all embankments and bridges as far as the first lock (ibid, 554).
59 BTHR/BCN 1/7B, 18 September 1801. It should be noted that a similar case, not completed until 1812, was the outcome of Pinkerton's association with the Barnsley Canal—cf W. N. Slatcher, 'The Barnsley Canal,' *Transport History*, I, no 1 (1968), 48–66.
60 Pinkerton, op cit, viii–x, xxv.
61 BTHR/BCN 1/42, 1 October 1802.
62 Someone of the same name was working under another contractor on the Thames and Medway and alone on the Yantlet Creek c 1823. If this was the same person he would have been about 80 at the time (Cf C. Hadfield: *The Canals of South and South-East England*, 92, 94n).

6
Managerial organization on the Caledonian canal, 1803–1822

Alastair Penfold

Proposals for constructing a canal through the Great Glen were made throughout the 18th century. In 1773 James Watt surveyed the line for the Commissioners of the Forfeited Estates, proposing a canal 10 ft deep at a cost of £164 000. John Knox pressed for its building in his survey of the highlands in 1784, as did local ministers in their contributions to the 'Statistical Account of Scotland' in the 1790s. Pressure came also from the British Fisheries Society, and John Rennie was commissioned to do further survey work in 1793. Continuing agitation from the Highland Society and the British Fisheries Society, combined with mounting anxiety over emigration, caused the Treasury to send Telford on a survey of the highlands in 1801. His involvement with Sir William Pulteney and the British Fisheries Society, undoubtedly helped him in this commission.

He was instructed by the Treasury to select the most suitable sites for fishing stations on the West Coast, to plan road and bridge communications on and between the mainland and islands, and to examine the possibility of constructing a canal through the Great Glen. Telford reported favourably to the Treasury and was sent back the following year. A select Committee was appointed in 1803 to consider his proposals, recommending that they be implemented as soon as possible. In the summer of 1803 Parliament approved the creation of a board of Commissioners to supervise the expenditure of an annual grant of public money for canal construction.

Acts of Parliament were obtained in 1803–04 and full scale canal construction commenced in the summer of 1804. The canal was opened in a very unfinished state in 1822, mainly to pacify mounting criticism over escalating costs. Telford's involvement in the project after that date was confined to compiling annual reports. The Board of Commissioners continued in existence until 1920.

Problems of managerial control were increased on the Caledonian Canal by distance, supply of raw materials and the unprecedented scale of the project. This paper will attempt to describe the managerial structure developed by Telford in response to the above problems, with special reference to the division of responsibility between Telford, Jessop, site engineers and contractors, and the contrast between survey and construction personnel.

The managerial team employed on the construction of the canal remained essentially unaltered throughout much of the construction period, both with regard to personnel and structure. To ensure clarity a brief diagramatic outline of the structure is given in Fig. 1.

Telford and Jessop

Telford and Jessop were both involved in the setting up and running of the managerial team and played a role in the overall project. In assessing their differing contributions it is necessary to answer the following points: who was responsible for the recruitment and, therefore, size of the managerial team; who decided their responsibilities; who assessed wage and contract rates; who decided what

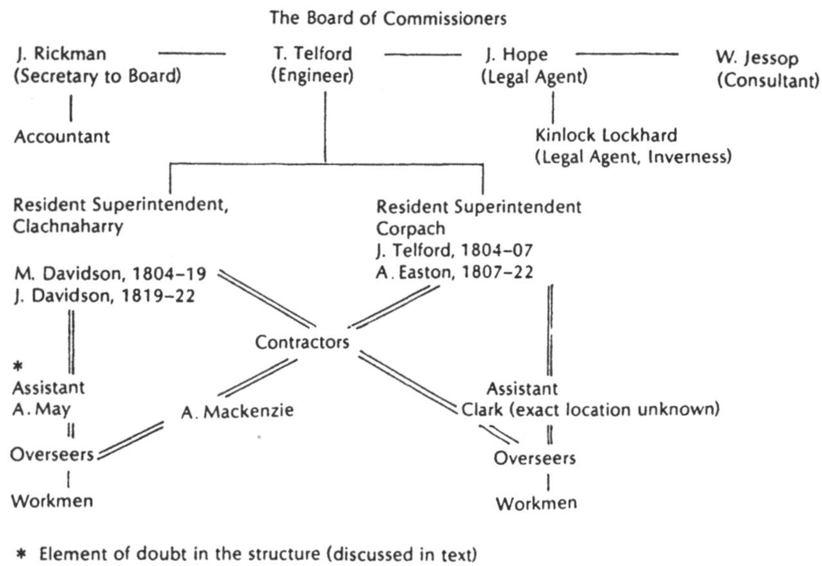

Fig. 1. Managerial structure

materials should be used; who appointed the contractors; and who implemented changes in design and line?

Jessop, together with Rennie, gave evidence before the select committee on the various Highland improvement schemes in June 1803, prior to the creation of the Board of Commissioners.[1] When asked to comment on Telford's proposals for the canal, he stated that he could not give 'an accurate judgement, not having seen the country'. However, he did provide an estimated cost of the whole project, of about £22 000/mile 'under the favourable circumstances of ground tolerably even, and the soil moderately good'.[2] As in the London Bridge enquiry, he was called upon by the Government to give his opinion on the feasibility of the project as one of the country's leading civil engineers. On 4 August, 1803, the newly-appointed Board of Commissioners took measures

> 'for obtaining the opinion and assistance of Mr William Jessop, another eminent and experienced engineer'.[3]

He was ordered to inspect the proposed line of canal with Telford,[4] which he did in October 1803, reporting back to the Board in February 1804.[5] A revised estimate was also submitted by him at this date. After 1804 he made a joint inspection with Telford in October or November of every year up to his retirement in 1813. He died in 1814.

Recruitment of managerial staff began in the early autumn of 1803 when Telford appointed John Wilson as Resident Engineer or Superintendent at Corpach, and A. May and W. Mackenzie as Superintendent/Pay Clerk at Clachnaharry.[6] These appointments were made before Jessop had become involved with the project.[7] It is probable that John Wilson was the same Wilson who had worked with Telford on Pontcysyllte Aqueduct as a contractor. Nothing is known about the background of Andrew May or William Mackenzie except that the latter was a land surveyor from Inverness. J. Smith was employed at Corpach to negotiate the purchase of country timber and arrange for its cutting and delivery.[8] A Mr Mason was employed to investigate the location of quarries. He was also involved in the preparation of a report on the entrances to the canal with Murdoch Downie.[9] A considerable number of surveying assistants was also taken on,[10] so that by the time of Jessop's visit in October 1803 a full-scale survey team was in operation.

Construction work commenced on a very limited scale in December 1803, but it was not until the passing of the Second Canal Act in June 1804 that moves were made to appoint permanent supervisory construction staff. Jessop was again excluded from the major decision-making over recruitment, simply endorsing Telford's appointments. Evidence for this belief comes from the Second Caledonian Canal Report which contains extracts from Telford and Jessop's correspondence over the new posts. The Commissioners reported that after having

> 'taken into consideration a representation made to us by Mr. Telford respecting the propriety of providing a constant superintendence on the spot, we appointed Mr. Mathew Davidson . . . (Clachnaharry District) . . . and Mr. John Telford . . . (Corpach District) . . . These persons were recommended to us by Mr. Thomas Telford, with the entire approbation and concurrence of Mr. Jessop, as men of tried ability and long experience.'[11]

Telford received authorization from the Board to appoint them.[12] He had written to Jessop on 8 June, 1804, suggesting John Telford and Davidson:

> 'The works being upon a scale of uncommon magnitude, and in a district of country unaccustomed to operations of this nature, I propose that such persons only shall be instructed with the chief superintendence and the execution of the principle works, as have to my own and your knowledge, for ten years past, been employed upon works of a similar nature whose abilities may be relied on and who are likely to enter with zeal into the spirit of the undertaking. . .'[13]

Telford had known Davidson virtually all his life, working with him as a stonemason in his native Eskdale before leaving for London in 1782. He had employed Davidson as site engineer on several bridge projects in Shropshire before appointing him Resident Engineer for Pontcysyllte Aqueduct.[14] Jessop had been Principal Engineer on the project and had become acquainted with Davidson. However, it is evident from the above correspondence that Telford and not Jessop appointed him. Similarly, John Telford was an ex-Telford Ellesmere Canal employee who had been fully versed in his working methods, acting as his personal assistant for some time.[15] Telford concluded his letter of 8 June by proposing that Davidson be provided with 'such assistants for counting the men and measuring the works as may from time-to-time appear necessary'.[16] For the middle district he proposed to employ 'persons of an inferior description' because 'the works there can be occasionally examined and checked by Mathew Davidson and John Telford'.[17] Jessop replied to Telford's proposals on 9 June, agreeing to all the appointments.[18] He made no amendments to the managerial team as suggested by Telford.

Wilson, May and Mackenzie were relieved of their supervisory posts during the ensuing months, Wilson re-emerging as one of the principal contractors, and May as assistant and pay clerk to Mathew Davidson. Mackenzie appears to have left the employment of the Commissioners altogether. Wilson's assistant continued to serve him in his new capacity,[19] (his name appears on wage certificates after 1804). It is not known how much Telford was involved in these reappointments, and details are extremely scarce of Telford and Jessop's involvement in the appointment of managerial staff below the level of superintendent. Jessop requested that Telford deal with the appointment of all assistants in June 1804.

> 'What Assistants they may want cannot at present be specified, they must from time to time be appointed as circumstances may call for them and this should be left to your discretion'.[20]

Telford and Hope decided upon the appointment of Kinlock Lockhart as the Commissioners' legal representative in Inverness in 1805.[21] Jessop does not appear to have been consulted. John Telford died suddenly in 1807 and was replaced by Alexander Easton, a former stone mason and Telford road inspector in Argyll.[22] Again Jessop does not appear to have been involved. No further managerial appointments appear to have been made before Jessop's retirement in 1813. Telford made Mathew Davidson the senior superintendent, appointing him as his deputy with responsibility for handling all financial matters for the canal when he left for Sweden in 1808.

It is clear that Telford, rather than Jessop, was responsible for managerial recruitment. He was more in touch with the day-to-day running of the canal and therefore better able to devise and appoint a managerial team for its supervision. In this instance at least, Jessop merely rubber stamped Telford's decisions.

Responsibilities

It was Telford, rather than Jessop, who decided upon the responsibilities of the managerial team. In his letter to Jessop of June 1804 Telford proposed that the superintendents fix contract rates in 'particular places' with the proviso that the 'state of such rates . . . be always immediately communicated to me and to be subject to my determination'.[23] Through such a system he undoubtedly hoped to control from a distance by financial manipulation. Telford requested that Jessop comment on the 'above proposal' and hoped that it gave 'a reasonable prospect of the works being carried on with economy, fidelity and success'.[24] No records survive of Jessop either opposing Telford's managerial team or suggesting improvements. It was, therefore, Telford's set of instructions which was adopted by the Board and used on all subsequent occasions for the control of site construction staff.

Jessop's participation in the fixing of contract and wage rates was considerable, as he accompanied Telford on the inspection tours when these matters were considered. He was not involved, however, in deciding the wage rates for the excavation of Corpach and Clachnaharry basins, which had been determined by Telford in 1803.[25] This work had presumably been performed by direct labour, as no contractors were appointed until June 1804 when Telford and Jessop received instructions to 'determine the rates of expense of the sundry works upon the line of canal . . . [and the price of] labour for the different sorts of work.'[26] Telford had contacted the Board of Commissioners in June 1804 requesting Jessop's assistance on the above points:

> 'It would be a great satisfaction to my mind, to have the assistance and advice of Mr. Jessop in Scotland'.[27]

Immediately after, Telford contacted Jessop, giving him a full breakdown of his proposals:

> 'Every part of the works which can be so managed, to be executed by measure of rates or prices, to be determined by you or myself after having maturely weighed every circumstance relative to the different works. All the cutting, puddlings and embankments to be let in small lots to different persons, . . . the general rules for prices to be determined by you or myself'.[28]

The various rates were fixed in August 1804 after a joint inspection. Telford and Jessop had previously been instructed to take careful consideration of local wage rates in deciding upon those for the canal.[29] This would suggest that those members of the managerial team who had been in the Highlands longest and therefore had a greater local knowledge (i.e. Telford) played the decisive role in the fixing of wages. This has to be counterbalanced, however, by the fact that Jessop had a far greater knowledge of wage rates throughout the country. Any substantial increase in the contract price was subject to approval by Telford and Jessop.[30] This policy was adhered to throughout Jessop's involvement in the project and beyond. Jessop's contribution to the fixing of prices would thus appear to have been considerable both in 1804 and after. It is not known how much he was influenced by Telford's fixing of wage rates in the autumn of 1803 or by his superior local knowledge.

Materials

In the location, purchase and transport of materials it would again appear that Telford provided the driving force. The survey team was fully involved in this department before Jessop set foot in the Highlands, Telford opening negotiations with Cameron of Lochiel for the purchase of country timber in late September 1803.[31] In the joint instructions of 1803 and 1804, Telford and Jessop were requested to pay special attention to the location of materials.[32] As in previous instances, however, the person with the most local knowledge had the greater say in deciding such matters.

By the summer of 1804 Telford had completed his third 'period of duty' in the Highlands; Jessop was just arriving for his second. Decisions regarding the purchase of timber may have been taken jointly, but it was Telford who handled all the initial enquiries and correspondence. Jessop was probably very influential in the purchase of ironwork, although here again it was Telford who handled the correspondence, shipping and erection. Contracts for ironwork were put out to tender, the offers made by Outram and Co. (Jessop's partner) and Hazledine being accepted. Hazledine had for some time been associated with Telford projects, providing the ironwork for the Pontcysyllte Aqueduct and many of Telford's bridges.

Machinery requirements were decided on at the joint inspection of August 1804 and orders were placed almost immediately afterwards for three Boulton and Watt pumping engines and a large amount of plateway.[33] The canal's first dredging machine appears to have been designed entirely by Jessop,[34] who was probably instrumental in the adoption of a steam dredger. Telford did not involve himself in the development of dredging machines after Jessop's departure, preferring to import specialized knowledge on the matter, in the person of Bryan Donkin.[35]

Contractors

The choice of construction contractors was made by Telford in June 1804. In that month he wrote to Jessop suggesting Simpson, Wilson and Cargill as the masonry contractors.

> 'In all matters relative to the execution of building I mean to employ John Simpson with whose abilities and character you have also been acquainted for upwards of ten years past. John Wilson and James Cargill are with him . . .'[36]

Jessop 'heartily concluded with Telford in appointing the contractors straight away rather than open the contract to competition [and] run the risque of getting very inferior men'. He also agreed to Telford's proposals for letting the earthwork in small lots to different persons, although he thought that 'the size of such lots [ought] to be proportioned to the ability of the undertakers' in the hope of finding some who 'may be able to manage a large contract, or a number of small lots'.[37]

Simpson, Wilson and Cargill had all been employed on the Ellesmere Canal, undertaking the masonry contract on the Pontcysyllte Aqueduct. Simpson had also been employed on virtually all of Telford's bridge contracts from 1790 onwards. The decision to bar competition on the masonry contract was thus understandable, given the wealth of experience offered by the successful contractors. Jessop's proposals over letting earth contracts in multiple units appears to have become standard practice, only half a dozen 'firms' being involved in the main excavation work between 1804 and 1822. Records of contractors are not numerous, only the names of Meak, Gillies, Ross, Davies and Hughes being recorded amongst the general contractors.

Design changes

There were several changes in the design of canal features and line alterations. Telford examined the line of the canal in 1801 and again in 1802, using James Watt's survey as a guide.[38] The second survey was comparatively detailed, listing the number of locks, bridges, aqueducts and culverts. This plan was accepted by the Parliamentary Select Committee and was included in the application to Parliament for obtaining the initial grant of £20 000.[39] In the Autumn of 1803 Telford was ordered to draw up a detailed plan of the canal suitable for reapplication to Parliament for a second more permanent Canal Act.[40] Jessop received instructions in late September to examine the line of canal.[41] In December 1803 he was ordered to prepare a report and estimate on the canal, using information supplied by Telford and his team of surveyors.[42] His report was submitted to the Board in February 1804, having first been endorsed by Telford.[43] It was then included in the papers used in obtaining the second Canal Act.[44]

The estimate of £474 531 which was included in Jessop's report was considerably more than Telford's 1802 figure (£349 617), mainly on account of the adoption of more expensive lock designs. The number of locks had been reduced by two to 23, and aqueducts reduced from twelve to seven, although bridges were up from twelve to 23.[45] As in other aspects of decision-making it is difficult to assess how much Jessop was influenced by Telford and his surveyors. Line changes had been decided upon before he even arrived in the Highlands, the position of the sea locks being altered after a joint report by Murdoch Downie and Mason.[46] General survey work proceeded well into 1804 and it was not until June of that year that Telford and Jessop received instructions to mark the exact position of the locks, bridges, weirs and culverts.[47] Many of the canal features were positioned during August 1804 although weirs and culverts were not definitely fixed, as their position depended on the raising of the fresh-water lochs, a task which was not to be performed until the completion of the project. Many of the weirs included on the original plans were not built until the 1840s.

The siting of locks and bridges in the middle district was also left open, as it was not intended to commence operations in that area until the two outer districts had been completed. Jessop was not involved in the detailed surveying of the middle district which did not start until the year of his departure in 1812.

Jessop appears to have been very much involved in the major design changes relating to lock construction and in the adoption of cast iron as a suitable material for bridge and lock gate construction. Telford had submitted turf-walled lock designs in his original 1801 survey. He argued that the slowness of operation would be counterbalanced by the cheapness of construction (at £5000/lock), and the accompanying estimate was based on the use of turf locks.[48] Jessop budgetted for more conventional masonry locks (at £7449 each) in his February 1804 estimate, which increased the overall cost of the project by £56 000. Telford, after considering the revised designs, gave his approval, though he stated he would not have accepted the revised plans if the cost for individual locks had exceeded £10 000. It is not known who was responsible for the design of the locks as built, although drawings of them are included in the Atlas of plates to Telford's 'Life'.[49]

Telford and Jessop decided to increase the size of the locks after a joint investigation into the type of vessel most likely to use the canal. They were assisted in this task by Sir William Rule, who supplied information on the size of Navy frigates and warships.[50] The decision to use cast iron for the lock gates appears to have been taken jointly, for Telford states in the 1814 Report that 'the high price of oak led Mr. Jessop and myself to adopt cast iron'.[51] The use of cast iron for accommodation bridges was also a joint decision, the basic design being based on swing bridges constructed in the West India Docks, a project which had strong Jessop connections.[52]

To conclude, Jessop's role in the project was essentially that of a monitor, rather than decision-maker. It was Telford who organized and carried out the initial survey work, it was Telford who recruited the managerial and contractual team and devised the procedure for their guidance and control, and it was Telford who handled all the problems of land purchase and finance. He was also responsible for the ordering and transportation of materials. Jessop was sent in by the Government to add weight to the project at a time when its future was uncertain; once the 1804 Canal Act had been obtained, his involvement in the project was limited to an annual joint inspection with Telford and to putting his signature to the latter's report to the Board of Commissioners. Telford continued to visit the construction site twice a year throughout the period of Jessop's association with the scheme, though this dropped to once a year after Jessop's departure.

Jessop's experience was undoubtedly a great help to Telford, who was, in comparison inexperienced in canal building, having been involved in the construction of only two canals (Ellesmere, as 'General agent' to Jessop, in 1793, and Shrewsbury in 1795) before his appointment as Engineer for the Caledonian in 1803. Jessop also had experience of inland lake and river navigation after work on the River Shannon. Apart from changing the design of locks and introducing cast iron, however, Jessop appears to have simply endorsed Telford's actions. He was also paid for his services, at least in the early years of the project, from Telford's own budget, which perhaps tells us much about his status vis-à-vis Telford and the canal.[53]

Resident and site engineers

It was through the next link in the managerial chain that Telford, and possibly Jessop, attempted to solve one of the major problems of the project – namely adequate control from a distance. An attempt is therefore made here to define the areas of responsibility allotted to the various site engineers and assistants with special regard to line and design changes and their overall contribution to the management and construction of the project.

As has already been noted, the evolution of the managerial structure was divided into two sections, the termination of survey work and the passing of the June 1804 Act marking the dividing point. Developments prior to June 1804 were concerned with controlling what was essentially a survey rather than a construction team. However, the basic procedure for the control of resident engineers engaged on construction work dates from this early period.

As previously stated, Telford had divided the project roughly in two, appointing Superintendents at each end, together with sundry assistants and surveyors. Provision was made for limited construction work in the two end basins which were marked out in the autumn of 1803. Fortunately, records of Telford's initial instructions to A. May, his Superintendent at Clachnaharry, have survived. They were written on 27 September, 1803, and ordered May to engage men for the excavation of Muirtown Basin and line of canal which extended from there to Clachnaharry. This line had been previously marked out by Telford and was to be excavated exactly to his directions. May was to ensure, if at all possible, that work was to 'be performed by the cubic yard, each man's work to be measured up monthly'.[54] Records of all expenditure were to be recorded in monthly paybills, which were to be dispatched, at regular intervals, to Telford, together with a journal of all that had occurred. Wages were to be fixed at 18*d* or 16*d*, according to quality of workmen.[55] Telford thus ensured full control over expenditure, accounts, type of labour to be used and line of canal to be excavated.

Similar instructions were sent to May's assistant, William Mackenzie, who was, when not engaged in recording the paybills, to visit the works and 'assist if necessary in measuring the work and in making agreements'.[56] A Mr Mason who was 'to manage the business about the quarries' appears to have had similar powers to May and Wilson, for Mackenzie was instructed 'to do the same with regard to Mr. Mason'.[57] John Wilson employed an assistant who dealt with the purchase and cutting of country timber.[58]

The early years of the project were thus very similar to other projects of the period, having a mixture of basic supervisory staff plus people allocated to one specific task, although on other projects this always appears to have been geared to the construction of one specific object or section, rather than to such matters as materials and finance, as appears to have been the case on the Caledonian. This degree of specialization does not appear to have been carried over into the post-1804 structure. During the months after September 1803, the individual members of the team were busily engaged on survey work, W. Hughes accompanying Jessop over the line at the end of October[58] and Mackenzie performing land valuations in the Corpach area.[59] The small number of workmen employed on excavation work was controlled by permanent overseers who presumably remained with the workmen on one particular site. The earliest reference to overseers is found in a Telford letter of February 1804.

> 'As well as engaging workmen and overseers it will be necessary to set them to work'.[60]

The essential elements of the managerial structure were therefore all present before the passing of the Canal Act in June 1804. The appointment of the two permanent Superintendents heralded the commencement of full-scale construction work. Telford was again responsible for drawing up their instructions, after receiving authorization from the Board for their appointment.[61] As before, the canal was divided up into two districts, each half having its own Superintendent, Mathew Davidson at Clachnaharry and John Telford at Corpach. No appointment was to be made for the middle district until the commencement of operations there.

Taken in conjunction with the earlier set of instructions, Telford's June 1804 directive provided the basic framework for the management of construction work up to the opening of the canal in 1822. As has been seen already, the superintendents were to let the cutting, puddling and embankments in small lots and 'where circumstances admit, a preference [was] to be given to letting several contracts to the same persons'.[62] They were also to let the masonry by contract. Telford was ordered by the Board to ensure that in 'no case whatever [was he] to allow the resident Superintendents upon the line of the canal to exceed the prices to be paid for labour, as previously settled by Mr. Jessop and himself, without specially reporting the same and receiving the sanction of the Board thereon'.[63] The Superintendents were also to ensure that the contractors kept in good repair any of the tools lent to them by the Commissioners. No mention was made of the recording of expenditure and procedure for the payment of workmen, which would suggest that they remained as before.

The instructions relating to the letting of masonry and general contracts appear to have been partially overruled by Telford and Jessop, as the line of canal and 'the mode in which the works are to be let' were decided in August 1804.[64] The Superintendents were allowed to fix rates 'applicable to particular places . . . according to the soil, situation and other circumstances,' subject to Telford's approval. In the event of the ground proving . . . .

> 'more difficult to work than the general appearance and trials already made had led us to conclude, a proportional allowance is to be made, but if any variation shall increase the expense above the rate of 6d. per yard on an average, every case of this kind is to be reported to Mr. Telford, and the agreement is only to be conditional until he has approved of the same.'[65]

As in the earlier 1803 instructions, Telford ensured rigid adherence to the prices fixed by him through the introduction of a clause giving him the right to reserve judgement on any increase. This feature was made more explicit in the June 1804 instructions.[66] Instances of increases in the contract prices were regularly recorded in Telford's annual reports, which would suggest that the above-mentioned safeguard worked efficiently.[67] Adherence to Telford's 'arrangements' by the Superintendents was given as one of the main reasons for the project's steady progress up to 1808.[68] No evidence exists to suggest that the Superintendents were not responsible for appointing general contractors. A very small number of individual firms was taken on, however, which would suggest that Telford and Jessop's desire that multiple units be let to single contractors was adhered to by Davidson and J. Telford.

On the occasions when design or line changes were implemented, as at Clachnaharry Sea Lock and Fort Augustus,[69] Telford and Jessop made all the major decisions, although much of the information must have been supplied by the Superintendents. Telford and Jessop also instructed the Superintendents as to where work should be concentrated, as in 1809 when they ordered Davidson to complete the section between Doughfour and Moulindour.[69] In addition to the 1804 provisions, the Superintendents played a vital role in the issue, transference and recording of finance and expenditure. From a comparatively early date, they were ordered to keep an exact record of all monies spent, in the form of a monthly pay-bill, which was sent to Telford together with all vouchers. It is also probable that the system of advanced monthly estimates introduced in 1804 was calculated by the Superintendents, as Telford made only two visits to the Highlands a year, although there is no documentary evidence for this assumption. The Superintendents received money direct from Telford rather than the Board.[70] It is not known if they handled all of the monthly allowance, although they did have control of the wages money, which was distributed by their subordinates. Since Telford was involved in all stages of ordering machinery it is conceivable that he paid the larger machinery and material suppliers directly from his account.

To conclude, it would appear that the Superintendents on the canal were kept in check by an elaborate system of monetary control and recording. Control of the money supply ensured control of contract price, wages and the amount of materials consumed. However, the Superintendents were responsible for the day-to-day running of the construction programme, receiving periodic visits from Telford and Jessop who decided any major issues on the spot. Their role in the managerial structure was thus second only to Telford. They do not appear to have had any say in what that role should be, the majority of decisions appertaining to this being taken before their appointment. The absence of detailed specifications for the project placed a greater emphasis on the skill and discretion of the Superintendents in carrying out Telford's wishes. Their very considerable achievement was marred by financial worries and inadequate supervision below Superintendent level, which resulted in structural failure and rebuilding in the 1840's.

Specialist staff

In addition to the permanent staff already mentioned, Telford employed specialists on particular topics, to advise on specific areas, usually relating to geographical features of the Highlands or machinery. In this first capacity, he employed Murdoch Downie, an Aberdeen sea captain, to survey the sea lochs and canal entrances, assisted by Mason.[71] Land valuation was carried out by George Brown of Elgin, probably the most experienced land valuer in the Highlands with A. Laughlands, another land surveyor and valuer, employed for the area around Fort William.[72] Numerous assistants and surveyors, including John Howell, and Barlow and Arrowsmith, were employed temporarily to carry out trial borings and to produce maps of the line of canal. Telford employed the Fyfe brothers to look after the pumping engines and at least one of the dredgers, and the Rhodes brothers of Hull on the design and construction of lock gates.[73] Bryan Donkin was consulted on an improved dredging machine.[74]

Lower management

Details of the lower managerial structure employed on the project are extremely scarce and vague, making it difficult to distinguish between those employed by the Commissioners and by the contractors. Immediately below the level of Superintendent came the post of General Assistant/Pay Clerk, which appears to have been occupied in the Clachnaharry district by Andrew May from 1804 to 1822.[74] He had previously been employed as temporary Superintendent before the appointment of Mathew Davidson. His main duties were the keeping of accounts and 'measuring the contents of the task work'.[75] The names of two other general assistants have been recorded, although their exact place of work is unknown. Presumably they were assigned to the western and middle districts. In pressing for a salary increase for the Assistants in 1813, Telford stated that they were paid £1.50 a week. He requested that this be increased to £100 per annum,[76] which gives some indication of the importance of the position.

The measurement of task or measure work, the main unit of labour on the canal, was a duty traditionally performed by assistant engineers, and it is not known if May and his colleagues worked on the actual line, measuring all the work as it was performed, or if they were based in site offices, receiving work records from overseers and possibly even Davidson and Easton. The large number of concurrent construction sites would tend to suggest that there were either more assistants/pay clerks whose existence was never recorded or, more probably, that it was the overseers who recorded the work performed on site. The small number of references to their being present on the construction site would suggest that they were concerned primarily with office work and that it was the overseers who controlled the project on the numerous construction sites. The number of overseers employed on the project varied from five to ten according to season, and was not affected by the number of workmen or sites.

May's knowledge of construction work was clearly very considerable, as he assumed the title of Superintendent for the Clachnaharry district after 1817.[77] It should be noted however that the main construction programme had moved to Fort Augustus by this period. It is not known if May continued in this capacity after the completion of the works at Fort Augustus. Reference is also made at this period to Clerk being employed at Fort Augustus as Superintendent.[78] Mathew Davidson had died earlier that year and had been replaced by his son, James, and Clerk's activities at Fort Augustus could possibly have been connected with these events.

The position of May and his colleagues was thus extremely ambiguous, appearing to require the qualities of engineer and accountant. There appears to have been considerable overlap between the roles of Superintendent, Assistant and Overseer, especially with regard to the measuring and recording of labour. There is even a remote possibility that the references to 'overseers and counters' recorded in the monthly employment figures, includes May and his colleagues, as no other record of their existence was kept, unless they were included under 'Mr. Telford's Clerks' in the annual management accounts.

The position of overseers in the managerial structure is also ambiguous. Overseers were included in the general employed figures, which recorded the number of carpenters, blacksmiths, masons and labourers. Presumably some of these workmen were hired directly by the Commissioners, but it would seem clear that the majority were technically employed by the contractors. Evidence of overseers coming within this second group is limited to isolated wage certificates and receipts. However, there is nothing to suggest that they were employed by the Commissioners. The names of at least two of the overseers are known, John Mackferson, Foreman of the Masons,[79] and Thomas Smith,[80] who had been employed by Telford prior to June 1804. The inclusion of overseers' signatures on all wage receipts would suggest that they kept records of work performed. It would also imply the formal handing over of money from one body to another.

The implications of this hypothesis are extremely serious as it would imply that the contractors had more agents on the construction site than the Commissioners and that the whole managerial team was dependent on co-operation between two groups whose interests sometimes conflicted, especially in the fields of finance and quality control. Detailed control of the project was left to the contractors (through their overseers), the Commissioners' agents providing overall supervision. There were possibly as few as five Government employees (the two Superintendents and three Assistants/Pay Clerks) on the project, if the above assumptions are correct. The duties of the overseers appear to correspond fairly closely to those of assistant engineers employed on contemporary civil engineering projects. They appear to have been responsible for one specific site, rather than large areas.

The importance of the contractors' overseers in the managerial structure would suggest that the contractors themselves played a key role. As has already been noted, the masonry contractors were well acquainted with Telford's working methods and the high standards demanded by him. Their permanent presence in the Highlands was virtually guaranteed by their heavy commitment to the Highland road and bridge scheme. They were provided with temporary accommodation at the commencement of the project but soon built themselves permanent dwellings at Inverness and later Fort Augustus.[81] One of them, John Wilson, had, as has already been noted, acted as Telford's temporary superintendent prior to June 1804. Telford appears to have been regularly accompanied by at least one of the masonry contractors on his annual tours of inspection. They were also involved in preparatory survey work, as can be seen from Telford and Wilson's investigation of Loch Oich in 1813.[82] They even acted as unofficial pay clerks, as Mathew Davidson reported in 1816:

> 'All is going well on the canal, I am in daily expectation of Mr. Telford here, Mr. Cargill and my son, James, are gone to pay the workmen at Fort Augustus.'[83]

To conclude, it would appear that the mysterious lack of information regarding Government supervisory staff under the level of assistant/pay clerk was due to the fact that there were none. The rest of the managerial team was made up of the contractors and their overseers. As a result, the contractors enjoyed a very considerable amount of freedom which resulted in shoddy workmanship and transgression from what was required. The problems were most apparent in the western and middle divisions, which contained a large proportion of the masonry works. Banavie Locks especially were not provided with adequate foundations or side walls, which resulted in rapid deterioration.[84]

It will never be known if Telford and his assistants were aware of these shortcomings at the time. May in his 1837 report certainly suggests that there was deliberate concealment:

> 'I have reason to believe that the contractor for these locks, while engaged in the actual execution of the works, was fully under the conviction (which was shared by many others at the time) that the navigation was a thing which was never to take effect and that his locks would consequently never require to come into actual operation'.
> '. . . That so imperfect a description of workmanship should have satisfied, or rather escaped the severe reprehension of Mr. Telford on his oc-

casional visitations to the canal, is surprising, and can only be accounted for on the supposition, which I believe to be a correct one, that the utmost pains were taken by the contractor to conceal, by a variety of arts, the true nature of his proceedings.'

Had the managerial structure been balanced in favour of the Commissioners rather than the contractors, shortcomings would have been corrected. In a time of high inflation the trust and freedom given to the contractors counted for nothing, and ironically, the rigid system of price control forced them into taking the obvious way out, namely bad workmanship. The problems of supervising so extensive a site, lack of capital, and eventually possibly lack of faith in the project all contributed to the further and continuing delays, so that by the time the Caledonian Canal had been completed it had been overtaken by events, and never fulfilled the hopes of those who had initiated the project.

Bibliography

Cameron A.D. *The Caledonian Canal.* Lavenham 1972, Terence Dalton Ltd.

Haldane A. R. B. *New ways through the Glens.* London, Thomas Nelson, 1962

Hadfield C. and Skempton A. W. *William Jessop.* Newton Abbot, David & Charles, 1979.

Rolt L. T. C. *Thomas Telford,* Longmans Green & Co., London, 1958

References

1. Third Report from the Committee on the Survey of the Coasts, etc. of Scotland. Caledonian Canal. June, 1803. Appendix 4.
2. *Ibid.*
3. 1st Caledonian Canal Report, 1804. (referred to subsequently as '1st Report').
4. Minutes of 2nd Caledonian Canal Meeting, 4 Aug., 1803.
5. Minutes of 11th Caledonian Canal Meeting, 6 Feb., 1804.
6. 1st Report 3. Accounts (Management) and Telford's instructions to May and Mackenzie, 27 September 1803.
7. Jessop to Rickman, 29 September, 1803. Jessop was Mayor of Newark at the time of Telford's departure for the Highlands in August 1803, and was unable to attend any canal business until October, requesting that instructions be sent direct to him.
8. Telford to Cameron of Lochiel, 3 October 1803. Scottish Record Office (SRO), MT1/1.

9. 1st Report, Appendix F. 'Report on the intended entrance from the Western Sea, at Corpach on Loch Eil by Messrs Downie and Mason, Fort William, 2 September 1803'.
10. *Ibid.*, Management and Survey Accounts.
11. 2nd Caledonian Canal Report, 1805 (referred to subsequently as '2nd Report').
12. Minutes of 19th Caledonian Canal Meeting, 11 June 1804; 2nd Report, Appendix C.
13. Telford to Jessop, 8 June, 1804.
14. See Rolt L. T. C. *Thomas Telford.* London, Longmans Green and Co., 1958, for general details of association between Jessop, Davidson and Telford.
15. *Ibid.* John Telford appears to have acted as Telford's personal draughtsman, preparing plans for Bridgnorth Church in the 1790s.
16. Telford to Jessop, 8 June, 1804.
17. *Ibid.* This proposal does not appear to have been implemented, Mathew Davidson assuming responsibility for the centre district.
18. 2nd Report, Appendix D; Jessop to Telford, 9 June, 1804.
19. *Ibid.*, Accounts (Management).
20. Jessop to Telford, 9 June, 1804.
21. 2nd Report.
22. 4th Caledonian Report, 1807; Also, Cameron, A. D. *The Caledonian Canal.* Laverham, Terence Dalton, Ltd., 1972, p.67.
23. Telford to Jessop, 8 June, 1804.
24. *Ibid.*
25. Telford to A. May, 27 Sept., 1803. SRO, MT1/1.
26. 2nd Report.
27. Telford to Rickman, 6 June, 1804. SRO MT1/1.
28. 2nd Report, Appendix C; Telford to Jessop, 8 June, 1804.
29. *Ibid.*, Main Report.
30. *Ibid.*, Appendix K. Report of Telford and Jessop, autumn 1804.
31. Telford to Cameron of Lochiel, 2 Sept. 1803, SRO MT1/1.
32. 2nd Report, Appendix B.
33. *Ibid.*, Main report and Machinery Accounts.
34. Skempton A. W. A history of the steam dredger, 1793–1830. *Trans. Newcomen Soc.*, 1974, 47, 103.
35. Donkin, S. B. Bryan Donkin, FRS, MICE, 1768–1855. *Trans. Newcomen Soc.*, 1949–51, 27,
36. Telford to Jessop, 8 June, 1804.
37. Jessop to Telford, 9 June, 1804.
38. *Op. cit.*, ref. 1, Appendix I.
39. *Ibid*, Main report.

A. Penfold 149

40. 1st Report 'Instructions to Mr Telford'.
41. *Ibid.*, Main report.
42. Minutes of 7th Caledonian Canal Meeting, 25 Nov. 1803.
43. 1st Report, and minutes of 11th Caledonian Canal Meeting.
44. Telford to Rickman, 15 Febr., 1804. SRO MT1/1.
45. 1st Report Jessop's report and estimate.
46. *Ibid.* See also refs. 7 and 9.
47. 2nd Report, Main report.
48. *Op. cit.* ref. 1, Appendix 1, Telford's 1801 report.
49. Rickman J. *Life of Thomas Telford.* London, 1838. *Atlas* Plate.
50. 2nd Report, Main report. Jessop's knowledge of Baltic shipping was non-existent, as he had told the June 1803 Select Committee.
51. 11th Caledonian Canal Report, 1814, Appendix C.
52. 3rd Caledonian Canal Report, 1806. It had originally been intended to construct timber bridges, similar to those on the Fortth and Clyde, but after consultations with Jessop, Telford decided to use cast iron.
53. Telford to Rickman, 24 March, 1804. SRO MT1/1.
54. A. May, 27 Sept., 1803. SRO MT1/1.
55. *Ibid.* The wage rates were listed only for day work.
56. Telford to W. Mackenzie, 27 Sept., 1803. SRO MT1/1.
57. *Ibid.* Reference was made in the 1805 accounts to Mackenzie as 'Superintendent' for the Clachnaharry district, although Telford's 1803 instructions make it clear that it was May who was given more responsibility. Mackenzie received £145 up to September 1804, including fees for land valuation.
58. Telford to Rickman, 31 Oct., 1803. SRO MT1/1.
59. Telford to Cameron of Lochiel, 22 Dec., 1803. SRO MT1/1.
60. Telford to Rickman, 18 Feb., 1804. SRO MT1/1.
61. Minutes of 19th Caledonian Canal Meeting, 11 June, 1804.
62. Jessop to Telford, 9 June, 1804.
63. 2nd Report, Appendix E.
64. *Ibid*, Main report.
65. *Ibid*, Appendix M. Prices of labour and workmanship, as determined by Messrs Jessop and Telford.
66. See Telford to Jessop, 8 June 1804, where constant reference is made to 'subject to my determination' and 'to be determined by you or myself' with regard to fixing of prices.
67. Telford and Jessop agreed to price changes in the contract for masonry in October 1806 (4th Caledonian Canal Report, 1807) and general excavation work performed by Meek in 1813 (11th Caledonian Canal Report, 1814).
68. 6th Caledonian Canal Report, 1809.

69. *Ibid.* 'Considerable' design changes were made at Clachnaharry sea lock during Telford and Jessop's 1807 autumn visit: the proposed coffer dam was scrapped and replaced by a solid embankment out of which the lock pit was to be eventually excavated.
70. *Ibid.* Money was paid into Telford's special bank account and then drawn out to pay for construction work: Presumably the Superintendants had some arrangement whereby they could draw money from this account in Telford's absence.
71. 1st Report.
72. 2nd Report.
73. Mitchell J. *Reminiscences of my life in the Highlands.* Vol. 1, 68.
74. 14th Caledonian Canal Report, 1817, Appendix E.
75. May was also responsible for handling early steamship business on the canal. See Haldane, A. R. B. *New ways through the glen.* London, Thomas Nelson, 1962.
76. Minutes of 63rd Caledonian Canal Meeting, 1813.
77. Paybill, Sept.–Oct., 1819. House of Lords Record Office (HLRO).
78. *Ibid.* Clerk received £7. 15s. 10d. a month, as did May at Clachnaharry.
79. Wage certificates, 1807–8. HLRO.
81. Cameron, *op.cit.*, ref. 22, p.55. Simpson and Cargill built houses in Telford Street, Inverness. Southey refers to Cargill having built himself a stone house at Fort Augustus.
82. *Ibid.*, p.78
83. Mathew Davidson to Hope, 12 Sept., 1816. HLRO.
84. Cameron, *op.cit.* ref.22, p.130. May had pointed out that the masonry in the Banavie flight was particularly bad.

7

Along the water: the genius theory. D'Alembert, Condorcet and Bossut and the Picardy canal controversy

Pietro Redondi

The manuscript published in this article is an important document that has up until now remained unrecognized with respect to the great technological controversy over the Picardy underground canal project, in the late eighteenth century. It actually consists of the report condemning this problematical innovation of canalisation techniques, supported by experiments carried out on the resistence of fluids on behalf of the d'Alembert, Condorcet and Bossut commission appointed by Turgot in 1775.

The reconstruction of the scientific, institutional and political motives surrounding this pertinent historical case, between the time of its conception under the Old Regime and its realization under the First Empire, reveals the intellectual and social factors that are indissociable from the institution of technology.

Le manuscrit publié dans cet article est une pièce importante et jusqu'ici méconnue de la grande controverse technologique issue du project du canal souterrain de Picardie à la fin du XVIIIe siècle. Il s'agit en effet du rapport condamnant cette innovation problématique de la technique de canalisation à l'appui des expériences sur la résistance des fluides de la commission D'Alembert, Condorcet et Bossut nommée par Turgot en 1775.

Entre la conception du projet sous l'Ancien Régime et sa réalisation sous le Premier Empire, la reconstitution des mobiles scientifiques, institutionnels et politiques de ce cas historique privilégié révèle les facteurs intellectuels et sociaux indissociables de l'institution de la technologie.

Your underground canal, especially, is an unprecedented chef d'oeuvre. Boileau said to Louis XIV, in that great century of style:

J'entends déjà frémir les deux mers étonnées de se voir réunir au pied des Pyrénées

Once his Successor has accomplished all his projects, the seas will no longer be astonished by anything; they will be well accustomed to such wonders.

The wonder, whose designer Voltaire[1] praised so effusively, was the project for an underground canal between Saint-Quentin and the Cambrai region. Its goal was to link up Picardy and Flandres, which would be the real key to an inland navigation system for France. This project was set up on the initiative of Minister Choiseul, through the genius of the renowned engineer of Picardy, Pierre-Joseph Laurent.

In congratulating this 'unprecedented chef d'oeuvre', Voltaire was contemplating a map of the canals criss-crossing the kingdom. To the south, there was the canal of the Two Seas, or Midi Canal, between Bordeaux and the Mediterranean. The Rhône and Saône rivers were navigable as far as Dijon. To the north, Dunkirk and the great port of Antwerp linked up with Cambrai. Here, in Picardy, the new Crozat Canal, completed in 1738, had joined up Saint-Quentin to the Seine river, by linking the Somme and the Oise rivers. The wonders of this navigation system included two new planned constructions in the south, the Burgundy Canal, to connect the Saône and the Yonne, which would allow navigation from the Mediterranean to the Seine via the Rhône; in the north, and underground canal cutting straight across the heights of the Vermandois above Saint-Quentin, which on the map appeared to oppose the link between the Paris Basin and Flandres, and between Northern ports and Holland[2] (Figure 1).

On the map, this small underground vein of considerable length (some 7,020 toises or 13,682 km), seemed to be a daring feat of French technology. The canal was, without doubt, the most vital and important one in France, both from an economic and a strategic point of view. According to Lalande, the 1770s saw 1,200 ships sail on the river Scheldt, between the port of Antwerp and Valenciennes.[3] Cloth, fabric and different types of wood and marble, ash and salted fish were the goods that travelled down from England towards Paris. Wines, oils and fruits were the products sent northward from the south and the Levant. The Scheldt was the traditional route for small boats carrying coal from the deposits in Flanders, whose new geo-

FIGURE 1 *Map intended to prove the utility of the Picardy Underground Canal allowing navigation from the North Sea, Holland and Flanders to the middle of France without going through the Channel.* From P.J. Laurent de Lionne, *Discours prononcé à la séance publique de l'Académie des sciences, belles-lettres et arts d'Amiens le 25 Août 1776*, Amiens 1783.

logical science made one assume the natural continuation from Liège to Valenciennes.

A canal penetrating into the hills of Saint-Quentin was also strategically important, given the disastrous experience of the Austrian

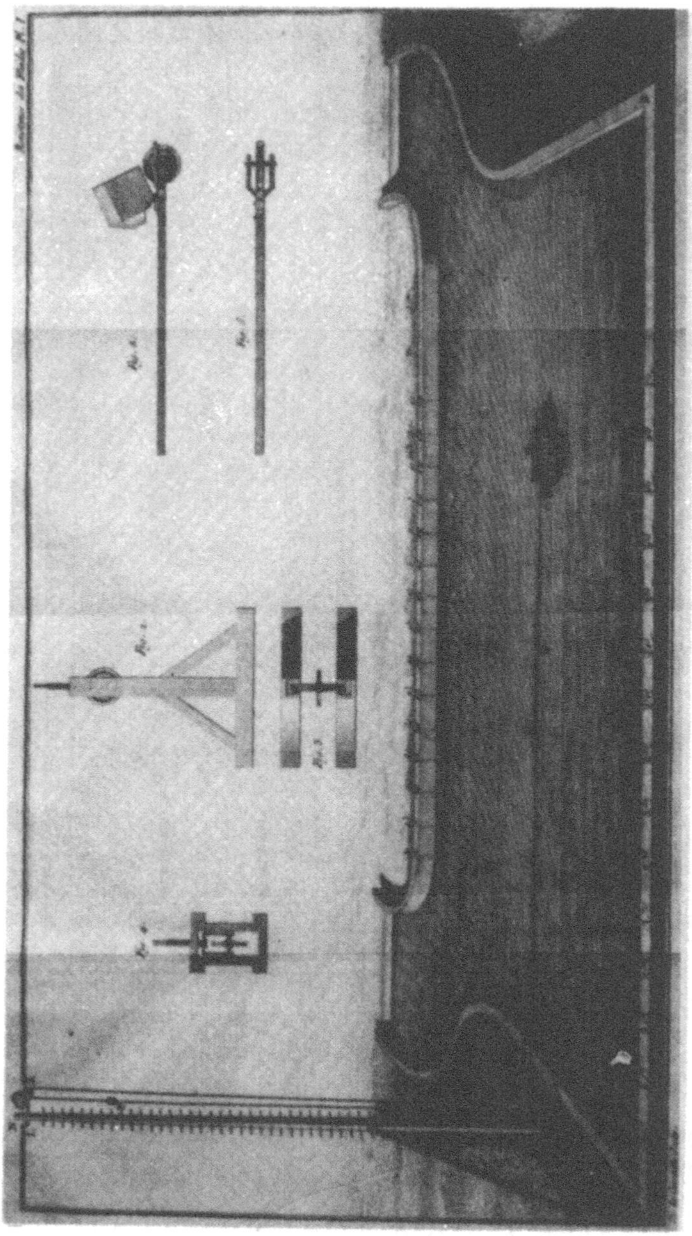

FIGURE 2 Experimental apparatus of the basin of the Ecole Militaire (summer 1775) to measure the resistance of water in the advance of a model boat. It consisted of a gravity dynamometer (R) and a cable (V) fastened to the model. The towing force was provided by a weight (P). Bossut's assistants were positioned at each stake along AB to determine the instantaneous passages of the model, "all ears" to a vocal computation of the half-second pendulum oscillations. Time was measured to ascertain the rate velocity/resistance of the model.

From d'Alembert, Condorcet, Bossut, *Nouvelles expériences sur la résistance des fluides*, Paris 1777, plate 1.

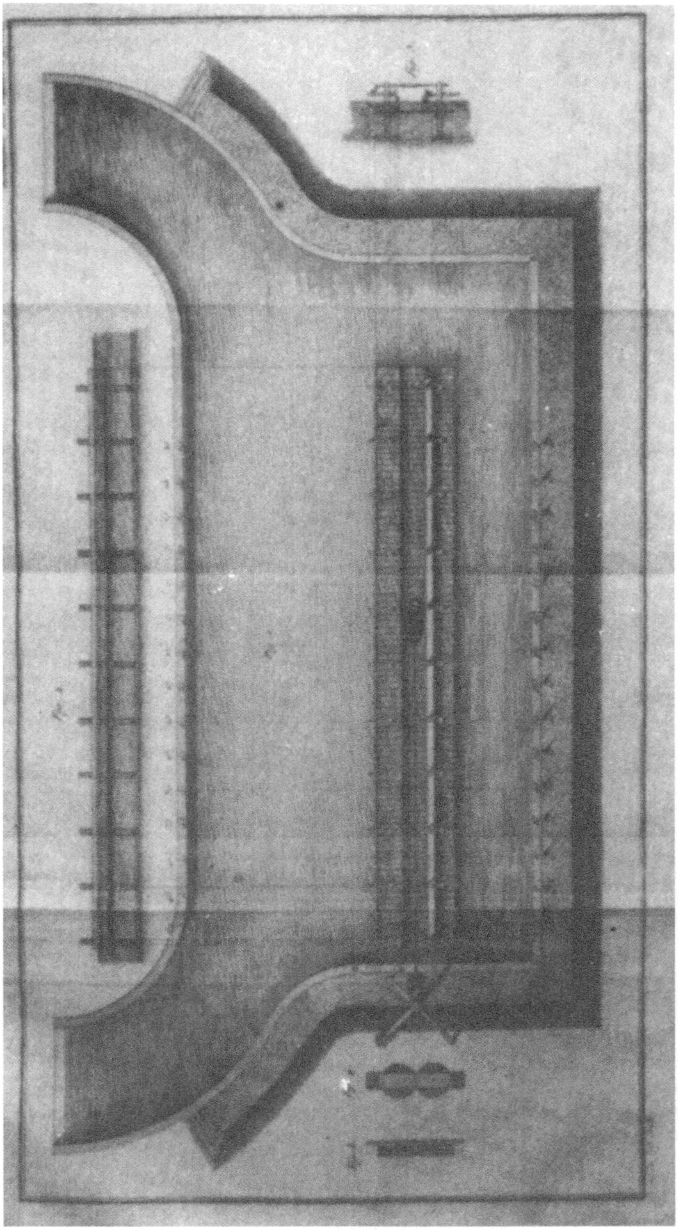

FIGURE 3 Vertical and horizontal section of the flume made in the basin of the Ecole Militaire to measure the resistance of water in a definited canal. From d'Alembert, Condorcet, Bossut, *Nouvelles expériences* . . ., plate 3.

FIGURE 4 The monumental northern gate designed by P.J. Laurent de Lionne in support of the Picardy underground canal. The small draft of the boats, the two tow-paths and the transversal bar allowed the hand-towing inside the tunnel, usually supplied by the families of the bargemen, including children. In 1858, the removal of one of the tow-paths necessitated the suppression of this system, also justified to stop "immoral acts" in the darkness of the tunnel. A horse-gear on the boats (1861), a steam towing-boat (1874) and then an electrified towing-boat (1927) provided new systems of towing by submerged chain. From *Carte topographique du canal souterrain de Picardie, ca* 1781, Archives Nationales F^{14}10127.

War of Succession. At that time, in 1744, the never-ending but costly corvées which were needed to resupply the army of Flandres with food and ammunition, had ruined the harvest. Even though, from that point on the English were the supreme rulers of the sea, should a new war break out, an inland system of navigation would allow men or artillery wagons from Flanders to be transported rapidly to Brittany or other provinces in danger. Even though Voltaire was right to mention the achievements of Sully and Colbert, he was not fully aware of either the technical difficulties or the scientific riddles and awesome costs (even in comparison with the famous Midi Canal) involved in digging such a long underground canal.

Under the Regency, in 1727, Saint-Quentin's chief engineer, De Vicq, had drawn up an initial plan, the proposals for which were revealed during the course of later discussions. The channeling of the river Scheldt which was extended practically to its source (near Vendhuile) was proposed first to reach the Somme by an underground canal of 3,550 toises (5,665 m) and then to cross the Bellenglise Valley and finally to connect with another underground canal of 700 toises (1,100 m) before reaching Saint-Quentin. This project, however, remained buried amidst the archives of the War Bureau, while a new project that called for a shorter link and a surer supply of water had been proposed. In 1767, when the Crozat Canal had been requisitioned by the State, the plan to link the Somme to the Scheldt by an underground canal held a certain national interest, especially since there was the possibility of greater financing.[4]

The Duke of Choiseul, Minister of War and of the Navy, as well as Minister of Finance, had entrusted the study of the project to Pierre-Joseph Laurent who was universally admired for his hydraulic experimentation in the region. Laurent was born in Bouchain, Flanders, in 1714, and was the son of the King's contractor, in Condé.[5] He, like De Vicq, had completed his apprenticeship with the defense contractor of Saint-Quentin. Both engineers, without doubt, owed their training as defense engineers to the inspiration which induced them to apply the tradition of military technology of underground passages to navigation projects. Laurent's technological inspiration for the construction of an underground canal came from other sources as well. In Rouen, since the seventeenth century, the water for fountains and ornamental lakes in the gardens of Château Legendre manoir was supplied by a famous underground network. Furthermore, the very same designer of the Midi Canal, the engineer Paul Rigaut, had carried out important underground projects at Castries, in Languedoc. Laurent, meanwhile, had built up his fortune and reputation as an engineer due to the success of the drainage of his silver mines in Brittany. 'A skilled engineer', Laurent owed his fame in the aristocratic circles to his invention of machines capable of raising the Postern gate at Valenciennes and also to the drainage channels of the lands of the Hainaut. More importantly, he owed his reputation to the system of manmade tributaries which had originally earned him Voltaire's first praises.[6] The Saint-Quentin canal, with its daring tunnel, was the result not only of the leveling carried out in the Vermandois valleys, which had been dismissed as too long and too costly to permit the construction of a surface canal, but also of Laurent's instinctive ability to appraise the strength and nature of the ground to be bored. The fundamental hypothesis underlying his project, that is the possibility of using the water bed as a supply for the underground canal, was the fruit of the 'inventive genius' of this engineer to whom Choiseul had entrusted, in 1767, the management of the canals in Picardy in order to link the Somme to the Scheldt. Short underground canals with narrow lateral towpaths which were used for towing the barges by hand, already existed in Holland, namely in Rotterdam and Amsterdam. Laurent had adopted this construction technique and planned to make his underground canal twenty feet wide (6.48m) including two lateral towpaths, two feet in width (0.648m)[7] (Figure 4). In England, in 1758, the engineer James Brindley had also built an underground canal almost five kilometers long, alongside the Duke of Bridgewater canal, near Manchester. Thanks to a convoy of small hand-pushed boats, the underground canal in England directly worked the coal mine. It was only three meters wide and was only partially vaulted.[8] Laurent too only intended to partially vault his big underground canal.

The excavation of the Laurent canal, begun in 1768, was initially based on the mining technique used for boring shafts; these shafts, situated at two hundred metres from each other, would then maintain the underground ventilation and water supply.

In fact, it was neither its impressive length, nor the number of successive shafts bored which contributed to the innovation of the canal technology involved in this grandiose project. Its novelty, in comparison with De Vicq's project, as well as with other canals built so far, went beyond the standard practice of traditional practical hydraulics. Such practice, set out for the canals at summit level in Bernard Forest de Bélidor's fourth book, *Architecture hydrauliqe* (Part 2), required that the water, necessary to maintain the movement of boats and locks, be obtained by storing up the winter surplus in large supply tanks.[9] De Vicq himself had followed these regulations when he planned to build a big and costly reservoir on the river Scheldt. Loyalty, however, to traditional doctrine, had no guarantee of success, as all the water in the reservoir would be soaked up by the underground water bed once it was released into the tunnel. Laurent's experience of soils must have prompted him to abandon De Vicq's plan and, with it, the classical method, in favour of a canal dug at a deeper level in order to harness the underground springs. Such an alternative was necessarily a lengthier one and it also had the disadvantage of necessitating the reduction in width of both the water basin and the towpaths for reasons of cost and reliability. The drawbacks that would result in terms of speed and navigation safety were unknown. A new technical and experimental problem was posed by their rational forecasting.

In any case, Laurent's project was justifiably valid, in that it conformed to the geophysical conditions which were exceptionally favourable for cutting an underground liaison between the Somme and the Scheldt; a summit level could in fact be supplied by the upper-valley of the Oise and by the Ardennes plateau dominating the Flanders' plain through which the Scheldt passes. Furthermore, the shafts dug by Laurent in the hills of Saint-Quentin showed the ease of excavation as well as the robustness of the stone beds which allowed the partial vaulting of this imposing underground canal.

After the total cost of the project had been estimated at four million livres by its designer, a royal decree of 300,000 livres, in 1769, raised in 1772 to half a million, guaranteed the accomplishment of the digging, the navvying and the earthworks. Under Laurent's management, the excavation work was vigorously undertaken by 1,500 workers, a work force swollen in ranks due to the poverty reigning in the countryside. The excavation of seventy shafts, each one to be dug by two men, had begun. The revetment of brickwork generally appeared superfluous. The underground water flowed as planned. In 1773, 5,000 toises (9,745 m), that is, two thirds the length of the tunnel, had been dug, and since 1772, an almost sixty-metre long section had been completed. This section was to become a source of alternate admiration and scepticism to its many illustrious visitors, a list of whom we possess:[10] men from the army, civil engineers from the Corps des Ponts et chaussées, local administrators, eminent figures from Flanders, foreign envoys and even Prince Doria Pamphili, the Papal Nuncio in France.

Not everybody was convinced, however, of the reliability of the Picardy Canal upon the completion of this section and a list of objections was anonymosly sent to the Finance Minister in August 1773.[11]

We were indirectly made aware of the nature of these objections when we read the report sent to Laurent by Choiseul's successor, abbé Terray; a difficult navigation as a result of the strong current of air, the weakness of the vault and the towpaths, the width of which was seen as insufficient for the towing of boats.[12] Laurent had time to reply by re-emphasizing the impossibility of a surface canal, and by proposing an increase in the size of the towpaths and the vaulting of a third of the underground canal at the most with bricks.

He died in Paris on 12 October 1773. On 17th January of the following year, abbé Terray appointed, by royal decree, Pierre-Joseph Laurent de Lionne, nephew and assistant to the project's designer, successor to the management of the work of the Picardy Canal.

In August 1774, however, Turgot, an apponent to the Choiseul party and the Queen, replaced Terray as Finance Minister and thus also became responsible for the Corps des Ponts et chaussées, the administration of which was controlled by Trudaine de Montigny (or Trudaine Junior). Turgot had, in fact, also visited the work site of the Picardy Canal, the cost of which had already risen to one million livres. The criticism which had arisen on the execution of the project, the worry about reducing the overall financial costs and the opposition to such megalomaniac spending, all combined to make Turgot aware of the major importance of inland navigation for the communication system and especially for the grain trade. Furthermore, the development of canals was in concert with the wish, strongly supported by Condorcet, to abolish the corvées and thus compensate for the lack of maintenance of the road system with the safety, capacity and continuity of transport on the waterways.[13]

In this general concern for public works, Turgot was so insensitive to the importance of military strategy, considering the policy of neutrality and peace that he was pursuing in order to successfully carry out his program of reforms, that he was always on the alert where proposals subject to examination and scientific application were concerned.

It was thus that Turgot entrusted Condorcet with the examination of two key areas in France's system of waterways: the Burgundy Canal and more importantly the Picardy Canal, a delicate affair from several technological and economic points of view.

The Corps des Ponts et chaussées was in charge of examining the canal projects. It included civil engineers, local inspectors, and, above all, Perronet, its pontiff and 'premier ingénieur du Roi' since 1763, and a member of the Academy of Science since 1765. The official opinion of the Corps des Ponts et chaussées, however, had not as yet been sought as far as Laurent's canal was concerned. Condorcet, the power behind the Turgot government, immensely disliked Perronet as well as his guild holder of the corvées system.[14]

> I have just returned from my trip to Flanders. I intend sending you presently a detailed report on the communication channels to be set up between the rivers in this part of France. I only ask you beforehand to urge Monsieur Trudaine to completely suspend, not only all the work on the underground canal, but also that on the river Somme ... In my opinion, there are some extremely worthwhile communications, which will run into even more difficulties than at the present if work is continued ... Monsieur Trudaine can be of no help to you in this case ... Perronet, who is at the head of this party, is an extremely ignorant and vain man, he has set up the Corps des Ponts et chaussées and he would like the kingdom to perish rather than strike a blow at such a fine establishment.

Interruption of work on the Picardy Canal followed ten days after this letter to Turgot, dated 10 October 1775. In using the words I did, Condorcet was perhaps a little harsh towards Perronet, the designer of the Pont de Neuilly (1774) who, with regard to hydraulic architecture and resistance of fluids, had practical knowledge far greater than that of any other *géomètre* in France. On the other hand, Condorcet was not exaggerating when he spoke of the danger for Turgot's new policy represented by the institution chaired by Perronet. Thus, a tug-of-war began between the Ministry and the administration of the Corps des Ponts et chaussées at the time of the abolition of the corvées. The power of the central and local authorities of the Ponts et chaussées, the interests shared by the contractors and the position of the civil engineers of the administration were consequently at stake. Distribution of funds, allocation of contracts, abuse of public funds, lack of control; it was an extravagant system of privileges and diffuse corruption in which contractors, concessionaires and Perronet's agents all floundered like fish in the sea. It is only the gravity of the repeated injunctions that makes one realise just how inefficient the latter must have been.[15] Turgot had decided to "drain the marsh dry." The examination of the Picardy Canal was left entirely to Condorcet, who would later write (as for the Burgundy Canal) ever controversial reports on the administration of the Corps des Ponts et chaussées:[16]

> I enclose for you my observations on the canal. I hope that you are satisfied with them. My visit only lasted two days. Perhaps it would be better from now on to survey all that which can support itself and all that will require masonry work. The extent of the survey will, however, be extremely difficult in the present circumstances and doubts will still exist about the two thousand toises which have yet to be dug. I will gladly take the responsibility for the different examinations which still have to be carried out, and I would prefer to go to such trouble rather than to leave everything to the civil engineers of the Ponts et chaussées, whose extravagance and harshness I fear, drawbacks which the administration by corvée has made them acquire.

We are not aware of the contents of the report accompanying the letter to Turgot, dated 15 October 1775, which followed Condorcet's visit, of 7 October, to Laurent's underground canal in the company of Laurent de Lionne and Lord Mansfield. The brevity of the visit, however, and the lack of leveling and measuring work, gives one the impression that it must have been a rough and hasty report. Condorcet viewed the whole issue of navigation in Picardy from an "encyclopaedic" and pragmatical angle, so he did not admire in the work an impressive aspect of grandeur worthy of the Roman Empire. He considered only the optimum goal for the State as contractor of public works. In fact, Condorcet appreciated the short-term financial and even ecological benefits, rather than the value of the intrinsic technological novelty or the long-term commercial profit.

Thus, instead of surveying, measuring or visiting other underground canals, Condorcet opted for an alternative, already considered by Bélidor forty years earlier, which would secure a liaison between the Paris Basin and Flanders. The solution was easier and longer, but would also be of benefit to the region (by lowering the death rate) as a result of the planned drainage.

Turgot had also sought out the advice of a prestigious ministerial commission, to take advantage of its competent and enlightened advice and to solve technological problems he encountered.[17]

You have had the kindness to consult me for information that you require for the Picardy Canal, and I have proposed the abbé Bossut as the géomètre most capable of answering all your queries. I have also considered that many other questions have arisen concerning inland navigation. It seems to me that it would be extremely useful to establish a permanent commission, for the examination of these questions, composed of géomètres from the Academy ... Monsieur d'Alembert's name would lend respectability to this commission, which would, at the same time, be a sort of tribute paid him by the government for his genius and his work on fluids for which he has developed the real theory. There is no doubt about abbé Bossut's practical knowledge ... Before limiting the proposed number to three, I considered other potential participants, but found no one who could contribute to the harmony and confidence of the commission.

When this suggestion was issued on 7 March 1775, d'Alembert, Condorcet and Bossut were appointed scientific attachés to the ministry of Turgot, without pay or title, to carry out "theoretical and experimental research on navigation canals for the purpose of trade."[18] It was to be a supercommission formed to examine the extraordinary problem posed by the Picardy Canal. Scientific genius came face to face with the inventive one, the enlightened spirit with the practice of art.

The Ecole des Ponts et chaussées had been completely short-circuited by the creation of the politically biased scientific commission. It did not take long for the criticisms to manifest themselves under the guise of complaints directed at the scandalous privileges granted to d'Alembert, who used his patronage in order to get the prestige of initiative.[19]

Perhaps you have been toid [wrote d'Alembert to Lagrange on April 14th] that I am in charge of the navigation canals, with a salary of 6,000 livres; this is not true. I have only decided, together with abbé Bossut and Monsieur Condorcet, to give Monsieur Turgot, the Finance Minister, my opinion on these canals, as he is a friend whom I have known for the last fifteen years. We have, however, refused the salary that he offered for the task. This undertaking has made it necessary for me to look again at Geometry and more so at fluids, topics on which I have a lot of material that has lain dormant for a long time.

In fact, d'Alembert, under direct invitation from Turgot, had taken seriously this cherished opportunity of carrying out a series of hydraulic experiments which he had been unable to do before, in 1752, at the time of his theory on fluids. The great scholar was without doubt aware of his own scientific problems.

The mathematical theory on its own had led to contradictory conclusions, such as the hydrodynamics paradox on the absence of pressures on a symmetric body submerged in a fluid, which d'Alembert had revealed in the fifth volume of his *Opuscules mathématiques* in 1768.[20]

Thus, at the very moment that the survey had been ordered in Picardy, the new ministerial commission received, on 11 April 1775 through Turgot, the financing for a large scale program of experiments on the resistance of fluids.[21]

These experiments would, in fact, be aimed at reproducing, under controlled conditions, the awesome hydrodynamic enigma presented by the navigation of boats in an inevitably narrow underground canal. To appreciate its importance, one only needs to visit the canal, which is still as it was when built at the beginning of the nineteenth century, yet only half as long as Laurent's original plans indicated. Due to the canal's one-way navigation system, the barges have to form convoys, and when a convoy enters the tunnel, it acts as a piston by stemming the water in front of it. This leads to a change of water level, which rises with the length of the oar and the draft of the boats. Until work carried out this century, which allowed the water to spread to the sides and deepened the canal, the convoys had to stop in the middle of the tunnel to allow the water to regain its balance by flowing alongside the boats. The whole operation took thirty minutes.

Even though the commission's experiments had been "particularly planned according to Monsieur d'Alembert's advice," these took place under the management of the abbé Bossut between July and September 1775, just before Condorcet's visit to Saint-Quentin's underground canal, on the basin of the gardens of the new Ecole Militaire in Paris.[22]

A canal made from wood was constructed in the basin which was 100 feet (32 m) long and 53 feet (17 m) wide to experimentally reproduce the conditions of an underground canal. Thirty model boats of various designs (some of which were taken from the naval collection gathered at the Louvre by Duhamel de Monceau), the collaboration of a team of engineers from the Ecole Militaire, amongst whom Monge and Legendre, all permitted abbé Bossut to carry out, with prompting from d'Alembert, this first of many hydraulic experiments using the naval basin technique (Figures 2 and 3).

There were three hundred experiments and with them a crop of scientific and technological results: assessment of the low tenacity and friction of water, of the approximate value of the law of the inverse square power, of the water resistance with respect to velocity, confirmation of the decrease in resistance in the case of total submersion of the driven body, formulae of oblique shocks, law of the resistance in relation to the displaced surface and, above all, of absolute imperative for the technology of canals, to make the latter as wide as possible, thereby avoiding underground canals.[23]

Neither Perronet nor other representatives of the Corps des Ponts et chaussées had been invited to participate any more than Borda or other members of the Navy. Instead, Turgot offered Bossut, in November, the magnificently paid Chair of Hydrodynamics, a new government institution, set up for the time being at the Oratoire, rue Saint-Honoré, and later annexed to the Ecole royale d'architecture at the Louvre — a relatively marginal institution, with a difficult future.[24]

In April 1776, abbé Bossut again presented to the Academy the *Nouvelles expériences sur la résistance des fluides*, signed by the three members of the ministerial commission. A new statistical method of appraisal, generalized by Condorcet, to obtain the average for the results of a large number of experiments was also published as a supplement of this volume. The *Discours préliminaire* of the reporters did not mention the delicate problem of the Picardy Canal, but it did contain statements the extremely harsh and controversial tone of which was a result of the prevailing tensions and rivalries.[25]

> Expect nothing from the narrow-minded practitioner lacking in principles and driven by a blind rage, who hastily puts together several facts and is then unable to explain their differences. There are no sciences without reasoning, nor sciences without theory.

Inventive genius and scientific theory came face to face and the latter demanded recognition under protection of the government. The following month, however, Turgot was dismissed. D'Alembert's personal grief at the loss of Mademoiselle de Lespinasse was also, most likely, a reason for the delay in the signing of the other document by the ministerial commission. This document, which was never published or made public at the time of its compilation, is the one which we publish here. It concerns the report titled *Observations sur le canal de Picardie*, some twelve pages long, signed at the Academy of Science by d'Alembert, Condorcet and Bossut (reporter), on 17 July 1776. We were made aware of this text as a result of four copies, two of which form separate files and two of which appear in registers, dated 8 October 1791. All these copies can be found attached to some miscellaneous files on the Picardy Canal affair which are kept at the Library of the Ecole des Ponts et chaussées. There appears to be no sign of either the original or other copies at the Academy of Science; any more so than there appears to be under the reference F14 (Public Works) of the National Archives, despite its voluminous files on the history of the underground canal. Moreover, there was a general discretion which surrounds the exact contents of this report. Both Lalande in his *Canaux de navigation* of 1778, and Laurent de Lionne in his apologetic replies, show an indirect knowledge of the contents of the report. Even Bossut and Condorcet, in intervening in the affair much later, would avoid explicitly taking into account the result of the report which remained confidential and somewhat mysterious. We can in fact today understand why this was so.

The report was delivered on 17 July 1776: the storm of controversy and the criticisms which followed Turgot's dismissal had still not died down. Deprived of power and political protection, the special commission headed by d'Alembert had practically fallen by that time. Even then, in addressing the report to Trudaine, the commissioners were only able to take advantage of the prestige of the body to which they belonged, the Academy of Science, with d'Alembert as its patron and Condorcet its new permanent secretary. Thus, it was no longer Minister Turgot, but the Academy which found itself involved in the affair through a commission which did not originate from within it but which had been appointed by Turgot and which outlived him by throwing the whole weight of academic prestige into a controversy requiring a very delicate decision.

It was all the more precarious, given that the report opposed Laurent's project and the Duke de Choiseul, and that it was controversial and hasty in its conclusions. The only direct confirmation of the whole affair, in fact, was to be found in the negative views expressed by one of the reporters who had visited the underground canal — namely the weakness of the vault, and the darkness that prevailed there. If these views, however, were those that Condorcet had sent to Turgot, we know that their exponent had been unfavourably influenced by the experience of the Ecole Militaire.

The basic contents of the report rejected the theoretical conclusions in condemning the underground canal from the point of view of resistance of water which "would be far greater than for an indefinite canal. This resistance could increase by a rate of 3 to 5, which is considerable quantity."

Beside this major defect, which would have made navigation of the canal long and difficult, in spite of its rectilinear layout, the system of towing seemed inefficient and dangerous. The shocks and movements resulting from the oblique movement of the boats submitted to impulses of unequal strength by the two men at each end of the transversal lever, would have damaged the lateral towpaths. Furthermore, the latter were considered too narrow for the movement of men, who in turn would have to be replaced by horses. Another danger was that of the solidity of the vault, which would require a revetment of bricks, further increasing the cost to such an extent that "it appears frightful."

While refraining from commenting on the key issue of the water supply, the report condemned as a luxury for the nation the Picardy Canal, as an impracticable project for reasons of construction and economics.

It should be pointed out that even De Vicq's project, of which the report initially took account, received the same condemnation resulting from the theory of the resistance of fluids. On the other hand, the three commissioners offered an alternative solution (and it is to this project that the last part of the text is devoted), which again called for a canal to be built entirely on the surface by a triple junction of the Oise, Sambre and Scheldt rivers;[26] one which would be longer, wider and safer.

Inspired without doubt by Condorcet's visit, this alternative, which adheres to the traditional technology of channeling rivers, appears to combine a whole series of advantages of a general nature, with ease of execution: the drainage of the marshes around the river Oise would reduce the region's death rate, the route through the Royal Forest of Mormal would automatically provide an income owing to the ease of transport of wood and therefore a sort of immediate source of self-financing for the canal, and even the material needed for the construction of the locks could be found on site.

Such global savings, however, so easily within hand's reach, were only apparent; the cost involved in the large number of necessary locks, and the length of the canal, meant that this alternative went against the declared main aim of establishing quick navigation between Paris and Flanders. The idea, when it came up much later, would still be rejected by canal engineers.

It is also important to point out that the report was dominated by a feeling of great responsibility for the exclusive role of the State as contractor. The offer by a private firm to take responsibility for public works of this nature, was explicitly rejected, because, in the case of failure: "the government would be held responsible for taking over the canal and repairing it and putting it back in order."

If, in the face of a long-standing tradition of scandals, such as the appalling system of corvées, this was the political feeling of the reporters, the attitute expressed in this document once more, rejected the proposals for an underground canal. This without either clear awareness or direct understanding of the hydrogeological advantages involved in this particular case, but rather, with a limited idea of the value of deduction used in theory and in scientific experiments. This, technically speaking, led in turn to a defence of the traditional methods which tended to consider canal technology from a conservative point of view.

The opening lines of the report also revealed a controversial contempt for the know-how of mechanics such as Laurent. "We are putting forward these views to counter the prejudice and ignorance of those who place practice above theory."

Perhaps the ministerial commission's hand had been forced by a series of rivalries and by purely economic and political factors. In order to fully develop their very recent scientific experiments and their new institutional role in the shadow of the authorities, the illustrious academics appointed by Turgot had not hesitated to adopt an absolute view of their knowledge, even though other experts had said they were wrong. The most technically valid solution which would enable a liaison to be established between the Somme and the Scheldt was without doubt a canal under the Vermandois plateau.

In August 1776, Trudaine de Montigny informed the contractor of his decision to definitely halt work on the project. We know about a text entitled *Réponse à l'extrait sommaire d'un projet envoyé par M. Trudaine à M. Laurent de Lionne* in August 1776; the contractor from Picardy proposed the technique used in the Dutch canals and small boats used in the Flanders. Where water resistance was concerned, he asked for "a large-scale experiment to try and solve the problem."[27] He also proposed a regular service of smaller and faster boats.

In a speech delivered to the Amiens Académie des sciences in August 1776, Laurent de Lionne presented himself as a victim of academic jealousy and defended the bringing together of local interest and foreign trade, by drawing on the praises of the most illustrious visitors to Saint-Quentin's underground canal. Among these admirers, one would later find Frederick the Great, who was to visit it incognito.[28]

Things would probably not have gone much further, had it not been for a dramatic, although foreseeable, event which placed the whole affair at the extremes of a politico-military scandal; great evidence of the urgent problem posed by the canal. In December 1779 Count de Byland's fleet carrying ammunition for the French military ports, was seized by Commodore Fielding and resulted in enormous losses for the State to which had to be added the cost of transporting arms and matting from Holland, first by land between Bouchain and Saint-Quentin and then by the more suitable inland navigation as far as Nantes.

An anonymous letter appeared in the *Mercure de France* in 1780 emphasizing that the presence of the "canal condemned by the members of the Academy of Science" would have prevented such a disaster.[29] For contemporaries, those who had decided to postpone the construction of the underground canal were nothing more than traitors who "sold themselves to England," as Condorcet would point out.[30]

The same accusations were to be revived in 1782 in a pamphlet published by the Amiens Academy of Science, signed by M. d'Agay, the province's administrator.[31] The Navy, through its physicist, Rigaud, also expressed itself in favour of the canal in the *Mercure de France* in 1780.[32]

The case was reopened in 1781 and its examination entrusted to the Corps des Ponts et chaussées, on behalf of which Chezy, associate director, and Perronet, in the same year, gave their direct expert opinion and a favourable account.[33] D'Alembert, Condorcet and Bossut's much debated report, however, appeared to remain confidential, as Perronet had to ask for it by written communication. It was perhaps as a result of this intervention by Perronet, that the report was handed over to the management of the Ecole des Ponts et chaussées, which wanted to sponsor and bring the project to completion.

D'Alembert died in 1783. In his eulogy, Condorcet made no mention whatsoever to the case. At the same time, however, he responded to the attacks by publishing an anonymous report entitled *Mémoire sur le canal de Picardie.* The report by d'Alembert's commission had in fact stirred up a scandal, and was most likely the cause of the discretion surrounding the document.[34]

The public is unaware of what was done by those [whom Condorcet does not even name] who were then responsible for the examination [of the canal]. They did not consider making any written statement against the insults that M. Laurent's admirers threw at them ... The view that one sought to spread, that the objections raised against the project were inspired by the envy that his genius had aroused, is absurd and can only appeal to the ignorant among us. Those involved in mechanics almost always complain about the scholars, but these technicians are only admired by society's other groups; the scholars alone can be their judges. It is not, therefore, surprising that they wish to challenge them. Thus, one becomes aware that it is the géomètres, the physicists and the engineers, and not the technicians or builders, whose opinion on the underground canal can have any influence on the enlightened.

Condorcet again mentioned the case in his *Eloge de Turgot,* published in London in 1786.[35]

The blame that they received for scorning the practical knowledge, for being jealous of the spirit of invention in the arts, and for holding to the views adopted by their Academy, only represented [for Turgot] the complaints of a charlatanism, which was indignant with a class of men trying to escape from its prestiges.

The controversy had, in effect, become a war of prestige and at the same time a scandal contesting the juridictional right of the Academy of Science over technical issues. On this important technological occasion, conflict arose between the social orders and the various State bodies concerning the old-age privilege practiced by the Academy.

In 1781 the Corps des Ponts et chaussées, considered the supreme organ, declared itself in favour of the Laurent Canal. It gave the same reply in 1791 to the Legislative Assembly.[36] The project was again passed by invoking a law of the first Brumaire of Year 3. It was, however, up to the government, set up on the 18th Brumaire, to reconsider the case.

Gaspard-Riche de Prony, the great engineer who had succeeded Perronet as head of the Ponts et chaussées, presented a new report, asked for by the First Consul, on 6 February 1801, at the Malmaison, which concluded with these words:[37]

One must regret that the doubts which led to the suspension of the canal work in 1775 may have deprived France of some of the great benefits which could have resulted from its completion. The rewards from this monument could have been reaped a long time ago. Both the State and Trade could already have made savings greater than the expenditure on the canal.

Three days later, Bonaparte went to Saint-Quentin and visited on horseback the old site, once more in the company of Laurent de Lionne and a string of engineers, civil servants and other noteworthy people — almost a mini-expedition to Egypt. At the underground canal opened at Tronquoy, which at one time delighted Frederick the Great, the ground had collapsed in twenty-five years and only the remains of the excavation were visible.[38] As a result of the visit, it was decided to let yet another commission examine the problem for the umpteenth time.[39]

Of all the projects, only those of Laurent and De Vicq were retained. The summer of 1801 was devoted to surveying and working on the soil. This new campaign at last showed the need for lowering the summit level of De Vicq's underground canal to a level similar to that planned by Laurent. The advantages and the cost evaluation of the two projects were submitted to the Assembly of the Ponts et chaussées on 4th February 1802. It declared itself by a large majority (21 votes to 9) in favour of Laurent's project.[40] Since the debate had been intense and the minority in the Assembly had published a harsh criticism of the majority's decision,[41] Bonaparte decided to submit yet another judgement on the controversy between Laurent's supporters and those of De Vicq; that of the reformed Academy of Science, whose great utilitarian value for the national economy Bonaparte wanted to assert through such an important opportunity.

It was thus that Cretet, a senior member of the Council of State, submitted to the First Class of the Institut the six voluminous bundles of documents concerning the Picardy Canal case, to present them to a commission made up of the following: Coulomb, Bossut, Laplace, Delambre, Legendre, Léveque and Louis Lefevre Gineau, reporter.

Charles-Bossut, now at the age of seventy-two, once more found himself amongst those judging the canal which he himself had condemned in 1776.[42] He was in fact the sole survivor of Turgot's commission and had not made a career for himself. Despite the republishing of his *Traité d'hydrodynamique* in 1786, the Convention decided to abolish his chair in 1788. This event took place even before the abolition of the Ecole royale d'architecture, of which he was a welcomed member.

Until then, it was thought that only institutional jealousies could have led to the destruction of Turgot's creation of technological pedagogy. These jealousies, however, also mirrored the case of the Picardy Canal at the time of the Revolution. If Revolution abolished Bossut's chair, it set up another one at the Ecole polytechnique, rightly entrusted to Prony, an engineer from the Corps des Ponts et chaussées, when Bossut was just an examiner at the Ecole polytechnique. One wonders just how far the serious controversy surrounding the Laurent Canal had influenced the projects to create, under the Revolution, a new scientific staff capable of controlling both the geometric and experimental knowledge and the know-how of the engineers. One must also question the role that this case must have played in the accusations launched at the Academy of Science for repressing the free inventive mind of the practitioners, accusations which must have influenced the decision to abolish the Academy.

The final report of the Institut's commission on the underground Picardy Canal was solemnly presented at the extraordinary session of the 28th Messidor of Year 10 (17 July 1802). All the main participants had assembled: Prony, Bossut, Chaptal and Monge, Bossut's former assistant during the experiments at the Ecole Militaire, who presided at the meeting. Bonaparte was also present. Having established the importance of a liaison between the Somme and the Scheldt as well as the viability of Laurent's and De Vicq's projects, Lefevre Gineau's report reached an ambiguous conclusion which echoed the earlier objections of the 1776 report:[43]

The very thought of an underground canal gives one the inkling of a great number of disadvantages, difficulties, obstacles and even foreseen dangers, and one cannot foresee, be it in the execution, solidity or navigation, whether in choosing between De Vicq's and Laurent's projects, your commissioners prefer De Vicq's canal because it is a shorter underground canal. They are, however, so struck by these disadvantages, that it is impossible for them not to express strongly the desire to see carried out, without resorting to underground navigation, the important junction from Belgium to France

FIGURE 5 Elevation and section of the southern gate of the Picardy underground canal at Riqueval, according to the executive design under the Consulate. The canal basin was now 5.20 meters wide and each tow-path 1.40 meters wide. The basin was considerably deepened and widened later, to reduce the effects of the resistance of water. From *Canal de St. Quentin. Plan, coupe, élévation de l'entrée du souterrain*, beginning of XIXth century, Archives Nationales, F^{14}10127-70.

and not to display to the government the wish that they get together so that the different projects that have been presented to them may be examined again.

A so never-ending postponement did not serve Bonaparte's purpose. However, the final opinion reverted to the First Class of the Institut which in fact officially opted for De Vicq's project without subscribing to the reasons and the final wish formulated in the commission's report. So, on 30 July 1802, Bonaparte and Chaptal decided in fact to carry out the project for the two underground canals between Saint-Quentin and Cambrai. Under the management of the engineer Gayant, the work would last seven years, without encountering any obstacles.

Napoleon inaugurated the canal on 28 April 1810, at the time of the new empress, Marie-Louise's honeymoon. The outline of the two underground canals was that proposed by Laurent, with a basin measuring 16 feet (5.20 m), but the towpaths had been widened. The towing was to be done by hand until the middle of the century. It was then done by horse, steampower and today electricity, in accordance with the constant increase in the number and tonnage of the boats using the underground canal.

A project of the Regency had become the completed work of the Empire. The case of the construction of the Picardy Canal had lasted almost one hundred years, constituting one of the longest and most complex cases in the history of the technology of canals (Figure 5).

No other mishap was to occur in the history of navigation of the longest underground canal in Europe. In 1810, however, on the day of its inauguration, the boats and their bargemen who had approached the dark underground canal did not dare enter it. A life-long exemption from the navigation duties had to be promised to the first bargeman to pass under the arch; the first boat to enter it was rechristened *Le Grand Souterrain.*

Observations sur le canal de Picardie

L'art de construire des canaux dépend absolument de la science de l'hydraulique. En effet, il faut s'assurer d'abord de la quantité d'eau que les sources, destinées à nourrir un canal, lui peuvent donner. Il faut ensuite fixer la position et les dimensions des Ecluses; les quantités d'eau qu'elles fournissent en un temps donné; la vitesse de cette eau; la force nécessaire pour mouveoir les bateaux; la résistance que ces bateux éprouvent dans des canaux de différentes largeurs, et dont les eaux sont plus ou moins profondes etc.

Toutes ces connoissances qui appartiennent à l'hydraulique sont nécessaires pour déterminer les dimensions d'un canal projeté et les travaux qu'il exige: et pour choisir entre plusieurs canaux celui qui mérite la préférence. La mécanique pratique ne peut, à cet égard offrir aucun secours. La seule machine qui soit employée dans les ouvrages dont il s'agit, est celle qu'on nomme Ecluse: et la construction en est connuë depuis long temps.

Il n'y a donc que ceux qui ont approfondi la Théorie de

l'hydraulique qui puissent prononcer en connoissance de cause sur le choix d'un Canal//et sur sa construction. Ces objets une fois decidés; l'exécution n'a plus que des[a] difficultés de pratique; et les Ingénieurs exercés dans ce genre de construction sont les personnes les plus propres à cet emploi.

Nous présentons ces réflexions générales contre le préjugé et l'ignorance, qui élèvent la pratique au dessus de la Théorie.

Le canal de Picardie n'a point pour objet, comme on l'a dit, de joindre l'Escaut à la Somme; mais l'Escaut à l'Oise; afin qu'il y ait une communication par eau, entre La Fere et Douai; objet important en tout genre; et, en général, une communication entre la Flandre et l'Interieur du Royaume.

Pour remplir cet objet, on pouvoit

1° Joindre l'Escaut à la Somme; parce que l'Oise est déjà unie à la Somme par le Canal Crozat.

2° La Sambre et l'Escaut à l'Oise.

De la jonction de l'Escaut à la Somme

Il y avoit plusieurs moyens de joindre l'Escaut à la Somme: on pouvoit y parvenir.

1° Par le Canal-souterrain de M. Laurent.

2° Par un Canal a ciel couvert [sic] qui // partiroit des environs du Catelet, pour joindre la Somme en suivant la vallée où coule l'Omignon.

3° Par un Canal qui partiroit de Marcoing sur l'Escaut, pour tomber dans la Somme un peu au-dessous de Peronne, à Bazincourt.

Le gouvernement a donné la préférence au canal souterrain de M. Laurent; mais comme ce canal tel que ce fameux machiniste le proposoit paroissoit à toutes les personnes éclairées, sujet à beaucoup de difficultés: que jamais aucun homme instruit dans la science de l'hydraulique n'avait été consulté sur ce projet:[b] qu'il n'existoit aucune comparaison entre ce projet et ceux sur les quels il avoit obtenu la préférence, Mons. Turgot a cru qu'il étoit convenable de suspendre les travaux de ce canal et d'en ordonner l'examen.

Mr. Laurent avoit entrepris en même temps de rendre la Somme navigable, depuis le point où le canal Crozat la quitte jusqu'à Amiens. Il s'est encore élevé des difficultés contre ce projet: et ces difficultés étoient faites par des hommes éclairés et sans intérêt; en conséquence on a ordonné également la suspension des ouvrages jusqu'après l'Examen du projet de Mons. Laurent. Il falloit donc résoudre les Questions suivantes.

1° Le canal souterrein tel que Mr. Laurent l'avoit conçu peut-il offrir une navigation sûre et commode?

2° Peut-on parvenir à ce but en changeant les dimensions de ce Canal, en le voûtant, en donnant une autre forme au tirage? et quels doivent êtres ces changements?//

3° Quelle seroit alors la dépense de ce canal souterrain et du canal à ciel ouvert qui se joindrait d'un côté à la Somme, à Saint-Quentin, et de l'autre à l'Escaut à Cambray?

4° Un canal à ciel ouvert par l'Omignon, est-il possible; et quelle en seroit la dépense?

5° Quelle seroit celle d'un canal par Marcoing et Bizancourt; et ce canal est-il praticable?

6° Combien en coûteroit-il pour rendre l'Oise navigable de La Fère à Boheries: joindre ensuite l'Oise à la Sambre par la Vallée de Noirieu, et la Sambre à l'Escaut, en faisant passer un canal dans la Forest de Mormal, et en suivant ensuite une des rivières qui, prenant leur source dans cette forêt, tombent dans l'Escaut?

7° Discuter laquelle de toutes ces communications seroit la plus avantageuse?

8° Quelle est la meilleure manière de rendre la Somme navigable, depuis St. Simon à Amiens?

Pour répondre à ces questions, il falloit avoir des nivellemens exacts de la Vallée de l'Escaut, depuis sa source jusqu'à Marcoing, de la vallée de l'Omignon, de celle de Bizancourt, de celle de Marcoing et des hauteurs qui séparent ces vallées, de la vallée de l'Oise, entre La Fère et Boheries, de la vallée de Noirieu et de celle de la Sambre jusqu'à Landrecy; de la forêt de Mormal et des vallées, où coulent les rivières qui vont de cette Forest dans l'Escaut, ainsi que des hauteurs qui séparent ces rivières: Il falloit enfin avoir nivellé le cours de la Somme.//

Mons. De Brie, Inspecteur des Ponts et Chaussées a été chargé de la première partie de ce travail.

Mons. de Chabaud, Ingénieur en chef, ordinaire du Roy à[c] Saint-Quentin, s'est chargé du nivellement de la Somme et de l'examen de la navigation de cette rivière.

Ces travaux préliminaires n'étant pas achevés, il est impossible de se décider avec certitude sur la meilleure communication entre l'Oise et l'Escaut; nous nous bornerons donc à discuter icy les questions relatives au canal souterrain, et à présenter quelques réflexions générales sur la communication par la Sambre, et sur la navigation de la Somme.

Du canal souterrein de Mr. Laurent

Ce canal doit avoir 16 pieds de largeur: les bateaux qui doivent y passer ayant 14 et jusqu'à 15 pieds de large: La voûte sera à plein cintre, taillée dans la pierre: Les banquettes pour le tirage seront de deux pieds de chaque côté non pavées et non revetuës. Ce tirage s'exécutera par des hommes qui pousseront de chaque côté du canal un baton de vingt pieds de long, fixé à un poteau placé à l'avant de chaque bateau.

Nous observerons:

1° que la résistance dans un canal si étroit sera beaucoup plus grande que dans un canal indéfini: cette résistandre pourra augmenter dans le//rapport de 3 à 5 ce qui est une quantité considérable.

2° Que la voûte seroit taillée dans une masse de craie où l'on trouve des cailloux ronds, articulés disséminés dans des bandes horizontales: Ces masses sont très communes en Champagne, en Picardie, en Normandie, dans l'Isle de France, quoique la craie ne soit pas toujours de le même dureté; ces masses sont de la même nature et ont la même disposition: partout elles se fendent dans tous les sens d'une manière irrégulière: il doit arriver souvent que des quartiers de pierres se détachent de la voûte.

L'un de nous a parcouru environ mille toises du canal de Picardie; partout il a observé ces fentes irrégulières que la filtration des eaux multiplie et qui sont d'autant plus sensibles,[d] que l'on est à une plus grande distance de la surface de la terre.

Dans la partie du canal qui conduit à l'endroit qu'on montre aux Etrangers; ces fentes paroissent moins: Elles ont été, on ne sait pour quel motif, bouchées avec du plâtre: ce qui prouve de moins, qu'elles auroient présenté une aspect éffrayant; quoique vraisemblablement on n'ait pas choisi l'endroit du canal le plus désavantage pour le montrer au public.

3° Que le frottement de l'eau détruiroit les parois du canal; que le marcher useroit promptement les banquettes et qu'ainsi il faudrait les revêtir et les paver.

4° Que le tirage proposé par Mons. Laurent est très incommode et très vicieux; en ce que les hommes appliqués//de part et d'autre aux extrémités du levier transversal qui pousse le bateau, ne peuvent pas s'égaler tellement en force et en vitesse, que le bateau marche toujours exactement dans la direction du canal, et qu'il ne se mette pas de biais, tantôt d'un côté, tantôt d'un côté opposé; Or tous ces mouvements latéraux s'exécutent aux dépens de la force motrice qui devroit être employée toute entière à faire marcher le bateau suivant le fil de l'eau; et ils pourront en certains cas absorber une grande partie de cette force destinée à vaincre la résistance considérable que le fluide lui oppose; sans compter que dans le cas où les deux forces seroient sensiblement inégales, le bateau choquerait par ses extrémité les parois du canal, et seroit exposé à se briser ou à s'endommager.

5° Qu'on seroit obligé de naviguer à la lueur des flambeaux; qu'ainsi la navigation par le canal de Mons. Laurent, seroit:

1° Dangereuse pour les hommes qui y navigueroient et même pour les bateaux que les éboulements pourroient écraser.

2° Qu'elle seroit interrompue par les éboulements.

3° Que le peu de largeur du canal augmentant la résistance, et le tirage étant très difficile sur la banquette, cette navigation seroit presqu'impraticable.

4° Que les éboulements qui rempliroient le canal et la destruction des parvis et des banquettes//occasionneroient des réparations immenses.

De toutes ces remarques, nous concluons que le canal de Mons. Laurent doit être abandoné ou corrigé.

Ce seroit mal répondre à ces difficultés que de proposer d'exécuter le canal à forfait et d'offrir des soumissions pour y faire le tirage: en effet lorsque le canal une fois terminé viendroit à être rendu impraticable par les éboulements et par la destruction des banquettes; lorsque les gens qui auroient entrepris le tirage se verroient obligés de l'abandonner, le Gouvernement se trouveroit forcé ou de reprendre le canal ainsi dégradé pour le réparer et le corriger, ou de renoncer à la communication entre l'Oise et l'Escaut.

On a proposé pour faciliter le mouvement des bateaux d'approfondir le canal, en laissant subsister la largeur déterminée par Mons. Laurent; mais ce moyen est très insuffisant. Il diminuera peu la résistance que le bateau éprouve à diviser le fluide; et le vice du tirage subsistera toujours en son entier: Si l'on veut corriger véritablement le projet de Mons. Laurent, il faut augmenter nécessairement la largeur du canal. Il faut la faire, par exemple, de 24 pieds: construire une large banquette d'un seul côté: substituer, pour le tirage, des chevaux aux hommes: voûter en entier le ciel du canal: revêtir les deux parois en pierre: paver la banquette: pratiquer d'espace en espace des garres, où un bateau auquel il arriveroit un accident, pourroit s'arrêter, décharger une partie de ses marchandises etc. Alors//la navigation seroit passablement commode, et n'auroit que l'inconvénient de ne se faire que par convoi et à des heures marquées: deux bateaux ne pouvant pas passer de front; et de ne se faire qu'avec des flambeaux; mais elle seroit toujours beaucoup moins avantageuse qu'une navigation à ciel ouvert. Reste à connoître, du moins à peu près l'augmentation de dépense que les corrections indiquées occasionneroient dans la construction du Canal.

Si la longueur du canal demeurant la même, on suppose que les dimensions des sections perpendiculaires à cette longueur augmentent proportionnellement; et que par conséquent les surfaces des mêmes section croissent, comme les quarrés de leurs côtés homologues; on verra, qu'en portant, la largeur du canal, de 16 à 24 pieds, le creux de l'excavation augmentera dans le rapport de 4 à 9. D'un autre côté, à mesure qu'on augmente la largeur du canal, les raisons de le voûter deviennent plus pressantes; et la dépense[e] pour la construction de la voûte devient plus considérable: elle paroit éffrayante.

On a dit et on répète tous les jours que le canal souterrein de Mons. Laurent seroit un bel ouvrage: un ouvrage comparable à ceux des Romains; ou, si l'on veut, aux Pyramides d'Egypte; mais ces ouvrages ne sont admirables que par leur solidité et leur grandeur: ils prouvent moins le génie de ceux qui les ont//executés, que la puissance de ceux qui les ont ordonnés: et c'est surtout par l'idée de ce qu'ils ont coûté qu'ils prouvent cette puissance.

On a déjà construit des canaux souterreins: et entre un canal de cent pieds de long; et un de cent lieues, creusé dans un terrein toujours le même; il n'y a d'autre difference qu'en ce que le dernier coûteroit environ douze-mille fois plus que le premier; mais tout est égal dans le mérite de l'invention et de l'exécution: Il est donc surprenant qu'on insiste sur l'honneur prétendu que le canal de Mons. Laurent feroit à la France qui n'a pas besoin de cette preuve de sa richesse; et qu'on imagine que des raisons de jalousie ayent pu en faire suspendre l'exécution.

De la communication par la Sambre

Cette communication seroit plus longue que par le canal souterrein: mais cette différence est très peu de chose; et elle est plus courte que toutes les autres.

Sans connoître précisément la quantité d'eau, il y a lieu de croire que la navigation pourroit s' y faire en tout temps: elle a des avantages considérables.

1° Toutes les denrées que le canal de Mons. Laurent pourroient amener, viendroient également par celui cy. On aurait de plus une immense quantité de bois, de//pierre bleuë, espèce de marbre grossier très utile pour les batiments et pour faire des vases et des ustencilles d'un usage commun.

2° On auroit, pour construire les écluses, l'avantage de se servir de ces mêmes pierres bleuës dont la chaux mêlée à la cendre de charbon de terre fait un excellent ciment.

3° On pourroit se procurer des moyens de construire le canal à peu de frais: en effet il passe à travers la forêt de Mormal, et il ouvre un débouché pour les bois de cette forêt: elle ne rapporte au Roy que 170.000 livres: mieux administrée et avec la facilité de faire aller les bois à Paris, elle doubleroit presque le prix; ainsi, en retenant sur les fonds destinés aux canaux 170.000 livres par an; et affectant les bois de cette forêt à la construction du canal, on diminueroit de près de moitié la dépense de cette construction.

Mons. Laurent se proposoit de faire passer le canal qui conduiroit de Saint-Simon à Amiens, sur la partie haute du vallon de la Somme: par là on employoit, soit pour le canal soit pour les contrefossés, des terrains utiles pour la culture. Le canal courait risque d'être desséché par les marais inférieurs: l'on manquoit une occasion précieuse de dessécher ces marais. Et l'on s'est assuré par des calculs exacts que des marais moins dangereux que ceux de la Somme abrégeoient d'un cinquième la vie moyenne des hommes.//

Telles sont les raisons qui ont déterminé à ordonner un nouvel examen.

On a répondu à ces objections, que quoique supérieur aux marais, le canal les dessécheroit: et c'étoit si bien là, le système de Mons. Laurent, que, dans le canal de jonction de la Scarpe à l'Escaut qui suit le vallon de La Censée, il se proposoit également de faire son canal supérieur à cette rivière.

On a dit encore que l'on pourroit dessécher les marais de la Somme par une opération particulière. Cela est vrai, mais pour cela il faudroit acheter des moulins et des étangs; au lieu qu'en faisant passer le canal dans le lit même de la rivière, on pourroit à la fois établir la navigation et produire le desséchement sans que la dépense fut augmentée autant, à beaucoup près. On a même dit que dans l'état actuel le vallon de la Somme rapportoit plus, que s'il étoit desséché; qu'après le desséchement, il deviendroit indéffrichable, que l'air des marais n'est pas mal sain etc.

De telles assertions ne méritent pas que l'on y reponde sérieusement, mais du moins ne suffisent-elles-pas pour prononcer sur un objet de cette importance, sans un examen fait par des personnes éclairées et à qui la Théorie de l'hydraulique et celle des constructions soyent familières.

A l'Académie ce 17 juillet 1776 signé D'Alembert, le Marquis de Condorcet et Bossut.

Notes

MS 233, tome 3d, FFs. 69-80 (22 × 34 cm.), Library of the Ecole Nationale des Ponts et chaussées (Paris): copy dated October 8, 1791 (in the upper margin). We publish this copy as the only dated and the only quoted under the names of the authors in the *Catalogue imprimé des manuscrits de la Bibliothèque de l'Ecole Nationale des Ponts et chaussées*, Paris.

Another very similar copy in MS Fol. 2784, FFs. 41-56. Two other copies — in the same library — in MS2560, no. 3, FFs. 3-12 (18 × 23 cm.) and in a separate cahier having the same format, in the same file. The few relevant variants of MS 233 compared with MS 2560 two copies will be pointed out in the notes.

a. des *instead of* les.
b. ce projet *is followed by these words* et ceux sur lesquels il avait obtenu la préférence.
c. Inspecteur ordinaire du Roy, en chef à *instead of* inspecteur en chef ordinare à.
d. *after* sensibles *MS 2560 says* et annoncent un éboulement d'autant plus prochain.
e. les dépenses *instead of* la dépense.

References

1. Letter from Voltaire to P.J. Laurent (Ferney, 6.12.1771), in P.J. Laurent de Lionne (1783), *Discours prononcé à la séance publique de l'Académie des sciences, belles-lettres et arts d'Amiens le 25 août 1776*, Amiens, I and sq. Cf. *Correspondence*, vol. 38, *The Complete Works of Voltaire*, ed. T. Bestermann, Banbury, Oxfordshire, 1975, 174-sq.
2. On the Picardy and Burgundy canals, see B. Forest de Bélidor (1753), *Architecture hydraulique*, Part Two, Book II, 362-373.
3. Cf. J. de Lalande (1778), *Des canaux de navigation*, Paris, 307.
4. Cf. "Renseignements et observations sur les moyens d'établir une navigation entre Valenciennes et Paris", manuscript Archives Nationales F14 616; also see the plans for De Vicq's project in the Archives Nationales F14 10127. Cf. also J. de Lalande (1778), 307-10; Ducros, "Mémoire historique et descriptif des projets ... pour la jonction de la Somme avec l'Escaut", manuscript 2566 Bibliothèque de l'Ecole Nationale des Ponts et chaussées, fol. 93. In general, the information in the book by the former director of the Saint-Quentin Canal: R. Macaigne (1934), *Le canal de Saint-Quentin*, Paris, 105-107.
5. See An. (1774), "Eloge de M. Laurent", *Le necrologe des hommes célébres de France*, Paris, 203-sq.; An. (1773), "P.J. Laurent", *Mercure de France*, November, 212; Michaud jr., *ad nominem* (1783), *Biographie Universelle*, vol. 23, 368.
6. See Voltaire to Delille (19.6.1761), *Correspondence*, ed. Bestermann, vol. 107, 267.
7. P.J. Laurent de Lionne, "Exemples de canaux souterrains comme celui de Picardie", manuscript 2566, Bibl. Ec. Nat. Ponts et chaussées, fol. 37.
8. Cf. J. de La Lande (1778), 504.
9. Bélidor (1753), Book IV, chap. V, 388.
10. Cf. the list appended to P.J. Laurent de Lionne (1783).
11. P.J. Laurent (1773?), "Canal de Picardie", manuscript 2566, Bibl. Ec. Nat. Ponts et chaussées, fol. 27.
12. An. (1773), "Observations addressées à feu Mr. Laurent par le Controleur gènéral ... le 15 août 1773", manuscript ibid., fol. 38.
13. The impulse given to inland navigation as an alternative to the difficulty of preventing the degradation of the roads for administrative and technical reasons is well illustrated in De la Fer de la Nouerre (1786), *La science des canaux navigables. De la possibilité de faciliter l'établissement général de la navigation intérierure du Royaume*, Paris, 2 vols. On the policy of transport and public works of Minister Turgot, P. Foncin (1872), *Essai sur le ministère Turgot*, Paris; but especially E.J.M. Vignon (1862), *Etudes historiques sur l'administration des voies publiques en France au XVIIe et XVIIIe siècles*, Paris, vol. II, particularly p. 34. See also J.M. Dotens (1829), *Histoire de la navigation intérieure de la France*,

Paris and B. Lepetit (1984), *Chemins de terre et voies d'eau*, Paris.
14. Ch. Henry (1882), *Correspondénce inédite de Condorcet et de Turgot*, Paris, 251-254, particularly 253. For Perronet's activity and the organization of the Ecole and the Corps des Ponts et chaussées, see C.C. Gillispie (1980), *Science and Polity in France at the End of the Old Regime*, Princeton, 479-498.
15. In 1740 the Dôle Bridge had collapsed as a result of the contravention of the rules of the art and of estimates. As regards to the injunctions, see especially the King's declaration of 20 September 1762, in A. Debauve (1893), *Les travaux publics et les ingénieurs des Ponts et chaussées depuis le XVIIe siècle*, Paris, 63.
16. Ch. Henry (1882), 258.
17. Ibid., 260-sq.
18. See L.P. Bachaumont (1781-1789), *Mémoires secrets*..., London, vol. VII, 305; P. Foncin (1877), 282; F. Arago (1847-49), *Biographie de Condorcet, Oeuvres de Condorcet*, Paris, vol. I, XXII.
19. *Oeuvres de Lagrange*, Book XIII, Paris 1882, 297.
20. Cf. d'Alembert (1768), "Paradoxe proposé aux géomètres sur la résistance des fluides", *Opuscules mathématiques*, Book V, Paris, 132-170. The *Paradoxe* had been introduced in d'Alembert (1752), *Essai d'une nouvelle théorie de la résistance des fluides*, Paris, 74-sq.
21. Cf. "Canal souterrain de Picardie, 25 frimaire an 10. Etat général des mémoires, notes documents et des cartes", manuscript 2639 Bibl. Ec. Nat. Ponts et chaussées.
22. See d'Alembert, Condorcet and Bossut (1777), *Nouvelles expériences sur la résistance des fluides* (read at the Academy of Science on 17 April 1776), Paris, 3.
23. Ibid., 192-sq.
24. See R. Hahn (1962), "The Chair of Hydrodynamics in Paris 1775-1791, a creation of Turgot", *Xème Congrès Int. d'Histoire des Sciences* (Ithaca 1962), Paris, vol. II, 751-754.
25. D'Alembert, Condorcet and Bossut (1777), 5.
26. Cf. Bélidor (1753), 370 and plate XLIII.
27. P.J. Laurent de Lionne (1776), "Réponse à l'extrait sommaire d'un projet envoyé par M. Trudaine", manuscript 2566 Bibl. Ec. Nat. Ponts et chaussées, fols. 3-12.
28. See d'Agay (1782), *Discours prononcé à la séance publique de l'Academie des sciences, belles-lettres et arts d'Amiens*, Amiens, 33-39 (BN Vp 247II).
29. (D'Agay?) (1780), "Lettre de M. d' ... à M. Rigaud, ingénieur de la Marine" (4 October 1780), *Mercure de France*, October, 44-sq.
30. *An.* (n. p., n. d.) [Condorcet, 1783], *Mémoire sur le Canal de Picardie*, republished in *Oeuvres de Condorcet*, Book XI, Paris 1847, 313-349.
31. D'Agay (1782), 38 and P.J. Laurent de Lionne (1783), 8-sq.
32. See "Lettre de Rigaud", *Mercure de France*, No. 50 (1780), next published in P.J. Laurent de Lionne (1783).
33. See A. Chezy (1781), "Mémoire" (3 March), manuscript 2950 and Perronet (1781), "Observations sur la partie de canal du Picardie qui doit être fait à couvert dans la longueur de 7020 toises ..." (9 January), ibid., Bibl. Ec. Nat. Ponts et chaussées.
34. Condorcet (1783), 7-sq.
35. Condorcet (1786), *Vie de Monsieur Turgot*, London, 93.
36. See the file *Pièces remises par le Comité d'agriculture et commerce à feu le citoyen Lessart ministre de l'Intérieur*, Arch. Nat. F14 616.
37. G.C. Prony, "Note sur les travaux ... pour la navigation intérieure de la France ... (11 Pluviôse Year 9), mémoire demandé au citoyen Prony par le Premier Consul ... objet de la conférence à la Malmaison du 16 pluviôse", manuscript 2566, No.

4 Bibl. Ec. Nat. Ponts et chaussées, 14. Another copy, manuscript Arch. Nat. F14 616.

38. Cf. R. Macaigne (1934), 117-119 for the local recollections.
39. The preliminary work by the engineers Viennois and Becquey, see Arch. Nat. F14 614; 616.
40. *Rapport du Conseiller d'Etat [Cretet] au Ministre [Chaptal] du 28 Germinal an 10*, manuscript Arch. Nat. F14 616, some of which was published in R. Macaigne (1934), 125-129.
41. *Opinion des ingénieurs composant la minorité de l'Assemblée des Ponts et chaussées dans la décision prise par elle le 15 ventôse an 10 sur le canal de jonction de la Somme à l'Escaut*, Paris 1802 (BN Vp 2387). The opponent document in complete: "Opinion de la majorité ..." (16 Pluviôse Year 10), manuscript Arch. Nat., F14 616. Joseph Laurent signed here as chief engineer of the canal.
42. See Lefevre Gineau's letter to Cretet (Dossier Lefevre Gineau) and the dossier of the extraordinary session of 17 July 1802 in the Archives of the Academy of Science.
43. See *Procès verbaux des séances tenues depuis la fondation de l'Institut jusqu'au mois d'août 1835*, vol. II, Hendaye 1912, 524-527.

8

Poverty, distress and disease: labour and the construction of the Rideau canal, 1826–1832

William N.T. Wylie

THE CONSTRUCTION OF THE Rideau Canal has usually been viewed as a British military engineering project. Imperial strategists intended the canal to provide a safe transportation route between Lake Ontario and Montreal by-passing the American border along the St. Lawrence River. While historians have differed on the strategic and financial viability of the project, they all have stressed the exploits of the engineers and entrepreneurs supervising construction.[1] This has encouraged a dramatic distortion of the construction process. Because of the pre-steam technology employed, the participation of thousands of labourers and artisans became crucial to the realization of the design. In most accounts, these workers are portrayed as quarrelsome, unreliable, and disruptive. In short, they are viewed as their masters saw them, primarily as instruments in the work-process.[2]

H. Clare Pentland advanced a revisionist portrait of canal workers generally, which has been expanded upon recently by Ruth Bleasdale. These writers have stressed the important role of labour in social development, the emergence of a capitalist labour market, and the incidence of class conflict between workers and employers.[3] Building on the studies by Pentland and

[1] The major studies are Robert W. Passfield, *Building the Rideau Canal: A Pictorial History* (Don Mills, Ont. 1982); Robert Legget, *Rideau Waterway*, revised edition (Toronto 1972); George Raudzens, *The British Ordnance Department and Canada's Canals, 1815-1855* (Waterloo, Ont. 1979).

[2] E.F. Bush summarizes the traditional accounts in *The Builders of the Rideau Canal, 1826-32*, Parks Canada, Manuscript Report Series No. 185, 1976, 18-25; Passfield's new study presents a somewhat more sympathetic view of the workers within the traditional context of the canal as an imperial project.

[3] H. Clare Pentland, *Labour and Capital in Canada, 1650-1860* (Toronto 1981); this is

Bleasdale, this paper will examine the lives of the workers on the Rideau during the construction period between 1826 and 1832. The analysis will consider the background of the labourers and then their employment including wages, conditions, and the problem of disease. The outbreak of disorder will also be examined though a study of this question is limited by the vagaries of the sources. Throughout the paper, the emphasis will be on explaining the difficult living and working conditions in the context of the structural forces of the economy which helped to produce them.

I. The Labour Market

IN SPITE OF BELIEFS at the time, there seems to have been no shortage of wage-earners seeking work in central British North America by the late 1820s. While craftsmen were still in demand, "common" labourers existed in Montreal and Quebec City in sufficient numbers to provide a plentiful supply for public works projects such as the Rideau.[4] This resulted from two factors, one international and the other regional. After the Napoleonic Wars a wave of British emigration began to North America. This migration originated in changes in the home economy which led to the displacement of farm workers from the land and artisans from their crafts. While some of these emigrants were English and Scottish, the majority were Irish, coming mainly from Ulster and its immediate environs, where the textile industry was in crisis. Probably few of these persons were paupers. Most possessed sufficient means to afford the cost of passage. Yet they often arrived in desperate condition. Exhausted by the unhealthy circumstances of their voyage, many were penniless as a result of their travel expenses. In British North America they sought wage labour to sustain their families and to accumulate the funds with which to establish themselves on farms or in shops.[5] Lieutenant-Colonel John By's description of the workers assembling at the Rideau Canal in spring 1827 probably reflected the situation generally:

---

a revised version of his Ph.D. thesis which was completed in 1960; Pentland, "The Lachine Strike of 1843," *Canadian Historical Review*, 29 (1948), 258-77; Pentland, "The Development of a Capitalistic Labour Market in Canada," *Canadian Journal of Economics and Political Science*, 25 (1959), 450-61; Ruth Bleasdale, "Irish Labourers on the Cornwall, Welland and Williamsburg Canals in the 1840s," M.A. Thesis, University of Western Ontario, 1975; Bleasdale, "Class Conflict on the Canals of Upper Canada in the 1840s," *Labour/Le Travailleur*, 7 (1981), 9-39.

[4] Pentland, *Labour and Capital*, 115-6; Fernand Ouellet, *Lower Canada 1791-1840; Social Change and Nationalism* (Toronto 1980), 157-8, 178.

[5] Donald H. Akenson, "Ontario: Whatever Happened to the Irish," *Canadian Papers in Rural History*, 3 (1982), 205-9, 218-9, 222-3; Gerald M. Craig, *Upper Canada, The Formative Years, 1784-1841* (Toronto 1963), 125-30; J.L. Runnalls, *The Irish on the Welland Canal* (St. Catharines, Ont. 1973), 4-5.

I therefore expect to collect a great number of persons on the Works by the first May and fear from the wretched condition of most of the emigrants applying to me for work, that it will be indispensably necessary to issue bedding to prevent sickness . . . at present the poor fellows lay with nothing but their rags to cover them, and their numbers are increasing, and the rainy season coming on, I dread the effects of Sickness and feel convinced that the distribution of bedding will be of the greatest importance.[6]

The labour market which the immigrants found on the St. Lawrence was already crowded with persons seeking jobs. Between 1790 and 1820 there had been a continual exodus of French Canadian farm workers to the cities in part because of the inability of the agricultural economy to support the growing population.[7] Some of these persons had been absorbed by the timber trade: by the late 1820s, about 2000 French Canadians were employed each winter in the timber shanties of the Ottawa Valley. Finding employment in Quebec and Montreal during the summers, however, was a constant struggle. In many years, there was a surplus of labour even before the arrival of the annual immigrant ships.[8]

The chief opportunity for employment was provided by the canal projects which proliferated throughout North America following the completion of the Erie Canal in 1825. In America, major canals were begun in Pennsylvania and Ohio as various seaboard centres struggled to tap the trade of the interior. In Upper Canada between 1826 and 1832, the Rideau and Welland Canal projects together may have employed 8-9,000 persons per year.[9] Compare this with the rate of immigration. During this period an average of more than 25,000 persons per year arrived at Montreal and Quebec City, with the largest numbers recorded in 1831 and 1832. Even allowing for the presence of women and children, the capacity of the public works was not enough to absorb the expanding labour force. This was reflected in the substantial outflow of recently-arrived persons to the United States which seems to have occurred during these years. On the Rideau the result of the labour surplus was that contractors were able to attract workers despite low wages and poor working conditions.[10]

[6] Public Archives of Canada (hereafter cited PAC), RG8, C, Vol. 43, p. 212, By to Durnford, 12 April 1827.

[7] The basis of the agricultural crisis is a source of contention. While F. Ouellet stresses declining crop yields, J.-P. Wallot and G. Paquet argue that the failure of the economy to adjust to changes in international demand was the major problem. The most recent contribution to the argument is R.M. McInnis, "A Reconsideration of the State of Agriculture in Lower Canada in the First Half of the Nineteenth Century," *Canadian Papers in Rural History,* 3 (1982), 9-49.

[8] Sandra Gillis, *The Timber Trade in the Ottawa Valley, 1806-54,* Parks Canada, Manuscript Report Series, No. 153, 1975, 115-7; Ouellet, *Lower Canada,* 178.

[9] Runnals, *Irish on the Welland,* 15; Bush, *Builders of the Rideau,* 18-9.

[10] These estimates are based on the data in Pentland, *Labour and Capital,* 81-4; the statistics probably err on the low side as Akenson argues in "Whatever Happened to the Irish?," 206-11.

II. Employment on the Rideau

WHILE THE RIDEAU CANAL was distinguished from other public works by the involvement of the British military, this made little difference in the treatment of the workers. This was partly because of the attitude of the military and particularly of Lieutenant-Colonel John By who was directly responsible for the project. By was a man of his class. As an officer and a "gentleman," he did not question the justice of a hierarchical social order or the inevitability of poverty. His main duty was to produce a canal of good quality as quickly and cheaply as possible. With these considerations in mind, the military decided to contract most of the construction to private entrepreneurs rather than undertake the work themselves. This was expected to result in lower costs. The contractors were paid only according to the work done, not for the expense of maintaining an establishment in a primitive environment during those seasons of the year when construction might be difficult.[11]

Thus, the establishment on the canal consisted of a small military component and a large number of private contractors. The military was stationed at Bytown where plans were drawn up and from which overseers were sent to supervise progress along the line of the works. Since different contractors tended to be employed for excavation jobs and construction of masonry locks and dams, there tended to be several entrepreneurs at each of the 23 stations. Moreover, many of these individuals sub-contracted part of their responsibilities to other businessmen. The result was a mélange of contractors with different backgrounds. Some of the most successful were Scottish immigrants from Montreal such as John Redpath, Thomas McKay, and Robert Drummond. Several were already residents in the area of the canal: Philemon Wright of Hull and John Haggart of Perth. Others were Americans eager to profit from the international boom in canals.[12] Most of the workers were hired by the various entrepreneurs though some were taken on directly by Lieutenant-Colonel By to do work not contracted out. Because of the number of employers and the scarcity of extant private papers, it is difficult to generalize about the condition of the work camps. Nonetheless, a common viewpoint was visible among employers. For purposes of the project, it was expedient to view the workers primarily as instruments of production required to facilitate the most economical completion of the project.

At the beginning of construction, the officers at Bytown took several steps to insure a supply of cheap labour. To attract workers, they advertised in the press that the government would furnish provisions, accommodations, and medical care at every site where they were not readily available. More notably,

[11] PAC, MG12, WO44, Vol. 18, pp. 72-3, Carmichael-Smyth to Mann, 14 March 1826.
[12] Bush, *Builders of the Rideau*, 14-17, 101-17; Queen's University Archives, Benjamin Tett Papers (hereafter cited Tett Papers), Business Records, Transfer Case 10, Day Book, 1828-1833.

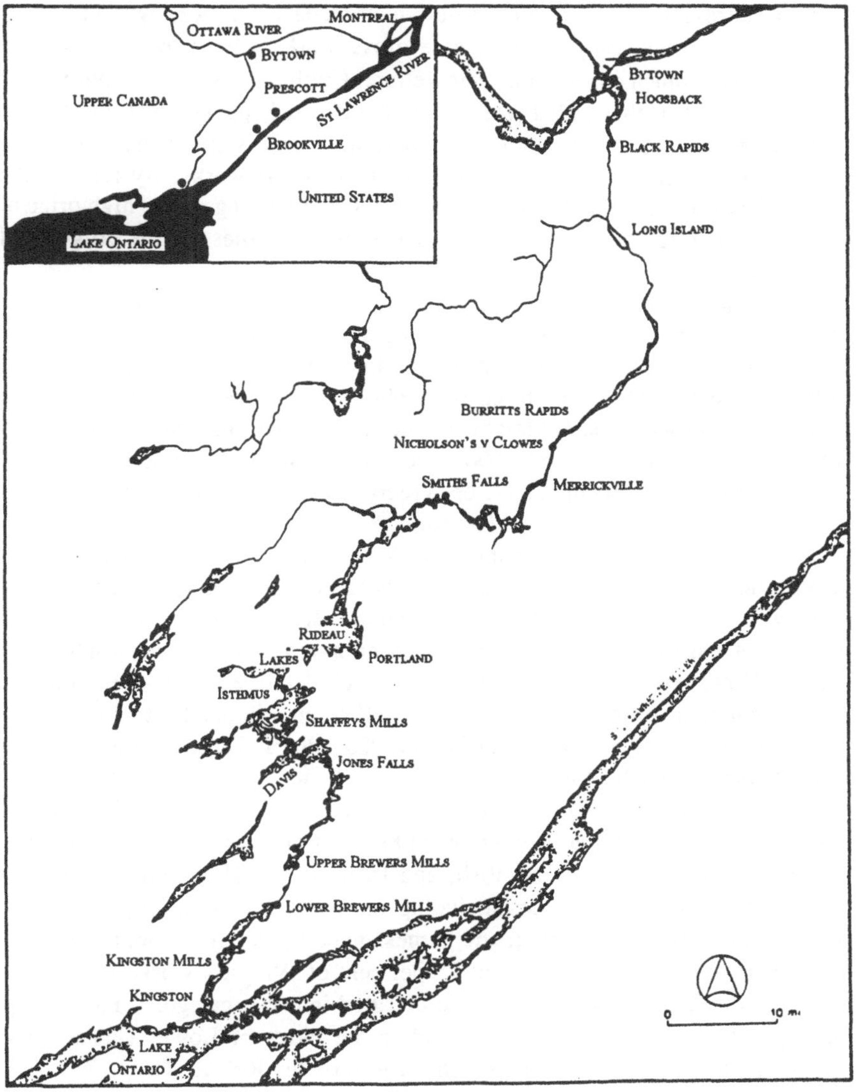

Map of the Rideau; drafted by Steve Epps, Parks Canada

to thwart an expected shortage of skilled labour, Lieutenant-Colonel By acquired the services of two companies of Royal Sappers and Miners comprising 162 soldiers skilled in such crafts as masonry, carpentry, and smithing.[13] In fact there was little difficulty in obtaining civilian labour. Not only was the country experiencing a great upsurge in immigration at the time, but there was

[13] John MacTaggart, *Three Years in Canada, An Account of the Actual State of the Country in 1826-7, 8 . . .* (London 1829), I, 160; Bush, *Builders of the Rideau*, 11-14.

also a depression in the timber trade which made even more workers available.[14] As a result, only when disease drove workers from the sites did the military show much concern for their welfare. Civilian artisans were allowed to complete most of the skilled work while the Royal Sappers and Miners were held in reserve for emergency construction situations, supervisory work, and above all as a quasi-police to encourage order on the works. By refused the offer of two additional companies of artificers commenting on the high price of their labour, since they must be paid year round regardless of the progress of the work.[15]

Because of lack of information, only a rough estimate is possible of the number of workers on the Rideau. During early spring 1827, observers were predicting that work would be provided that year for 1,000 craftsmen and 4,000 labourers. The work force probably varied between 2,000 and 5,000 during the following years. Major sites such as Bytown, the Isthmus, and Jones Falls employed approximately 500. The smaller stations may have averaged about 50 each. Most seem to have been recruited from Montreal and Quebec or nearer the works at Bytown and Kingston.[16]

The ethnic composition of these workers is the subject of some dispute. Traditionally, the skilled tradesmen have been described as English and Scottish with a smattering of French Canadians. Because of the comments of persons at the time, the labourers have been assumed to be overwhelmingly Irish.[17] However, the two enterprises for which detailed information is available, Philemon Wright and Sons and McKay and Redpath, seem to have employed a very large population of French Canadians. These contractors were responsible for much of the work at Bytown, Dows Swamp, Hogsback, Burritts Rapids, and Jones Falls, which included several of the major sites. About 50 per cent of their skilled workers were French Canadians with the rest being distributed among English, Scottish, and Irish names. This mixture seems to have held for all of the trades including masonry, stone-cutting, quarrying, smithing, carpentry, coopering, and harness-making. A large proportion of the unskilled were also French. About 75 per cent of Wright's workers and two-thirds of the Redpath-McKay work force fitted into this category. Faced with unemployment in the timber trade and in the cities of Lower Canada, the French joined the British in the search for jobs on the Rideau.[18]

[14] A.R.M. Lower, *Great Britain's Woodyard: British America and the Timber Trade, 1763-1867* (Montreal 1973), 67-76; Michael Cross, "The Lumber Community of Upper Canada, 1815-67," *Ontario History,* 52 (1960), 214.

[15] PAC, RG5, Al, By to Mann, 10 June 1828, cited in Pentland, *Labour and Capital,* 222 n. 166.

[16] Bush, *Builders of the Rideau,* 18-19.

[17] These descriptions are reviewed in *ibid.*

[18] PAC, MG24, D8, Wright Papers (hereafter cited Wright Papers), Vol. 103, Oxford Snye Day Book, 1828-9; *ibid.*, Vol. 127, pp. 67, 505-81; Montreal, McCord Museum, Redpath Papers (hereafter cited Redpath Papers), Item 5, Workers' Salaries, 1827-8; *ibid.*, Item 6, Ledger, 1826-9; *ibid*, Item 7, Day Book, 1828-32.

The contribution of the French was greater at some sites than others. It seems possible that they were preferred over the Irish by contractors experienced in working in Lower Canada. While the latter had a reputation for rowdiness, the French were considered more reliable and more familiar with the conditions of the country. Both the Wrights and Redpath and McKay seem to have maintained continuing relationships with some of their workers. Wright, who was involved in the Ottawa timber trade, may have shifted some of his labourers to construction during the depression of 1826-7 and afterwards moved them freely between his Gatineau farming operations and the works at Dows, Hogsback, and Burritts Rapids. McKay and Redpath were already employing considerable numbers of workers in their masonry and building activities in Montreal and agreed to send portions of the paycheques of their workers on the Rideau to their wives in Montreal.[19]

Old Saw Mill at Kingston Mills; labourers working on timbers probably to be used at lock site visible in the background; Burrowes Sketch, Archives of Ontario

The proportion of Irish was probably exaggerated by commentators at the time because the migrants were so highly visible in their poverty and behaviour. Nonetheless, they may have played a greater role at sites where the contractors were not from Lower Canada. For instance, at Chaffeys Lock where the principal contractor was John Haggart from Perth, the workers seem to have been almost exclusively British and perhaps mostly Irish.[20] These immigrants seem to have come directly from the St. Lawrence and not over-

[19] Wright Papers, Vol. 127; Redpath Papers, Item 5, Workers' Salaries, 1827-8.
[20] Tett Papers, Business Records, Transfer Case 10, Day Book, 1828-33.

land from American ports. The sources agree that persons in the United States at this time could see no advantage in travelling north when there was work closer at hand on the American canals.[21]

Was employment on the Rideau profitable? In the absence of careful studies of the colonial standard of living, a rough comparison of wages and costs may be attempted. The general picture sketched in both Canadian and American literature seems applicable to the Rideau. A distinction was made between craftsmen and the less skilled. The former were in short supply and were treated with some respect. The latter were plentiful and regarded chiefly as commodities. The labourer on most North American canals existed near the subsistence level and frequently was unable to support a family.[22]

Workers on the Rideau usually were hired by the month and paid a daily rate. There was little piece work with the exception of some stone cutters and quarry men and some common labourers hired directly by the military. The latter were described as of inferior abilities to the average workmen and thus were paid according to the amount of work done.[23] Skilled labour seems to have been paid between five and seven shillings a day. Lieutenant Edward Frome, who participated in the project, noted the following levels: carpenters, 5/ to 5/6; masons, 5/ to 6/6; stone cutters, 6/ to 7/; sawyers, 5/ to 6/ ; and smiths, 5/ to 6/.

While the average labourer's salary was 2/6, there were variations in the rate between 2/ and 3/.[24] Levels were of course lower if room and board were provided: Philemon Wright who seems to have operated in this way paid 1/3, 1/6, and 1/10.[25] The variations were perhaps dependent in part on the type of work performed. The contractors themselves received different rates for the excavation of soil and rock, for puddling and embanking, and other activities. Wages, however, also varied according to the season and the year. They were generally higher in the summer than in the winter when there were fewer jobs to be filled. The general wage levels in the eastern part of the province were driven down in 1827 when too many workers arrived on the Rideau to be employed. There may have been some recovery in 1828-9 when it became necessary to raise wages as disease drove men from the works.[26]

[21] MacTaggart, *Three Years in Canada*, I, 329.

[22] Pentland, *Labour and Capital*, 186-9, 196; Bleasdale, "Class Conflict on the Canals," 15-7; G.R. Taylor, *The Transportation Revolution, 1815-1860* (New York 1968), 289-90.

[23] Redpath Papers, Item 6, Ledger, 1826-9; PAC, MG12, WO44, Vol. 15, p. 81, Sir John Webb to Byham, 1829.

[24] Frome's Paper in *Professional Papers on the Royal Engineers* 1 (1837), cited in PAC, MG29, A24, Drummond Papers, Vol. 1; Redpath Papers, Item 5, Workers' Salaries, 1827-8; Bush, *Builders of the Rideau*, 20.

[25] Wright Papers, Ruggles Wright, Lists of Men Engaged, 1828; *ibid.*, Vol. 127, p. 67, 432, memorial of P. Wright and Sons to Lord Aylmer, 1832; *ibid.*, p. 67, 472-3, List of Men and Supplies, n.d.

[26] Pentland, *Labour and Capital*, 116.

The costs of living in the camps would have taken most of the earnings of a common labourer. Expenses were high not only because of the difficulties of transporting provisions over long distances, but because of the determination of many contractors to make a profit on supplies as well as construction. Even in the Wright camps where board was provided, the company made substantial profits on whiskey which they purchased from local settlers at 2/ a gallon and sold in large quantities to the workers at 2/9. The situation was worse when all necessities had to be purchased at the company store. The Redpath and McKay workers seem to have been constantly dependent on the company for credit and continually in debt.[27]

The plight of the workers was exacerbated by the insecurity of employment. Many contractors came to the works with insufficient resources to fulfill their responsibilities. On some occasions they were late in paying wages; at other times they gave up their contracts entirely plunging their crews into unemployment.[28] Conditions were worse in winter when some sites shut down altogether, the price of provisions rose, and the climate aggravated health problems. Describing the labourers returning from the Rideau to Kingston in October 1827, one commentator noted that he had never seen such distress: "There is scarcely a hut or log-house here but is filled with sick and needy, who are suffering, not only from Disease, but also from Hunger, and from almost every other misery concomitant upon the want of the common necessaries of life."[29]

III. The Working Life of the Rideau Labourer

THE CONSTRUCTION OF THE RIDEAU was conducted mainly by hand with the aid of small tools and animals for heavy hauling. The shortage of mechanical labour-saving devices is striking. Short tramroads were constructed across the sites at Hogsback and Jones Falls probably for the use of horse-drawn cars. A horse pump was placed at Bytown and a steam pump apparently existed at Kingston Mills to cope with leakage into the lock pits. Plows, scrapers, and dredges used for clearing the land on the Erie and Welland Canals were not visible on the Rideau probably because the rocky terrain of much of the region would have rendered them ineffective.[30] The failure to make more extensive use of steam power to facilitate construction was a feature of the canal projects

[27] Wright Papers, Vol. 103, Oxford Snye Day Book 1828-9; Redpath Papers, Item 6, Ledger, 1826-9; *ibid.*, Item 7, Day Book 1828-32.
[28] PAC, MG24, E6, MacTaggart Papers, pamphlet "Notices on the Rideau Canal," (Kingston 1832), 9.
[29] *Ibid.*, RG5, A1, R.W. Tunney to Hillier, 27 October 1827.
[30] Wright Papers, Vol. 127, p. 67, 510, November 1828; PAC, RG8, C, Vol. 47, p. 244b, Statement of the Progress of the Works, 1 March 1829; Alvin F. Harlow, *Old Towpaths: The Story of the American Canal Era* (New York 1926), 53; Runnalls, *Irish on the Welland*, 13.

generally in North America and also of the early railway camps in Britain during these years. Experiments with new equipment seem to have been discouraged by the abundance of cheap labour.[31]

Construction at Kingston Mills; work on dam in progress in centre background; the steamboat 'Pumper' at the far left was used to drain water flooding the lock pits; Burrowes Sketch, Archives of Ontario

Some tools and implements were provided by the labourer, while others were furnished by the contractor. Picks and shovels, for instance, were usually supplied by the former, while wheelbarrows were the property of the latter. Drivers with teams of oxen and horses were usually recruited from the thin ranks of nearby settlers or from farmers around Hull and Kingston. Being in short supply, these carters tended to receive good wages.[32]

Work on the Rideau included the clearing and excavation of earth and rock, masonry quarrying and construction, the erection of wooden weirs and bridges, and the building of rubble embankments and dams. A description of several of these activities will serve to underline their essential character. The first task was the clearing or grubbing of the land. This involved the extensive use of the axe to remove brush and trees and was work with which the French Canadians especially were already familiar from their experience on the farms and in the bush. Large timbers would be hauled away by oxen, but the stumps presented a

[31] Raphael Samuel, "The Workshop of the World: Steam Power and Hand Technology in mid-Victorian Britain," *History Workshop*, 3 (1977), 28-9; Terry Coleman, *The Railway Navvies; A History of the Men Who Made the Railways*, revised edition (London 1981), 53-4.

[32] Wright Papers, Vol. 101, Waste Book 1828; *ibid.*, Vol. 103, Oxford Snye Day Book, 1828-29; *ibid.*, Vol. 127, p. 67, 505, Work and Supplies at Hogsback, 1828; Tett Papers, Transfer Case 1, Item 2, H. Magregor to Benjamin Tett, 20 October 1832; *ibid.*, Business Records, Transfer Case 10, Day Book, October 1828.

more formidable problem. They had to be undermined by "pooling in" (digging under them) or occasionally by blasting. Workers not familiar with the techniques were sometimes injured by falling trees or collapsing stumps.[33]

Large volumes of earth and rock were moved at the sites to produce canal cuts or embankments. This involved pick and shovel work which was the forte of the Irish, some of whom may have worked on roads and canals in the British Isles before emigrating. The earth was usually loaded into wheelbarrows or horse-drawn carts for transport. The removal of these loads from lock pits and cuts was accomplished by the construction of barrow runs, composed of planks laid up the sides of the embankment and up which the barrows were wheeled (see drawing). The running was performed by the strongest of the men who frequently were aided by ropes attached to the barrow and to their belts which then ran up the side of the slope and round a pulley at the top.[34]

Brewers Lower Mill; excavations of canal and construction of locks in progress; note the barrow run on the left side of the picture; Burrowes Sketch, Archives of Ontario

The quarrying of stone masonry required special skills. For this reason and because of the fact that the quarries were often isolated from the rest of the works, separate camps were provided for some of these gangs. The work was conducted partly in winter when one of the methods for splitting rock was to insert water into drill holes which would then expand with freezing. Blasting, however, was unavoidable. This was probably the most dangerous aspect of an accident-prone environment. The process involved the drilling of holes with

[33] MacTaggart, *Three Years in Canada*, II, 103-4.
[34] Harry Hanson, *Canal People* (London 1978), 18-9; Coleman, *The Railway Navvies*, 46-7.

hand equipment, the insertion and packing of powder and its detonation. There were injuries firstly because of the so-called "merchant powder" which was composed of three parts "nitre" to one part sulphur and charcoal.[35] Even an experienced quarry man was not able to anticipate the size of the resulting explosion. The danger was compounded by the inexperience of the workers and the carelessness of management. Seeking to blame the Irish labourers whom he reviled, John MacTaggart, the Clerk of the Works, commented that some would take quarrying work "never thinking that they did not understand the business. Of course, many of them were blasted to pieces by their own shots, others killed by stones falling on them."[36] Yet he also revealed that elementary safety precautions had not always been taken, for only "at length... [was] the matter so systematized, that a number of shots were always prepared to be fired at once... so that all the quarriers got out of the quarry to a respectable distance before the mine was sprung."[37]

After being split, much of the further preparation of the stone was done by masons working with chisels and other tools at the quarry. The blocks were then hoisted on to sledges or carts with the aid of hand winches and transported to the lock pits. At the sites, the blocks were raised into position by hauling them up inclined planks by means of a block and tackle. Crowbars and hard wooden mallets were used to manoeuvre them into position. The joints between the stones were customarily grouted with a mixture of cement and clean white sand. The cement was obtained from Hull which provided a better quality than that imported. The backs of the walls were packed with water-resistant puddle which was a mixture of clay and water. Carpenters and smiths were then employed in completing the lock chambers. The large gates were constructed from oak obtained in the region. Crabs and other pieces of cast iron were furnished by the forges at St. Maurice in Lower Canada. Flat iron, on the other hand, was English and available in large quantities in Montreal.[38]

The working life of the labourer was arduous. During the summer, he worked fourteen to sixteen hours, six days a week. While some sites were not active in the winter, those operated by P. Wright and Sons and by McKay and Redpath were, meaning that work went on in extreme climatic conditions.[39] By their nature, the tasks were at the same time repetitive, boring, and dangerous. There is some evidence that the employee approached the job with the expecta-

[35] Legget, *Rideau Waterway*, 46-7.

[36] MacTaggart, *Three Years in Canada*, II, 245; it is regrettable that in analyzing worker behaviour we have to rely on the descriptions of MacTaggart and a few other commentators, who, almost without exception, subscribed to the racist stereotypes of the Irish then current in Britain.

[37] *Ibid.*, II, 246.

[38] *Ibid.*, I, 340-1; Karen Price, *Construction History of the Rideau Canal*, Parks Canada, Manuscript Report Series No. 193, 1976, 37-8.

[39] Wright Papers, Vol. 127, pp. 67, 536-69, Work Sheets, 1827-29; Redpath Papers, Items 4-7, Day Books and Workers' Salaries, 1826-32.

tion of injury. MacTaggart wrote of the foolhardy disregard of the labourers for their own safety. As he put it, "it is vain for overseers to warn them of their own danger, for they will pay no attention."[40] If such carelessness existed, it was born of the economic necessity to keep working regardless of the risks. A similar attitude was reported among the timber workers in British North America in this period, and also among the railway navvies in Britain. They exhibited a bravado amounting to a kind of group pride in their ability to endure in the face of daily danger.[41]

IV. Living Conditions

TWO KINDS OF COMMUNITIES sprang up during construction, the work camps which were scattered along the line of the canal, and the urban shanty towns which emerged in the centres of Kingston and Bytown. In the camps, the primary problem was supplying the physical wants of a large labouring force in a wilderness environment. To attract workers, contractors felt compelled to import provisions, but always with an eye to minimizing costs or generating profits where possible. They also made what arrangements for accommodation as seemed necessary to keep workers on the sites. Their activities varied with the isolation of the station and the size of the camp.

Most provisions were ordered from Montreal, though quantities were also assembled at Kingston, Brockville, and Hull. Montreal was preferred in part because it was the major regional entrepot. It was also the home base of influential contractors such as McKay, Redpath and Thomas Phillips and the location of the military supply depots of the Ordnance Board. During 1827, the military shipped large amounts of supplies up the Ottawa to Hull to insure that there would be sufficient quantities for contractors to keep the work going. In ensuing years, the Merrickville section received goods overland from Prescott on the St. Lawrence. For the works near the Rideau Lakes, roads between Brockville and Oliver's Ferry and Portland were used. The stations between Davis Mills and Kingston Mills were supplied from Kingston. Because of the limited development of the region along the canal, local settlers were mainly limited to providing hay and oats for the livestock at the works.[42]

The rudiments of a healthy diet were made available at most sites to keep the labourers fit for work. As previously described, the organization of food distribution varied. While Wright provided room and board, Redpath, Haggart and others kept stores where large profits could be taken. Nonetheless, they all

[40] MacTaggart, *Three Years in Canada*, II, 245-6.
[41] *Ibid.*, II, 171; Coleman, *The Railway Navvies*, 71-9.
[42] V.A. George, "The Rideau Corridor: The Effect of a Canal System on a Frontier Region, 1832-95," M.A. Thesis, Queen's University, 1972, 58; Redpath Papers, Item 7, Day Book 1828-32; Wright Papers, Vol. 103, Oxford Snye Day Book, 1828-29; Tett Papers, Transfer Case 1, Item 2, H. MacGregor to Benjamin Tett, 30 November 1832; *ibid.*, Business Records, Transfer Case 10, Accounts 1830-54.

provided certain staples including beef and pork, flour which was baked into bread at the camps, and potatoes. These were supplemented with peas, beans, corn, tea, sugar, and salt and also by small amounts of local butter, eggs, and fish in season. The volume of consumption seems comparable to that in the timber camps and in the railway construction camps of Britain. The workers there were voracious eaters, frequently consuming four to five pounds of food and drink per day.[43]

Whiskey and tobacco were available to relieve the drudgery of the daily routine. Free supplies of grog were sometimes distributed to encourage the men to stay on the works in winter or during periods of disease. More commonly, both commodities were sold in quantities which suggest a high rate of consumption. It was not unusual for the individual to purchase a half pound of tobacco and a gallon of whiskey at weekly intervals.[44]

Some arrangements for shelter were made by the contractors at many sites. Where it was possible, skilled workers were boarded with local settlers. At small sites, such as Black Rapids, several large log houses might be constructed. These buildings were more numerous at larger sites such as Jones Falls where Redpath provided lodgings for 200 and one building was described as 30 feet by 40 feet and of log construction. At the Isthmus in 1830, the military felt compelled to build what amounted to a small village of shanties to lure back workers frightened away by disease. The most ambitious undertaking was that at Bytown where the basis of a permanent settlement was laid. A military encampment was constructed on the west side of the locks and a series of large houses, described as barracks, for the civilians on the east side. Here as elsewhere, shanties erected by the labourers existed side by side with those provided by the employers.[45] The shanties were probably similar in most respects to those in French Canadian timber camps and at British construction sites. These buildings housed the workforce in a communal setting composed of a single room with beds arranged around the walls. If MacTaggart's description of Upper Canada may be relied upon, these beds may have consisted of little more than a plank, perhaps cushioned with straw stuffing, on which several men may have struggled to sleep at one time. The shanty was also a centre for eating. It usually included a central fireplace with an opening in the roof to permit the escape of smoke.[46]

[43] Wright Papers, Vol. 103, Oxford Snye Day Book, 1828-29; Tett Papers, Business Records, Transfer Case 10, Day Book 1828-33; Coleman, *The Railway Navvies*, 86; PAC, Mossom Boyd Company Papers, Unaccessioned material box A, file 10, Food Consumption, Haliburton, 1883-4 (I am indebted to Chris Curtis for this reference).

[44] Legget, *Rideau Waterway*, 40; Wright Papers, Vol. 103, Oxford Snye Day Book, 1828-9.

[45] Tett Papers, Business Records, Transfer Case 10, Day Book 1828; PAC, MG24, H12, John Burrows Sketch Book, 1827; Wright Papers, Vol. 127, p. 67, 429, memorial of P. Wright and Sons to Lord Aylmer, 19 September 1832; PAC, RG8, C, Vol. 45, p. 22, Lieutenant-Colonel By to General Mann, 23 January 1828; Bush, *Builders of the Rideau*, 23; MacTaggart, *Three Years in Canada*, II, 243.

In matters of dress of course, the workers were primarily dependent on their own resources. The French Canadians and many of the British immigrants seem to have been appropriately clothed for the environment. The typical dress included trousers, shirt, and various combinations of vests, frocks, jackets, and headgear. While some boots and shoes were purchased in the camps, moccasins were favoured because of their comfort by the French Canadians and other long-time inhabitants of the country. Shirts seem usually to have been of linen. Sometimes they arrived in the camps ready-made and sometimes as materials to be made up, presumably by the wives of local settlers employed for the purpose. Some French Canadians purchased the colourful sashes for which they are well known. Winter clothing included mittens, perhaps the French Canadian capote (blanket coat) though there is no direct evidence of this, and a few sheepskins and buffalo robes which some of the artisans could afford to buy.[47]

The labouring Irish, arriving with whatever had been on their backs at home, were particularly unprepared for the rigours of work in the bush. MacTaggart reported many appearing in "breeches that bind at the knee and stockings." This was unusual in North America where trousers had been common since the end of the eighteenth century. More importantly, as Lieutenant-Colonel By noted, the thin clothing of the Irish did not leave them prepared for winter. Though he provided 1,000 blankets the expense of which was to be deducted from their wages, these deficiencies probably contributed in part to the illness which was almost endemic among the Irish on the canal.[48]

The canal labourers have traditionally been portrayed as living in isolation from their families in remote camps far from established communities. Ruth Bleasdale has recently qualified this image by pointing to the existence of large numbers of families near the works on the St. Lawrence and Welland Canals in the 1840s.[49] This does not seem to have been the case on the Rideau where most camps were located farther from settlement. In this situation, the French Canadian women preferred to remain at home in Lower Canada while many of the immigrant families gathered in shanty towns at Bytown and Kingston close to sources of food and supplies.[50]

[46] Gillis, *The Timber Trade*, 313-8; Samuel Strickland, *Twenty-seven Years in Canada West or the Experience of an Early Settler* (1853; Edmonton 1970), II, 285-6; MacTaggart, *Three Years in Canada*, I, 292; Coleman, *The Railway Navvies*, 80-2.

[47] Wright Papers, Oxford Snye Day Book, 1828-29; Tett Papers, Business Records, Transfer Case 10, Day Book 1830; MacTaggart, *Three Years in Canada*, II, 211; Provincial Archives of Ontario, Costume Society of Ontario, Notes on Men's Occupational Clothing of the 19th Century; I am indebted to Tina Bates for information on early nineteenth-century occupational clothing.

[48] MacTaggart, *Three Years in Canada*, I, 290; PAC, RG8, C, Vol. 43, p. 212, By to Durnford, 12 April 1827.

[49] Bleasdale, "Class Conflict on the Canals," 18-9.

[50] Redpath Papers, Item 5, Workers' Salaries, 1827-8.

Axeman in dress typical of Irish labourer at Long Island, c. 1830; drawing by J.P. Cockburn; Royal Ontario Museum

The impressionistic picture available of these predominantly Irish communities stresses their deprivation. The employer's interest in supplying his workers with food and accommodation did not extend to their families. Forced to rely on their own meagre resources, the residents of Corktown at the north end of the canal constructed mud huts. Without proper ventilation or sanitation, these shanties were frequently the centres of disease. In the midst of the squalor, the immigrant women struggled to maintain the organization of their families while the men were away. The inhabitants clustered tightly together and demonstrated a high level of mutual aid in the fight to survive. They also drew on the Irish tradition of solidarity in opposition to outside authority, in this case reacting with hostility to the attempts of British soldiers to police their community.[51]

V. Illness and Health Care

THE MOST SERIOUS THREAT to the well-being of the workers and their families was disease. This was in fact the case at canal projects across the continent where fevers were frequently reported, especially in swampy areas. Variously described as ague or swamp fever at the time, this illness was usually malaria according to leading medical historians.[52]

The disease was apparently brought to the continent before 1800 and spread by the anopheles mosquito which was abundant in swamps and bushy areas. By biting infected persons, the mosquito absorbed the micro-organism responsible for the sickness and then transmitted it to other inhabitants.

The symptoms were described vividly by MacTaggart who was himself a victim:

> The *Fever and Ague* of Canada are different, I am told, from those of other countries: they generally come on with an attack of bilious fever, dreadful vomiting, pains in the back and loins, general debility, loss of appetite, so that one cannot even take tea, a thing that can be endured by the stomach in England when nothing else can be suffered. After being in this state for eight or ten days, the yellow jaundice is likely to ensue, and then fits of trembling — these come on some time in the afternoon, mostly, with all. For two or three hours before they arrive, we feel so cold that nothing will warm us; the greatest heat that can be applied is perfectly unfelt; the skin gets dry, and then the shaking begins. Our very bones ache, teeth chatter, and the ribs are sore, continuing thus in great agony for about an hour and a half; we then commonly have a vomit, the trembling ends, and a profuse sweat ensues, which lasts for two hours longer. This over, we find the malady has run one of its rounds, and start out of the bed in a feeble state, sometimes unable to stand, and entirely dependent on our friends (if we have any) to lift us on to some seat or other.[53]

On the Rideau, the illness was most prevalent in the marshy portions of the canal during the periods of hot muggy weather in late summer. In three consecutive years, 1828-30, it stopped work in August-September on much of the southern portion of the canal between the Isthmus and Kingston Mills. The

[51] PAC, RG5, A1, R.W. Tunney to Hillier, 24 October 1827; MacTaggart, *Three Years in Canada*, II, 243-4.

[52] Taylor, *Transportation Revolution*, 289-90; Harlow, *Old Towpaths*, 53; Bush, *Builders of the Rideau*, 25.

[53] MacTaggart, *Three Years in Canada*, II, 17-8.

mortality rate was relatively low, 4.1 per cent during the peak period in 1830, yet this represented 55 deaths. The proportion of persons infected was quite high, 59.8 per cent at this time. Since the sickness left the worker debilitated for several months, it posed a serious threat to the schedule and the cost estimates of the project.[54]

The military took several steps to minimize the effect of the disease on the process of construction. The prevailing theory was that ague was caused by bad air. (This was the original meaning of the word "malaria.") Hence, Lieutenant-Colonel By acted to establish a 250 foot corridor along each side of the canal to improve air circulation. Land purchases were made for this purpose at Clowes Quarry, Old Slys, First Rapids, Chaffeys, Upper and Lower Brewers, and Kingston Mills. The policy was actually implemented at the Isthmus where a substantial amount of ground was cleared around the works. The canal design was also changed in several places to minimize the labour required to complete the project. Once again, the alterations made at the Isthmus were the most striking. Here the depth of the canal-cut was reduced by four feet over a distance of almost a mile by raising the water level with the assistance of locks at the entrance of Mud Lake and at the Narrows on Rideau Lake.[55]

Specific provisions were required for health care. The military had anticipated this need at the beginning of construction. Writing in the press in winter 1826, MacTaggart had promised that, "Surgeons shall be engaged, and furnished with medicines for the benefit of the sick; as the swamp wilderness and swampy water, may sometimes create distempers."[56] The only steps taken, however, were intended for the permanent employees of the military including the Royal Engineers, the Royal Sappers and Miners, and a handful of civilian functionaries including the Clerk of the Works. For these persons, a system of wage stoppages was adopted which amounted to a form of health insurance. Deductions were made from the pay of each individual in return for treatment in the event of illness. To provide care, a hospital was constructed at Bytown under the supervision of Dr. M.H. Tuthill, an Ordnance surgeon.[57]

The military response to the epidemics was to appoint a second surgeon and to construct a temporary hospital at the Isthmus where the Royal Engineers took over direct supervision of construction from the private contractors. Medical care was provided to the labourers along the line on an *ad hoc* basis. While By proposed extending a scheme of insurance to them by deducting sums from the crown payments to the contractors, this was never approved. Since the original contracts had contained no reference to this, it would have invited

[54] Dianne Patychuk, "Malaria on the Rideau: A Case Study in the History of Health Care in Canada," revised draft, B.A. Hons. Thesis, Queen's University, 1979, copy on file Parks Canada, Cornwall; this source forms the basis for much of the following discussion.

[55] *Ibid.*, 25-6.

[56] MacTaggart, *Three Years in Canada*, I, 160.

[57] Bush, *Builders of the Rideau*, 28-9.

litigation from the entrepreneurs. More importantly, such a scheme would have entailed a much larger medical establishment on the Rideau than the military were prepared to provide.[58]

Consequently, the contractors were deemed responsible for the health care of their workers and no attempt was evident to establish a system of insurance analogous to that of the military. The majority of workers and their families received attention only when they could afford it and when persons with some medical knowledge were in the region. These were sometimes the military surgeons and sometimes local physicians. The costs of visits and medicine were beyond the means of most labourers. The typical medical visit cost at least one pound at Burritts Rapids and perhaps much more at other places. The most effective medicine was quinine — from the extract of cinchona bark — which had been in use since the eighteenth century, but which was in short supply. MacTaggart observed that, "it seems to be a very potent medicine, but being very dear, poor people are at a loss to procure it."[59]

The burden of malaria rested most heavily on those least able to bear it, the common labourer and the destitute immigrant, who were often the same person. The disease was more devastating to those with lower health resistance. Living in crowded quarters and working exhausting hours, the common labourer was probably more vulnerable than the artisan and manager whose conditions tended to be much better. Moreover, in delayed fashion, the Irish immigrant was now reaping the bitter fruits of his dislocation and miserable existence en route to the new world. Because of his wretched physical state, he was likely more susceptible than the French Canadian. The outbreak of disease drove many of the labourers away from the works and employment. It meant that many others returned to their families in Kingston or Bytown sick and unable to work, if they returned at all.

VI. Labour Unrest

THE INFORMATION REGARDING LABOUR conflict on the Rideau is sketchy. Nonetheless, an analysis reiterates the arguments made by Pentland and Bleasdale for early nineteenth-century British North America. They link labour unrest primarily to economic factors but also to ethnicity. Particularly they emphasize the importance of the Irish tradition of direct action against the property and persons of their enemies, regardless of the boundaries of the law. Reacting against the inequities of their homeland, the Irish expressed their frustration in terms of conflict with one another — especially between Protes-

[58] Patychuk, "Malaria on the Rideau," 74-7.
[59] Wright Papers, Vol. 103, Oxford Snye Day Book, 1828-29; Patychuk, "Malaria on the Rideau," 12; MacTaggart, *Three Years in Canada*, II, 18.

tants and Catholics — and by opposing those in positions of authority — especially representatives of the British government.[60]

Violence and disorder appeared on the Rideau soon after the arrival of the workers. In spring 1827, a prominent Bytown citizen reported to the Civil Secretary:

> It is with deep regret, that I have the honour to acquaint you, — for the information of His Excellency the Lieutenant-Governor, that the public peace is most dreadfully disturbed — and the lives and property of the inhabitants in danger day and night in this new Town, by drunken, riotous, persons employed on the works of the Rideau Canal, — as there is not an evening passes — not even the Sabbath excepted — wherein there is not a riot and general fight.[61]

These conflicts often involved Irish workers. Sometimes they quarreled with one another and sometimes with persons outside of the Irish community. The latter incidents tended to occur when traditional ethnic, occupational, or geographical boundaries were crossed. The outbreaks were typically limited in scope, rarely outlasting the particular confrontation. For instance, when the local gentry held a village fair at Bytown in 1829, brawling broke out between Irish workers and farmers who had assembled from the surrounding area.[62]

On a few occasions, the traditional division between Catholics and Protestants was manifested. In Ireland this cleavage was grounded not only in religious and ethnic differences, but also in economic circumstances. Protestants and Catholics in Ulster struggled over land and jobs.[63] Members of both factions were present on the Rideau. While the actual numbers are unknown, the proportions of Catholics and Protestants may have been similar to those recently computed by Donald Akenson for the whole period 1815-46. According to his calculations, approximately two Irish Protestants had arrived in Upper Canada for each Irish Catholic.[64] Only one major confrontation occurred on the Rideau. In December 1829, a battle took place between Orange and Green at the Isthmus in which more than 400 partisans were involved. This was likely precipitated by a struggle for jobs. Construction had been shut down earlier in the year because of disease and the difficulties of excavating through rocky and spring-filled terrain. Workers had flocked back to the site in late fall for the resumption of construction.[65]

[60] Pentland, *Labour and Capital*, 113-4; Bleasdale, "Class Conflict on the Canals," 19-31; see also Kenneth Duncan, "Irish Famine Immigration and the Social Structure of Canada West," in Michiel Horn and Ronald Sabourin, eds., *Studies in Canadian Social History* (Toronto 1974), 158-9.

[61] PAC, RG5, A1, W. Wilson to Hillier, 18 May 1827.

[62] Michael Cross, "Violence and Authority: The Case of Bytown," in D.J. Bercuson and L.A. Knafla, eds. *Law and Society in Canada in Historical Perspective* (Calgary 1979), 7-8, 10.

[63] Cecil J. Houston and William J. Smyth, *The Sash Canada Wore; An Historical Geography of the Orange Order in Canada* (Toronto 1980), 9-10.

[64] Akenson, "Whatever Happened to the Irish?," 220-1.

[65] *Independent Examiner and Bathurst District Advertiser*, 1 January 1830; Price,

Given the circumstances it is surprising that there were not more instances of religious confrontation. During the building of the canal, the membership of the rabidly anti-Catholic Orange Order spread through the communities touched by the work-force.[66] Catholics and Protestants were in almost daily contact on the works. But instead of fighting each other, the two factions seem to have directed their energies primarily towards survival in a hostile environment. This struggle tended to bring them into conflict with third parties.

Much Irish lawlessness was clearly economic in origin. Concerning Bytown, Lieutenant-Colonel By wrote during winter 1826:

> I have had some trouble with the lower class of Irish, and I think it will be indispensably necessary, when the stores are set up in January to have a Sergeant and twelve men stationed under my orders on the Rideau to protect the stores and buildings, and to preserve the trees, there being no civil authority in the neighbourhood of Hull.[67]

Arriving penniless, the immigrants required food, shelter, and firewood to survive the winter. Without other means, they tended to take what they needed without regard to the legalities of property rights.

The "Battle of Merrickville," which actually seems to have taken place at nearby Clowes in 1829, was another example of this. Trespassing on the farm of Benjamin Mosier, an immigrant canal labourer by the name of Thomas Foley cut timber which he probably intended to use for firewood. When the farmer called on the law, Foley's friends first repulsed a constable and later a sheriff and his deputy. Finally, Captain Daniel Burritt and a company of the Grenville militia took three of the labourers into custody and transported them to Brockville where they were acquitted after the authorities could get none of the Irish to testify against the others.[68]

Low wages and poor working conditions presented another occasion for solidarity among the Irish. In spring 1827, the labourers staged at least three strikes at Bytown. Little is known about these struggles. While the ethnic cohesion of the Irish probably facilitated organization, there is no evidence of the secret societies visible in the strikes on the Welland and St. Lawrence canals in the 1840s. The strikes at Bytown apparently involved all the workers of the area, not only the Catholic and Protestant Irish, but also the French Canadians in the neighbouring timber trade. This set these incidents apart from the friction between French and Irish workers typical in the timber trade, the ship yards, and in canal construction in the first half of the nineteenth century. This strife was usually occasioned by competition for jobs. On the Rideau, the

---

Construction History of the Rideau Canal, 246-8; I am also indebted to Kathy McKenna for her information on these events.

66 Houston and Smyth, *The Sash Canada Wore*, 18-9, 22.

67 PAC, RG8, C, Vol. 43, p. 11, quoted in *ibid.*, MG29, A24, Vol. 1, Andrew Drummond Manuscript History of the Construction of the Rideau Canal, 18.

68 *Independent Examiner and Bathurst District Advertiser*, 11 September 1829; Legget, *Rideau Waterway*, 157-8.

Jones Falls c. 1838; Boatloads, possibly of immigrants, passing through the canal; watercolour by P.J. Bainbrigge; PAC, c11835

oversupply of workers at the beginning of construction seems to have led to unexpectedly poor terms of employment which united the workforce against their employers.[69]

During the construction of the Rideau, the strike actions of 1827 were isolated incidents. Most labour unrest seems to have focused on the acquisition of the immediate necessities for survival such as food and firewood, and to have been directed against individual property owners and contractors who stood in the way of their provision. Why were there not more instances of concerted actions against contractors? Ethnic and religious divisions within the workforce were certainly a factor. Probably more important was the sustained opposition of the military. The civilian magistrates and constables scattered among the adjoining settlements were no match for the potential might of groups of workers. In the absence of an effective civil authority, Lieutenant-Colonel By mobilized the military to suppress strikes and maintain order. After the arrival of the first company of Royal Sappers and Miners in 1827, there was no further systematic opposition to authority. Small detachments of soldiers were stationed as needed along the line of the canal to "check the disorderly

[69] Pentland, *Labour and Capital*, 190; Cross, "Violence and Authority," 10; Bleasdale, "Class Conflict on the Canals," 22-4, 28-9.

conduct of the labourers." In so doing, By had established an important precedent in the history of Canadian labour relations. This use of troops to suppress labour unrest was one of the first instances of a practice which became quite common in Canada during the next century.[70]

VII. Conclusion

THE RIDEAU AND THE WELLAND CANALS were the first projects in Upper Canada to attract wage-earners in the 1000s. In many respects the experience of the workers on the Rideau may be considered a preview of what labourers generally could expect in Upper Canada in the 1830s and 40s. Because of an unfavourable labour market, workers were forced to endure difficult and often dangerous working and living conditions. They responded with a militant, although sometimes fragmented, resistance. This in turn was met by provincial authorities with powerful new methods of social control. Because the Rideau was an explicitly military project, troops were more readily available there than they would be elsewhere. But during the 1830s efforts were made to establish more effective civil institutions such as the police, the provincial penitentiary, and prisons to suppress labour unrest. The economic future of the workers was bleak. The evidence suggests that few wage-earners on the Rideau were able to save much from their earnings. While some may have become struggling farmers, artisans, or storekeepers, the majority had to continue competing for jobs with successive annual waves of immigrants. Some of these persons became members of an embryonic international proletariat seeking work in the timber trade of the Ottawa, on the public works projects north and south of the American border, and finally in the factories which began to proliferate in the 1850s. In this sense, the history of the workers on the Rideau was one episode in a continuing story of poverty, transience, and fruitless pursuit of the elusive opportunity.

The research for this article was conducted under the auspices of Parks Canada. I would like to thank my colleagues for their helpful criticisms.

[70] PAC, RG5, A1, By to Mann, 10 June 1828; Pentland, *Labour and Capital*, 190.

9

Hugh McIntosh (1768–1840), national contractor

Mike Chrimes

'the principal contractor for public works in England', James Walker (1843)[94]
'a man of property' John Rennie (1821)[84]
'He has the reputation of being worth £1,000,000 of money', D. Pollock (1834)[94a]

INTRODUCTION

Published historical studies of the civil engineering contractor are rare and the majority are based on eulogistic accounts of the Victorian and Edwardian contractors.[1,2] From such accounts it would appear that the industry progressed to maturity from nothing within the lifespan and careers of Thomas Brassey and Sir Samuel Morton Peto. This would seem a simplistic and unlikely version of events. To come up with an alternative is, however, no easy task. Most railway histories give no details of contractors, and precious few about construction.[3] Charles Hadfield's canal histories are better, but it remains very difficult to build up a picture of civil engineering contracting in the canal age. Common sense would suggest that as the volume of work increased at this time, an increasing number of individuals would find it possible to make a living not just as a canal cutter or mason but, with some capital resources, as a small-scale contractor. Available evidence suggests that not only is this true, but that by 1810 there were several contractors in existence with many years experience and considerable resources. It is regrettable that at present only one of these firms is at all well-known: Jolliffe and Banks,[4] and the only other about which much has been published—the Pinkertons—are associated with such levels of incompetence as to almost defy belief.[5] Biographies of Telford mention Gibb, Cargill, Simpson, Hughes, Hazledine and Wilson among early contractors, but details are lacking about much of their careers.[6,7]

These problems were highlighted recently by the discovery of the papers of William Mackenzie (1794–1851) who, in partnership with Thomas Brassey, established the British overseas civil engineering contracting industry in the 1840s.[8] It was difficult to evaluate the importance of the archive and Mackenzie's importance as a contractor in the absence of comparative information about Mackenzie's contemporaries and precursors. To facilitate this, information about early contractors and their known contracts was compiled from a variety of sources, and summarised as a table (Table 1). This highlighted the scale of the contracts carried out by Mackenzie and Brassey in the 1840s, but also revealed how little had been published about early contractors of importance. The most glaring omission must be a study of the McIntosh family, most notably Hugh McIntosh, who would appear to have been the first contractor who had a national organisation, and was possibly the first to operate overseas.

HUGH McINTOSH (1768–1840), NATIONAL CONTRACTOR

TABLE 1

Leading public works contractors prior to 1850 (arranged chronologically)

Name	*Dates active*	*Number of contracts before 1850 (1840)*	*Number of Railway contracts 1850 (1840)*	*Mileage*	*Approximate value*
Dyson (J., J. jnr and T.)	c.1767–1820	29			
Pinkertons	1767–1831	35			£0.34m
Nowells[1]	1780–1850	71(58)	36	100(45)	£1m
Jolliffe and Banks	1789–1834	102	1	4	£4m
H. McIntosh	c.1790–1840	120	11	30	£3.5m
D. McIntosh	c.1831–1841	29	24	100	£1.5m
W. Hazledine[2]	c.1790–1840	36			
J. Simpson	c.1790–1820	41	1		
Bough, Hughes and Mills	c.1794–1830	22			£0.25m
T. Townshend and Son	c.1800–1851	18(15)	9	40	£1m
Gill, Hodges & Co.	1817–1829	45			£0.076m
W. Dargan	1819–1856	39(12)	17(2)	400(14)	£4m
Pritchard & Hoof (& Hill)	1820–1855	20(11)	12(3)	60(8)	£0.5
J. Stephenson[3]	c.1820–1848	46(13)	46(13)	600(30)	£9m (£0.3m)
T. Nicholson	c.1820–1861	9(2)	9(2)		£0.3
Tredwells	c.1823–1870	33(14)	27(10)	250(35)	£1.5m
Jackson and Bean	c.1826–1880	26(9)	19(3)	150(20)	£1.1m
Grissell and Peto	c.1830–1846	54(15)	32(6)	235	£5.5m (£0.4m)
William Mackenzie[4]	c.1833–1851	17(12)	9(9)	(41.5)	£0.8m (£0.6m)
T. Brassey[5]	c.1834–1871	18(7)	17(6)	350(83)	£3.25 (£0.8m)
G. Wythes	c.1837–1875	18(2)	17(2)	150(10)	£1.5m (£0.1m)
Fox Henderson[2]	1841–1856	24	1		
Mackenzie and Brassey[6]	1841–1850	48	45	1300	£17m
M B and Stephenson[7]	1844–1851	21	21	422	£7.5m

1. The Nowells, in association with Hattersley, Hemingway, etc., continued as contractors for most of the century; this includes the known work of Jonathan (1757–1810), Joseph (1784–1836), Benjamin (1785–1823), John Willans (1806–1851), Jonathan Willans (1808–1846) and James (1793–1859) Nowell.
2. Primarily ironwork contractors. Ironwork contractors have generally been excluded. Butterleys are alleged to have supplied iron rails for over 100 railway lines, mostly mineral lines, by 1830.
3. Includes 21 contracts of Mackenzie, Brassey and Stephenson
4. Excludes contracts with Brassey and Stephenson.
5. Brassey's works include some contracts in which Mackenzie had an interest, but which Brassey completed.
6. Includes contracts with John Stephenson.
7. Excludes contracts of Mackenzie and Brassey.

All figures should be multiplied by 50 to give approximate present day values.

EARLY CAREER

Hugh McIntosh (1768–1840) was born on 4 December 1768 at Milntown, Kildrummie, Nairn, the second son of David McIntosh (c.1730–90) and Margaret Tolmie. He came from an agricultural background, and following some education, probably at Inverness Academy, moved south in pursuit of work, with Alexander Mackenzie (1769–1836), father of William.[9] His career began as a navvy on the Forth-Clyde Canal, then in its final phase of construction under Robert Whitworth (1739–1799).[10] As work on the canal was completed many of the Scottish workforce moved to England to work on Whitworth's other projects such as the Leeds–Liverpool Canal. Alexander Mackenzie spent most of his career there,[11] but the first contracts which McIntosh took in his own name were on the Lancaster Canal, where the engineer was John Rennie.[12] McIntosh was to work regularly with Rennie until the latter's death in 1821.[13]

In 1797 McIntosh began work on a succession of contracts on the Grand Trunk Canal system. Soon after he married Mary Cross (1773–1833) the daughter of William Cross (1744–1827), a farmer from Kings Bromley and an agent of the Canal Company, at Cheddleton, Staffordshire, on 10 December 1798. Here their only surviving child David (1799–1856) was born. Their first home was described by the contractor Thomas Nicholson as a contractor's hut built of stone excavated for the canal works, with an earth floor and straw bedding.[14] It is possible that marriage brought additional capital, or perhaps the Grand Trunk contracts were very profitable. At all events in 1803 the family moved to Poplar, to enable Hugh to supervise directly the massive East India Dock excavations. Work continued on the Grand Trunk down to 1816, and the seeds of a national organisation had been sown.[15]

DEVELOPMENT OF THE FIRM

The current whereabouts of McIntosh's business records are unknown. What follows has therefore had to be put together from what is known about other contractors of the time, and evidence which survives elsewhere. After the completion of the East India Docks contracts Hugh had an office in Bloomsbury, initially at 19 Charlotte Street until 1821, and subsequently at 38, and then, from 1823, 39 Bloomsbury Square.[16] At this, the head office, he would have had clerks and book-keepers. Ledgers and account books would have been kept including a daily journal of transactions, account books for individual contracts detailing payments and receipts, and further ledgers detailing accounts with clients and suppliers. Each transaction would probably have been recorded and copied into four different ledgers. Site agents would have kept receipts, pay records, and records of progress, etc., passing bundles on to head office when the final accounts were settled. Such practices are supported by surviving records of William Mackenzie, who was trained by McIntosh, and those of Thomas Townsend.[17] They are also indicated by references to the detailed books presented as evidence in the court proceedings between the family and the Great Western Railway. Detailed record keeping was essential to a contractor's profits, particularly if claims needed to be substantiated. Hugh McIntosh also appears to have early appreciated the need for legal advice. He made use of the services of Edward Tyrrell, Remembrancer of the City Corporation, from 1805 and regularly made use of his son Timothy's services later in his career.[18] His reputation was such that in 1833 Sir John Rennie described him as having the 'character of being a very litigious person'.[19]

The whole enterprise relied upon able subordinates, particularly the site agents, who had considerable independence—on occasion too much. William Betts (snr) exceeded his and in 1835 had to withdraw tenders for work on the Grand Junction Railway, as they had been made without Hugh McIntosh's knowledge.[20] The first agent we know of was Hugh's youngest brother James (1776–1855), who probably joined him in Staffordshire, and later married Mary Cross's sister Eleanor in 1811.[21] The most important associate was Hugh's son David.[22]

David was educated at Mr. Stock's school in Poplar before attending Glasgow University (1814–17). This, and Lawrence's unfinished portrait of him, are a reflection of Hugh's prosperity at the time, and the decision for David to abandon his studies and start to work for his father on the Edinburgh and Glasgow Union Canal must have been due to pressure of work, rather than shortage of money. David was almost certainly the first contractor to receive a university education, and poor health is probably the chief reason why his career did not eclipse that of his father. Unlike many nineteenth

Fig. 1. David McIntosh, c. 1813. Oil painting by Sir Thomas Lawrence. Its present location is unknown.

century contractors, the McIntoshes do not appear to have had partners to share the risk and generally seem to have financed their projects from their own capital resources.

It was not however entirely a family firm. Many of McIntosh's agents became well-known in their own right, and made substantial fortunes from their connection with McIntosh.

CANAL CONTRACTS

McIntosh's contracts are too numerous to detail.[23] His early reputation was based on his skill in earthworks, and it was only from around 1820 that he became significantly involved in other kinds of contracting.[24] Canal contracts were where he first gained his experience. For the Grand Trunk he built the Leek canal, the Burslem Branch, Lawton Branch, the Dane River feeder to Rudyard reservoir and its completion, Caldon Low railway, Lam End railway, Foxley railway, Lawton Middlewich and Stone cuts, repairs near Warrington and Hanley, and a road near Burslem.[23,25–27] While the Grand Trunk Canal works were in progress he took other canal contracts on the Kennet and Avon,[23,28] Croydon (1801–1809),[23,29] Thames and Medway (1800–1801),[23,30] and Regent's canals (1812–1820).[23,31] On the latter John Aird senior's father, Robert, died in a landslip, although most of the company's problems were financial rather than technical. With David he took contracts on the Edinburgh and Glasgow Union Canal (1817–1822),[32] Thames and Severn canal (1820–1822),[33] Sharpness Docks and the Gloucester–Berkeley Ship Canal (1822–1827),[34–36] and lots 6–8 on the Great Western Canal (1831–35).[37] Among his final works were contracts on the Aire and Calder Navigation (1834–1841), which included a cut for the Stainforth and Keady Canal, Stanley Ferry Aqueduct,[38] and Goole Docks.[39] James Abernethy, then a young assistant engineer, described how McIntosh, by then a blind old man, inspected the works with the help of his agents.[40] Problems encountered there included a cofferdam failure. For the Grand Junction Canal (1837–1840) McIntosh built the Wilstone extension, including Tringford pumping station.[41] The growth of McIntosh's enterprises can be judged from the fact that whereas the early lots on the Grand Trunk were for a few hundred pounds,[42] on the Union Canal he offered to take on all the advertised lots offering a surety of £10,000.[43,44]

LONDON DOCKS AND ASSOCIATED WORKS

He was heavily involved in the construction of new docks in London, carrying out works for almost all the dock companies.[45] These included East India docks where he was troubled by press gangs (1803–1807,[46,47] and 1816–1817),[48] Brunswick wharf (1832–34), Greenland,[23] Grand Surrey (1811–1812),[49] and Commercial docks (1815–1821),[50] Regents Canal basin (1819–20), London Docks and Shadwell basin (1826–1833)[51] and West India Docks (1806, 1832–1833).[53,54] Associated with these were roads to East India Docks, Mile End Road, Barking Road and Tilbury Road.[23] Hugh purchased property in the area, most notably from the East India Dock company, and began some housing development.[55] The profitability of his contracts is reflected in his ability to buy almost 70 acres of land in 1813 for £14,913. The scale of his purchases meant that much of the property was not developed until after his death by David, and later by a nephew, also called David, who inherited much of his property.[56] Hugh had a wharf at Trinity Street, Rotherhithe, in the late 1830s for access to the London–Greenwich Railway works.

ROADS

McIntosh built other toll roads[23] outside Docklands. These included Highgate Archway (1810–1813) and Vauxhall Bridge Road (1816) in London. He also built roads near Manchester, Tewkesbury and at Clayton Hill, Brighton (1807).

GAS AND WATERWORKS CONTRACTS

McIntosh became an important contractor for gas and waterworks companies, laying pipes and building retorts, etc., for the Chartered Gas Light and Coke Company (1815–)[23,57] in the London area and also Bristol gasworks (1817).[23,58] He worked for the Grand Junction (1811–12, 1821), East London, Chelsea, and Kent waterworks (1809–) companies.[23,59,60] He also laid sewers around London at Nine Elms, Southwark and Deptford.[61] John Aird snr established his reputation under McIntosh on this kind of work before going to work for the Phoenix Gas Company.[62]

GOVERNMENT CONTRACTS

Hugh McIntosh's ability was rewarded by substantial government contracts. John Nash remarked, when explaining why no other contractors had been invited to tender at Buckingham Palace: "The work was not offered to any person but Mr McIntosh—his known probity, his great means of performing the work in the short time allowed him, and my own experience as to his means of executing great works on the most moderate terms were considered by the Board and myself as sufficient achievements to employ him without other competition, and his proposals were accepted seeing that they were within my estimate.[63]"

He was responsible for enormous works at Pembroke (1813–1816, 1830–32)[64] Plymouth (1823–1832)[65] and Portsmouth (1824–32) dockyards.[66]

Fig. 2. Mill and bakery at the Royal William Victualling Yard. *Photograph, A. B. George, PHEW Collection, Institution of Civil Engineers*

The Plymouth contracts involved work for the Royal William Victualling Yard at Cremill Point, carried out under a series of thirty-one contracts awarded between 1824 and 1832 to a combined value of more than £300,000. The Victualling Board, succeeded by the Admiralty, began to get concerned in 1832 about payments to McIntosh, and by 1834 the two sides were in dispute with claims of nearly £75,000. In August 1835 Sir John Rennie was replaced as arbitrator by George Ledwell Taylor, breaking the tradition established by his father of the chief engineer being sole arbitrator. Taylor blamed the problems on the Resident Engineer, Richards, although they may have been aggravated by the fact payments were being made on account to enable McIntosh to continue working, rather than by strict certification of measured work.[66a]

McIntosh carried out much work connected with the Royal parks and palaces, including ornamental ponds, roads etc for the Commissioners of Woods and Forests in Regent's Park (c.1812, 1827–1832).[23,67] He laid water pipes for Windsor Castle, Hampton Court, Kensington Palace and Brighton Pavilion.[23] He carried out excavating work at Buckingham Palace (1825–1826), Hampton Court Water Gallery (1829), the British Museum (1823–1828) and the State Paper Office (1829–1831).[68] He was the contractor at Chetney Hill (1801–1818) near Sheerness, for works associated with the Lazarette (quarantine establishment), chiefly the excavations of a canal and dock. This, the only British example, was designed by Samuel Wyatt, with John Rennie acting as engineer. It was abandoned by 1824.[69,70] McIntosh was also possibly the first British contractor to work overseas when he undertook the demolition work at Flushing for Lord Liverpool (1809).[23] He built the North Hyde Canal and mounds (1814),[23] and was contractor for Melville Hospital, Chatham (1827), where William Betts was agent.[70a]

DOCK AND HARBOUR WORKS

His harbour works included the first railway owned harbour at Whitstable (1831–1832),[71] Shoreham harbour (1817–1821),[72] Dover harbour (1812–1815, 1833–1836),[73] Hull Junction Docks (1826–1829),[74] and a timber pier at Southampton (1832–1833) and its stone replacement (1837–1838). Here William Betts was again his agent.[75] McIntosh also began work on Brighton Chain pier but gave it up due to the ground conditions.[76] He was responsible for various sea defence works, most notably at Dymchurch (1820–32),[77,78] and in the Gillingham area.[23] David built Black Point lighthouse near Beaumaris (1835–1838).[79]

BRIDGE CONTRACTS

McIntosh was never trained as a mason and as a result he was never a specialist bridge contractor.[80] However, as his business grew, his firm gained the prestige and expertise necessary for such work. He built bridges at Harmondsworth (1824–1825), Brentford (1824–1825)[81] and South Bridge, Northampton (1815–1816),[82] widened Clopton Bridge, Stratford-upon-Avon (1813–16, 1818)[83] and repaired Rochester bridge (1821).[84] He was contractor for Telford's Mythe Bridge, Tewkesbury (1823–1826)[85] and Folly Bridge Oxford (1827). He built floating bridges over the Itchen (1834–36)[86] and at Portsmouth (1838–1840).[87,88] Financial problems led to McIntosh running the Itchen bridge company for a time. He gave evidence to the House of Commons on London Bridge (1821)[89] and carried out site investigations for the new London Bridge in 1823;[90] David provided evidence for the report on the new bridge's stability in 1831[91] and also repaired Blackfriars Bridge (1834–42).[92–94]

RAILWAY CONTRACTS

In many ways the most interesting aspect of the McIntosh story is the way in which McIntosh embraced the opportunities offered by the railway boom of the 1830s. Although some contractors developed with the early railways, like John Stephenson and Thomas Nicholson, and other contractors with previous contracting experience like the Nowells, and Tredwells moved into railway contracting, few had the McIntoshs' experience of managing large capital projects. Although large capital contracts for railways emerged in the 1840s with the rise of Mackenzie and Brassey, the relatively small lots preferred by the early railway engineers like the Stephensons were inappropriate for the McIntoshes, who since 1810 had been taking on work simultaneously to the value of several hundred thousand pounds.[94a] McIntosh had been involved in the early railways associated with canal and dock works, and helped construct the Plymouth–Dartmoor Railway (1819–1922).[94b] The scale of work they were

Fig. 4. Clopton Bridge, Stratford upon Avon, view of CI footpath, and masonry widening by Hugh McIntosh.

capable of is indicated by the London & Greenwich Railway (1833–1841), which was built in its entirety by Hugh McIntosh. It has been suggested that the engineer George Landmann drew up the plans in McIntosh's office. McIntosh himself had a financial stake in the line. In most of the other early railway lines for which they tendered the McIntoshes took a significant proportion of the work, and their railway contracts exceeded £1m in contemporary value.

Fig. 5. London and Greenwich Railway arches over the Surrey Canal.

The London and Greenwich suffered from a shortage of capital, and as early as 1836 it was reported the London and Greenwich company owed Hugh £10,000. At a Special General Meeting on 21 July 1837 it was reported there were large sums owing to the contractors and McIntosh's solicitor Tyrell became involved in obtaining a further Act to authorise more capital. An arrangement was made between February and June 1838 to resolve the problem, and McIntosh received some payment in bonds. The line suffered labour problems, and English and Irish navvies had to be separately billeted. The total value of contracts was in excess of £140,000.[95]

Surprisingly the firm did not tender for work on the London & Birmingham Railway although approached about the Tring cutting, which it was felt required an experienced contractor;[96] given that many of the original contractors had problems this may well have been a wise decision. David was responsible for extensive works on the London and South Western Railway, including the section

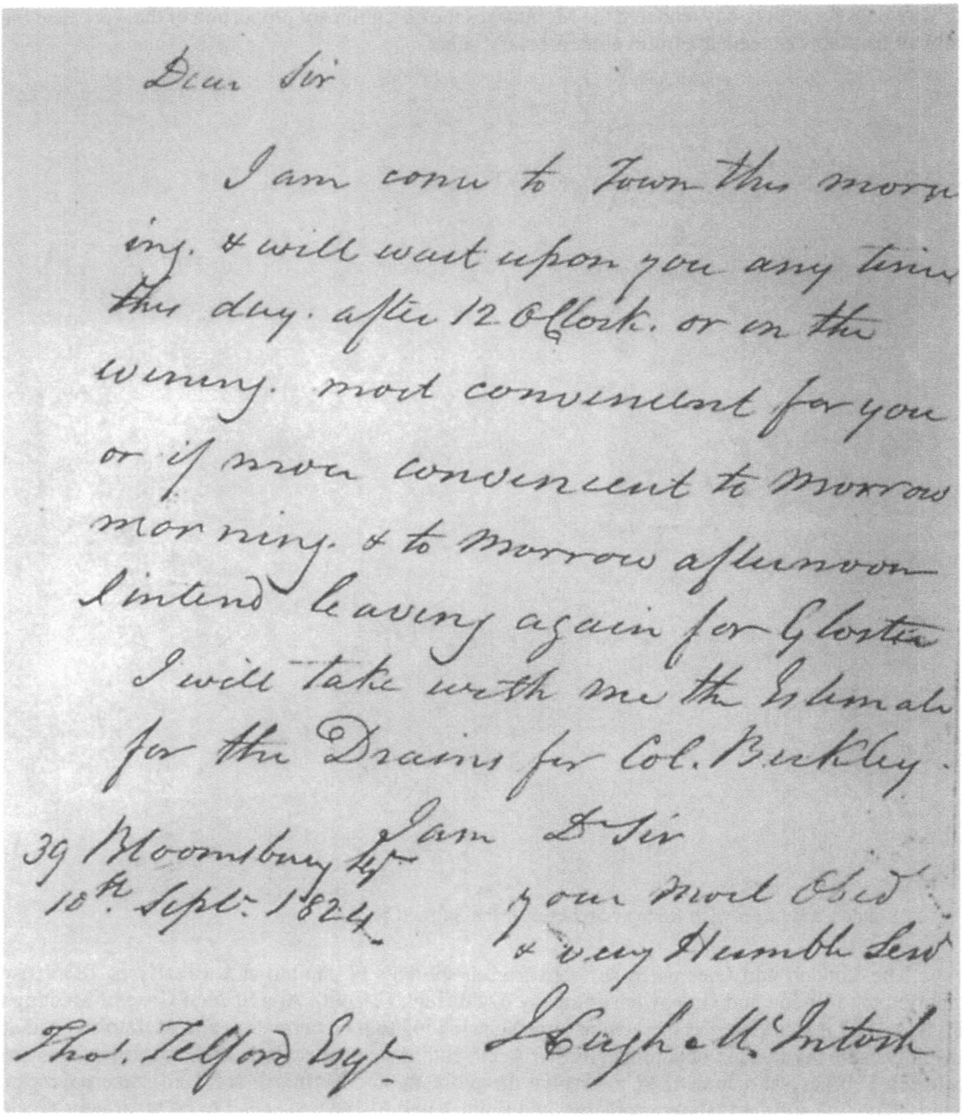

Dear Sir

I am come to Town this morning. & will wait upon you any time this day. after 12 O'Clock. or on the evening. most convenient for you or if more convenient to morrow morning. & to morrow afternoon I intend leaving again for Gloster I will take with me the Estimate for the Drains for Col. Berkley.

I am Dr Sir
your most Obedt
& very Humble Servt
Hugh McIntosh

39 Bloomsbury Sqr
10th Septr 1824

Thos. Telford Esqr

Fig. 6. Letter from Hugh McIntosh to Thomas Telford, 10 September 1824 relating to drainage work for Colonel Berkeley. *Telford Mss, Institution of Civil Engineers Library*

between Southampton and Winchester, and masonry works at the London end.[97] On the Midland Counties Railway, where William Betts was the agent, David had all the contracts between Rugby and Leicester.[98] At Dutton Viaduct on the Grand Junction David's agent was the young Edward Ladd Betts.[99,100,101] After inspecting works on the viaduct the young David Stevenson remarked: 'they have a great deal of fine machinery and tackling and it is altogether one of the best conducted works I have seen'.[102] David had smaller contracts on the Northern and Eastern,[103] and Birmingham and Derby Junction railways.[104]

The McIntoshes took eight contracts on the Great Western Railway with a combined tender value of £188,903, and payments totalled £221,148.[105,106,107] Their claims of £265,557 were turned down by I. K. Brunel. The legal case was pursued through Chancery, and not finally settled in their favour until 1866, long after the death of the original litigants. It was not an episode which reflected well on either Brunel or the GWR. It also highlighted the problem of determining responsibility for a specific contract, as when Hugh endeavoured to obtain payment for final accounts on his contract, Brunel linked it to outstanding disputes on David's contracts. The McIntoshes also had several contracts on the North Midland (1837–1840)[108] and Manchester and Leeds Railway.[109,110] It was while meeting his agents for these lines at Wakefield that Hugh died of apoplexy on 31 August 1840.[111,112]

DEATH AND BURIAL

Following Hugh's death in Wakefield his body was transferred to London where he was buried in the family vault at St. Matthias Church, Poplar. His brother James is buried in the same tomb.

McIntosh suffered an eye infection while working on the Gloucester and Berkeley Canal in 1826, which resulted in blindness.[113] This may in part explain why many of the later contracts, particularly the railway contracts, were taken in David's name. Hugh remained active until his death in 1840, and kept tight control on site visits by asking detailed questions. At his death the family were the greatest contractors in the country, and David was ideally positioned to capitalise on the railway mania. Unfortunately he had a weak heart. He retired shortly after his father's death, and died in 1856, when he was buried at Kensal Green cemetery. Hugh had left a considerable fortune of £300,000, more than any civil engineer to that date. Aside from his principal office at 39 Bloomsbury Square, and his property portfolio in the East end he owned an estate at Marshalls, Romford, Essex, which he purchased in 1828 but never made use of.

McINTOSH'S LEGACY

Hugh McIntosh was one of the key individuals in developing the British civil engineering industry. By the early nineteenth century he had built up the expertise and capital resources to provide the sort of reliability engineers like John Rennie and Thomas Telford needed to ensure their works were built to satisfactory standards for their clients. Remarks by Telford on progress on the Union Canal are typical in this respect: "In regard to Mr McIntoshes work, great progress has been made since my last inspection and much of it has been brought into a regular state, his having the command of capital, and being a person of uncommon activities. I have no doubt, but if he applies himself to the utmost extent possible, that the sundry works, under his contract may be accomplished in due time, but not small exertions will be required for that purpose."[114] Although Hugh relied in part on his family, chiefly his brother James, and son David, to manage his works, a large number of other famous contractors and engineers were trained under him—the Aird family,[62] William[115] and Edward Ladd Betts,[101] Hugh[116] and Alexander Mackenzie Ross,[117] William Radford[118] and his son,[119] William Henderson,[120] James Leishman[121] and William Mackenzie.[8,122] They enabled McIntosh to establish himself as the first contractor with a nationwide organisation. He instilled in them a knowledge of sound practices, as witnessed by Joseph Horne after McIntosh's death. He praised McIntosh's management of horses, which he rarely overworked, and was thus able to use them for 5 or 6 years without a day's illness.[123]

HUGH McINTOSH (1768–1840), NATIONAL CONTRACTOR

ACKNOWLEDGEMENTS

I would like to thank Professor A. W. Skempton for allowing me access to his files on early contractors, and Dr. David Brooke for his helpful advice. Thanks also to Paul Parkes for his secretarial assistance. My greatest debt is due to the late Frank D. Smith, whose genealogical and other research into contractors has been left to the Institution of Civil Engineers.

NOTES AND REFERENCES

1. See for example R. K. Middlemas, *The Master builders*, (Hutchinson, London, 1967).
2. R. S. Joby, *The Railway builders: lives and works of the Victorian railway contractors*, (David & Charles, Newton Abbot, 1983) is better, but is still very thin on the earlier period.
3. An exception is J. Marshall, *Lancashire and Yorkshire Railway*, (David & Charles, Newton Abbot, 1969–1972), 3 vols.
4. H. W. Dickinson, Jolliffe and Banks, contractors. *Transactions of the Newcomen Society*, vol. 12, 1931, pp. 1–6. This draws heavily on S. Wells, *The History of the drainage of the . . . Bedford level*, vol. 1, 1830, pp. 773–779.
5. S. A. Broadbridge, The Pinkertons and the Birmingham Canal, *Transport history*, vol. 4, 1971, pp. 33–49.

   The Pinkertons were approached by Brunel about the supply of labour and bricks for the Thames Tunnel.
6. A. Gibb, *The Story of Telford*, (Maclehose, London, 1935).
7. A. Penfold, *Managerial organisation of the Caledonian Canal*, in A. Penfold ed., *Thomas Telford, engineer*, (Thomas Telford Ltd., London, 1980), pp. 129–150.
8. M. M. Chrimes and others, *William Mackenzie: giant of the railways*, (ICE, London, 1994).
9. A. Mackenzie, *History and genealogies of the Mackenzies*, (1894), p. 621, and related correspondence in *Mackenzie collection*, ICE Archives.
10. A search through Scottish Records Office files on the canal revealed nothing about this stage of McIntosh's career.
11. No evidence has been found to link McIntosh with the Leeds–Liverpool Canal; Alexander Mackenzie seems to have made his career on it. (see *PRO/RAIL/848/8*)
12. See for example *Rennie reports, Lancaster Canal*, ICE Archives.
13. *Rennie reports*, ICE Archives.
14. T. Nicholson, *Strictures on a pamphlet published at the request of the Manchester Statistical Society*, (J. Gadsby, Manchester, 1846), p. 17.
15. It is possible members of the Cross family acted as Hugh's agents in the Midlands. A Mr. Cross was working for McIntosh in 1831 (*Telford mss.* LO/107, ICE archives).
16. James Walker, *Autobiography*, (mss, ICE archives). p. 74; Letters in ICE archives (*Telford mss, Edinburgh and Glasgow Union Canal*); Letter from 38 Bloomsbury Sq., 9 Dec 1822, (*Telford collection*, Ironbridge Gorge Museum, *T/GB/73*); Letter from D. McIntosh, 39 Bloomsbury Sq., to T. Telford, 12 March 1823, (*Telford mss.*, ICE archives, *LO/63*).
17. *Mackenzie* and *Townshend collections*, ICE archives.
18. Timothy Tyrell (junior), was one of Hugh's executors and David's heir. It was he who pursued the GWR through Chancery. He died in 1865.
19. Quoted in S. Porter ed. Survey of London, vol. xliii: Poplar, Blackwall and the Isle of Dogs (1994). p. 277. By this time Rennie's view had been soured by experiences at Plymouth. On 3 January 1832 he had recommended McIntosh to the Dock Company (*Reports*, vol. 8, p. 23).
20. Grand Junction Railway Minutes, 8 August 1835, (PRO/RAIL/). It is unclear which Betts made the original tender. William Betts snr (1790–1867) is most likely but it was his son Edward Ladd Betts who was working on the Grand Junction Railway at the time (see 101 and 115 below).
21. James McIntosh was one of Hugh's executors. He seems to have retired by the 1840's. He was involved with much of the government work in the 1820s and 1830s (see 68 below). From 1844 he was living at 4 Richmond Terrace, Islington.

HUGH McINTOSH (1768–1840), NATIONAL CONTRACTOR

22. Obituary notice, *Minutes of Proceedings of the Institution of Civil Engineers*, vol. 16, 1857, pp. 162–163.
23. McIntosh listed his early important contracts in a letter detailing his experience for the Edinburgh and Glasgow Union Canal Company, from 19 Charlotte St., Bloomsbury, dated 3 November 1817. (*Telford mss., Edinburgh and Glasgow Union Canal*, ICE archives).
24. John Rennie, in his report to the bridge trustees on tenders for repairs to the Rochester Bridge, notes McIntosh has 'done a great deal of work under me, though the principal part of it has been earthwork'. J. Rennie, *Reports*, vol. 11, 1821 (ICE archives), p. 418.
25. J. Lindsay, *The Trent and Mersey Canal*, (David & Charles, Newton Abbot, 1979).
26. P. Lead, *The Caldon canal and tramroads* (Blandford, Oakwood Press, 1979).
27. *Rennie reports*, ICE archives, *passim*.
28. *Rennie reports*, ICE archives, *passim*.

    In Rennie's report to the Committee of Management of the Grand Trunk Canal, 15 April 1801, McIntosh is the only contractor whose performance is not severely criticised. His account totalled c. £2800. (*Reports*, vol. 2, pp. 203–208).
28. K. R. Clew, *The Kennet & Avon Canal* (David & Charles, Newton Abbot, 1968).
29. C. Hadfield, *The Canals of South and South East England* (David & Charles, Newton Abbot, 1968), pp. 109–116.
30. C. Hadfield, *The Canals of South and South East England* (David & Charles, Newton Abbot, 1968), pp. 81–93.
31. *Gentleman's magazine*, 14 October 1812; H. Spencer, *London's canal* (Lund Humphries, London, 1976), describes a battle between McIntosh's men and William Agar's servants when the contractors attempted to enter Agar's property on April 15 1815.
32. J. Lindsay, *The Canals of Scotland* (David & Charles, Newton Abbot, 1968), pp. 66–79.
33. H. Household, *Thames and Severn Canal*, (David & Charles, Newton Abbot, 1969). David reported, on behalf of his father, on the state of the canal on 28 June 1820 (Glos. Records Office, *Ref TS/240*), and the following year the firm implemented some of his recommendations.
34. G. N. Crawford. Manuscripts of the Telford Collection (Ironbridge), *Journal of Gloucester Society for Industrial Archaeology*, 1981, pp. 9–26.
35. *Telford mss*, ICE archives, *ref T/GB*. These include estimates for land drainage works for Colonel Berkeley (£895/3/9).
36. *Telford collection*, Ironbridge Gorge Museum, *ref T/GB; T/GB/72* refers to Mackenzie and McIntosh's estimates for the works (£102,883/14/3; payments exceeded £126,000). (Agreement PRO/RAIL/829/18). Progress on the original canal had been poor, and Telford had been asked to advise. McIntosh was invited to submit a quotation for the completion of the work, and in the autumn of 1822 visited the works with William Mackenzie before submitting his estimates. Work began in 1823, with George Barratt and Hugh Rose (Ross) as agents at Purton and Slimbridge, and William Henderson at Gloucester Docks. Despite the large number of men (523) and horses (100) on site progress was slow. In May 1825 David McIntosh was temporarily put in charge, but progress remained unsatisfactory, and claims were made. In August 1826 McIntosh began to transfer men to Plymouth due to pressure of work there. Alexander Mitchell was an assistant on these works. (*Minutes of Proceedings of the Institution of Civil Engineers*, vol. 9, 1849, pp. 103–104).
37. *PRO/RAIL/829/18*. Problems had affected the original canal, built 1810–1814, and following reports by James Green it was decided to proceed with further construction, McIntosh submitting tenders in 1831. Payments were c. £16,000.
38. *Journal of the Railway and Canal Historical Society*, vol. 21, 3, 1975, p. 55. Work began on the aqueduct in July 1836 and it was opened in August 1839. McIntosh allegedly lost heavily on the aqueduct contract, and was awarded £2000 in claims in 1840. Payments for these works exceeded £174,806.
39. C. Hadfield, *Canals of Yorkshire and North East England*, (David & Charles, Newton Abbot, 1972), vol. 1, pp. 150, 161.
40. J. S. Abernethy, *The Life and times of James Abernethy*, (1897), pp. 64–65.
41. A. Faulkner, *Grand Junction Canal*, (David & Charles, Newton Abbot, 1972), p. 137. The Agent was probably William Radford junior. (*ICE Candidates circular* 993, 1849). Payments for the works at Tring were £12,235.

HUGH McINTOSH (1768–1840), NATIONAL CONTRACTOR

42. *Rennie reports* (ICE archives), vol. 2, 1801, p. 208.
43. Letter from H. McIntosh, 3 Nov. 1817; Letter from G. Moncrieff to Telford, 5 Feb. 1818, (*Telford mss, Edinburgh & Glasgow Union Canal*, ICE archives). McIntosh tendered £37,100 for five lots, through Alexander Mackenzie, and eventually took on lots 1–8 and 23 and 24, for c. £43,000.
44. *Edinburgh and Glasgow Union Canal Minute Book*, 10 March 1818, (Scottish Record Office, *SRO/BR/EGU/1*).
45. A. W. Skempton. Engineering in the port of London, 1789–1808. *Transactions of the Newcomen Society*, vol. 50, 1978–1979, pp. 87–108; Engineering in the port of London, 1808–1834. *Transactions of the Newcomen Society*, vol. 53, 1981–1982, pp. 73–96.
46. *East India Dock Minutes*, (Museum of London in Docklands) September 1803; October 1803 mentions the press gang.
47. J. Rennie (& R. Walker) *Reports: East India Docks* (ICE archives). McIntosh excavated c. 857,000 cubic yards of earth in the Import Dock, and payments exceeded £33,000.
48. *East India Dock Minutes*, 30 August 1816. Tender for raising mud from basin into barges 6d. per cubic yard with engines; for use of barge 3d. per cubic yard.
49. *Grand Surrey Dock Minutes*, (Museum of London in Docklands) 2 October 1806, Hugh McIntosh's tender for excavating basin accepted; 20 Feb. 1807, record of payments.
50. *Commercial Dock Minutes*, (Museum of London in Docklands) July–October 1815, contain the tender and a record of payments for the excavation of the timber pond.
51. London Docks Committee Minutes (Museum of London in Docklands).
52. E. F. Clark, *George Parker Bidder* (KSL, Bedford, 1983), pp. 28, 40, 80, 107, 170–174, 419; E. F. Clark, G. P. Bidder at the London Dock extension works 1826–1828, *Transactions of the Newcomen Society*, vol. 55, 1983–1984, pp. 177–209. Agents were Messrs Ross, McInnis (McKennis), and Wright.
53. The value of the work for excavating a timber pond, installing a sewer and a cast iron bridge was c. £4500. Agents were John Fraser and A. M. Ross. Sir J. Rennie. Reports (ICE Library), vol. 8 (1831–1833), pp. 23; 67–70; 177–200; 217; 219; 246; 271; 324; 348; 353; vol. 9, 1833, pp. 49, 52, 53, 61, 63, 72, 154; 197, 214.
54. S. Porter ed., *Survey of London vol. xliii: Poplar, Blackwall and the Isle of Dogs*, (Athlone Press, London, 1994), p. 277, 311, 331–332, 576–585, 591–592.
55. S. Porter ed., *Survey of London vol. xliii: Poplar, Blackwall and the Isle of Dogs*, (1994), p. 74, 123–124, 144, 149, 151, 188–189, 576.
56. David (1815–1881) was the son of Hugh's brother John, It was he rather than his cousin who resided at Hugh's properties near Romford, and developed Havering Park.
57. S. Everard, *History of the Gas Light & Coke Company*, 1812–1849. (London, 1949), p. 69.
58. H. Nabb. The Bristol gas industry, 1815–1949, *Historical Association, Bristol Branch, pamphlet* 67, p. 47.
59. E. Boult, *The Goodliest place in Middlesex*, p. 204. There was a payment in 1812 of £2780 for the reservoir at Ruislip. The Consulting engineer was John Rennie, and the engineer William Anderson. Possibly William Radford junior was agent on later works (*ICE Candidates circular 993*).
60. G. M. Saul, John Rennie (1761–1821): one of his outstanding contributions to waterworks technology, *Transactions of the Newcomen Society*, vol. 59, (1987–1988), pp. 3–18; see pp. 8, 10. McIntosh was paid 5/6 per yard for laying the 42 in mains.
61. *SKCS/121, 124, 133* (Greater London Record Office).
62. R. K. Middlemas, *The Master builders*, (1963), p. 121; published obituary of John Aird in *Engineering*, vol. 21, 12 May, 1876, p. 382 (see also article by the author for the *New Dictionary of National Biography*). I am uncertain of the relationship, if any, between J. P. Munro, who worked for McIntosh in 1831, and the Munro who was involved with Phoenix Gasworks.
63. H. M. Colvin, *The History of the King's Work* (HMSO, London, 1973), vol. vi, p. 139.
64. J. Rennie, *Reports* (ICE archives), vol. 8, (1814–1816), p. 299, 411–412; vol. 9, 1817–1818, p. 101, 245, 250; vol. 11, 1820, pp. 120–126; vol. 12, 1821, p. 187; Sir John Rennie *Reports* (ICE archives), vol. 2, 1824, pp. 23–24. McIntosh's work exceeded £130,000 in value.
    Agents at Pembroke were William Radford snr (1813–1816), and William Henderson, who continued to work there after McIntosh's death. (S. Peters. *The History of Pembroke Dock* (E. Stock, London, 1905), pp. 7–9).

HUGH McINTOSH (1768–1840), NATIONAL CONTRACTOR

65. Work began on the reservoir for the Victualling Establishment at Plymouth Dockyard with a tender of £5501/17/6, and Sir John Rennie's original overall estimate was for £291,512. The Agents were William Radford assisted by his son. Sir John Rennie, *Reports* (ICE Archives), vol. 1, 1824, p. 419; vol. 2, 1824–1826, pp. 13, 35–36, 58–59, 64, 88, 184–189, 210–213, 228, 232, 236, 297, 301, 310, 320–321, 395, 426–428, 438; vol. 3, 1826–1827, pp. 27, 35, 45–46, 64, 105, 174–175, 180, 351, 356, 373, 378, 407, 418; vol. 4, 1827, pp. 71–76, 109, 110, 112, 116, 156–157, 165–8, 192, 229, 296, 298, 300, 303, 306, 343, 366, 368, 375, 394; vol. 5, 1828–1829, pp. 31, 36, 51, 77, 117, 140–141, 162, 201–202, 208, 209, 232, 340–341, 353–373, 377–396; vol. 6, 1829–1830, pp. 10, 16, 22, 34, 38, 97, 100, 104–109, 137, 138, 140, 141, 149, 164, 166, 170, 260, 283, 285, 317, 331–337, 355, 368–387, 391–392; vol. 7, 1830–1831, pp. 11–27, 47, 49–54, 64–65, 115, 120, 132–133, 152–157, 160–166, 185–186, 254–255; 294–295, 367–369, 383–384, 407, 411–415, 423–424, 434–435; vol. 8, 1831–1833, pp. 11, 20, 33, 57, 72, 88–91, 100, 103, 104, 143–145, 147, 169–173, 209, 210, 211, 213–215, 215, 218, 220, 232–234, 261–265, 272–284, 286–395, 298–301, 304–306, 355–358, 364, 367–368, 370–371, 375–377, 386–399, 412–418, 421, 424, 432; vol. 9, 1833–1834, p. 1, 17, 24, 43, 54, 57–60, 62. 64, 67–68, 71, 73, 73, 94, 95, 97, 98, 143, 147, 152, 153, 184, 209, 216–220, 244, 291, 294, 301, 344, 359, 409, 410, 412, 417, 430; vol. 10, 1834–1835, pp. 27–30, 50, 51, 75, 78, 80, 83, 112, 114, 129, 131, 163, 202, 21–218, 220, 272–273, 276, 357, 362, 364, 367–368, 388, 407–413; vol. 11, 1836, p. 323.

66. J. Coad, *Historic architecture of the Royal Navy*, (Victor Gollancz, London, 1983), pp. 123–124; Sir J. Rennie, *Reports*, vol. 7, 1831, p. 409, *Reports*, vol. 8, 1832, pp. 222–223; The value of the works was approximately £200,000.

66a. G. L. Taylor, *The Autobiography of an octogenarian architect*, (London: Longman, 1870), vol. 1, pp. 164–165.

67. PRO/WORKS/1/19 Letter 22 Nov. 1831 from B. C. Stephenson requesting payment of £7500 to McIntosh; PRO/WORKS/5/120 refers to total payments of £11,706/19/– to Hugh McIntosh, PRO/WORKS/16/27/2 contains Hugh McIntosh's tender for works around St. James Park of £8300 (11 June 1827); PRO/WORKS/16/290 contains Hugh McIntosh's estimate for clearing out St. James' Park Lake for £336/17/6. James Leishman was agent on these works.

68. H. M. Colvin, *The History of the Kings Works*, (HMSO, London, 1973), vol. VI, *passim*. McIntosh tendered unsuccessfully for the excavations for the Houses of Parliament in 1837. The value of the contract for the Hampton Court water gallery was £785.

69. J. Rennie, Reports (ICE archives), vol. 4, 1806–1807, pp. 80–83, 110–111, 120, 167–169, 274–275, 348, 390–391; vol. 5, 1808, pp. 219–221, 294–295; vol. 6, 1811, pp. 423; vol. 7, 1813, pp. 314, 329–330, 362–363; vol. 8, 1815–1816, pp. 324, 383–384, 391–392; vol. 9, 1818, pp. 73–74, 391–393, 400. McIntosh experienced problems in gaining access to property and placed a succession of claims. He also fell seriously ill in 1811, which hampered progress. His payments exceeded £36,000, and his men were among the last to leave the site in 1818, after construction supervision had been abandoned.

70. P. Froggatt, The Chetney Hill Lazaret, *Archaeologia Cantiana*, vol. 79 (1964), pp. 1–15.

70a. G. L. Taylor, *The Autobiography of an octogenarian architect*, (1870), pp. 173–174.

71. R. B. Fellows, *History of the Canterbury and Whitstable Railway* (Canterbury, 1930). David's tender of £10,000 was accepted in May 1831.

72. A. W. Skempton, William Chapman (1749–1832) civil engineer, *Transactions of the Newcomen Society*, vol. 46, (1973–1974), pp. 70–73.

73. *Report from the Select Committee on Dover Harbour with the Minutes of evidence* (London, 1836), p. 51. McIntosh's tender for work on the west pier was £19,300. William Betts was agent. He was living nearby in 1815 when his son Edward was born, and McIntosh probably had a contract there then, on an initial phase of work which followed Ralph Walker's report of 1812.

74. M. W. Baldwin, The Engineering history of Hull's earliest docks, *Transactions of the Newcomen Society*, vol. 46, (1973–1974), p. 9. Value of contract £95,000.

75. The original contract for £9,275 was agreed by William Betts on December 24 1832. The pier repairs in 1838 cost c. £3,100.

76. J. G. Bishop, *The Brighton Chain Pier* (Brighton, 1897).

HUGH McINTOSH (1768–1840), NATIONAL CONTRACTOR

77. J. Elliott. Account of the Dymchurch wall which forms the sea defences of Romney Marsh. *Minutes of Proceedings of the Institution of Civil Engineers*, vol. 6, 1847, pp. 466–481; E. F. Clark, *George Parker Bidder* (1983), pp. 204–5; 431–432. McIntosh is known to have worked on the sluice, but I am unclear about his role on the sea wall, where over 1000 men were employed at times in the 1820's.

78. James Leishman was agent for at least some of these works. He acquired property in the area he leased to his wife's family, the Waddells, marrying Sarah Waddell (1811–1894) at Dymchurch Parish Church on 24 March 1829.

79. J. Walker, *Autobiography*, (mss, ICE Library), p. 313.

    Edward Ladd Betts was agent. There is a drawing of the lighthouse at Trinity House, dated 1835, signed by James Walker and David McIntosh.

80. It is possible McIntosh carried out bridge repairs at Weymouth in 1823–24, when Mr. Betts was asked to report on the state of the bridge. William Betts' third son George was born there.

81. *Report of the Committee of Magistrates appointed to make enquiry respecting the public bridges in the County of Middlesex*, (1826), p. 221 (Brentford), p. 252 (Harmondsworth); *GLRO: MA/D/Br/12 & 46.*

82. *Northamptonshire archaeology*, vol. 15, (1980), pp. 150–151.

    The Engineer was William Anderson, engineer to the Grand Junction Waterworks, who often worked with McIntosh, and was familiar with his work at Rochester and London Bridges.

83. *Stratford upon Avon Corporation Records. BRT* 5/9/12; 5/9/14; 5/8/12; 5/8/21.

84. *Report from the Select Committee of the House of Commons to inquire into the State of London Bridge*, (1821), p. 123, minutes of evidence of Hugh McIntosh; J. Rennie. *Reports*, (ICE archives), vol. 11, 1821, p. 418. Tender £8578/10/8.

85. W. Mackenzie, Account of the bridge over the Severn, near the Town of Tewkesbury, *Transactions of the Institution of the Civil Engineers*, vol. 2, 1838, p. 1–14. The manuscript version of this paper (*ICE OC 285*) contains unpublished material on this project. Value of the contract c. £10,000 (J. Rickman ed., *Life of Thomas Telford* (1838)).

86. B. C. Jones, Crossing the Itchen, *Southampton papers*, 1. It appears that McIntosh was owed c. £19,500 at the end of 1838, and he offered to pay for the Parliamentary expenses for a further Act to enable the tolls to be raised and debts cleared. In 1849 the Company was virtually bankrupt again and William Betts (snr.) was involved in reviving the scheme, and replacing the bridge in 1854.

87. J. Finch, *The Port of Portsmouth floating bridge* 1838–1961, (unpublished typescript, Portsmouth City Records Office, c. 1971).

88. (First) *Annual report of the Portsmouth Floating Bridge Company* (1840), (Portsmouth City R.O.), p. 3. D. McIntosh paid £6,807/10/–.

89. *Report from the Select Committee of the House of Commons to inquire into the state of London Bridge*, (Corporation of London, 1821), Minutes of evidence of Hugh McIntosh, pp. 123–125.

90. T. Telford, *Report. . . on the effects which will be produced on the river Thames by the rebuilding of London Bridge*, (London, 1823). According to Telford papers at ICE, McIntosh's foreman was Mr. Kitching. (*Telford mss*, ICE archives, *LO/63*).

91. T. Telford and J. Walker, *Reports on the present state of the new London Bridge*, (London, 1831–1832); *LO/107*, (*Telford mss*, ICE archives).

92. J. Walker, in, *Report to the Common Council, from the Committee appointed in relation to Blackfriars Bridge*, (Corporation of London, 1833).

93. J. Walker, in *Report to the Common Council, from the Committee appointed in relation to Blackfriars Bridge*, (Corporation of London, 1840), pp. 13–14. Principal agent William Radford jnr. reported on sources of stone for the new parapet.

94. J. Walker in *Report to the Common Council, from the Committee appointed in relation to Blackfriars Bridge*, (Corporation of London, 1843), pp. 7, 9–11. Total value of contract £95,075/18/10.

94a. The unsuitability of the methods employed on the Liverpool and Manchester Railway, and the advantages of employing a large contractor are discussed with specific reference to McIntosh by Joseph Locke when giving evidence on the London and Southampton Railway. *Minutes of evidence taken before the Lords Committees (sic) . . . Railway from London to Southampton* (1834), pp. 258–269.

94b. H. G. Kendall, *Plymouth and Dartmouth Railway*, (1968).

95. R. H. G. Thomas, *London's first railway, the London and Greenwich* (Batsford, London, 1972). PRO/Rail/389/1 details McIntosh's tenders and payments. One agent was J. Rowland, assisted by James Hodges of Montreal Bridge fame. Agent between Deptford and Greenwich was Charles Nixon, (*Select Committee on Westminster Bridge, Report and Minutes of Evidence*, London, 1844. (HC 477, 1844), p. 73). The story about the plans is referred to in F. R. Conder *The men who built railways* (Thomas Telford, London, 1983).

96. PRO/rail/384/277; *Minutes of London Committee, London and Birmingham Railway*, 23 July 1834; 17 September 1834.

97. *PRO/rail/412/7* and */412/8*. Payments in excess of £177,000. Contracts were for Southampton–Bartholomew Hyde, agreed 13.5.1836; brickwork London–Wey Navigation, agreed 9.9.1836; and works near Winchester, agreed 9.9.1836. William Betts was agent at the Southampton end, and Radford, snr. at the northern end.

98. *PRO/rail 490/25*. The value of the tenders for the 6 contracts (9–14) was £258,629. The Minutes of the South of Trent Committee refer to various delays. *Herapaths journal*, 4 Feb. 1843, refers to an outstanding claim for £7,500. The case was reported in *Railway and canal cases*, 3, 1842–1846, pp. 780–790. I am grateful to Elton Engineering Books for a copy of this.

99. *PRO/rail/220/20*. Tender price £54,440, 11 February 1834.

100. N. Webster, *Britain's first trunk line*, (Adams & Dart, Bath, 1972), p. 42.

101. Obituary notice of E. L. Betts, *Minutes of Proceedings of the Institution of Civil Engineers*, vol. 36, 1873, pp. 285–288.

102. D. Stevenson, Diary, 1 July 1835, (National Library of Scotland).

103. *PRO/Rail/541/3*. James Walker, the engineer for the line, was keen on having McIntosh as the contractor, and on 5 December 1837 suggested McIntosh might carry out the works for a lower price; in October 1839 he approached David McIntosh about the contract between Broxbourne and Roydon. Payments c. £26,000. (£23,262 according to *Railway times* 2, 1837).

104. *PRO/rail/36*.

105. D. Brooke, The Equity suit of McIntosh v the Great Western railway: the Jarndyce of railway litigation. *Journal of transport history* (forthcoming).

106. Hugh's contract was 9L, James Leishman acting for him; David's were initially 3L, 3B, 3B extension, 6B, and, following the removal of William Ranger from his contracts, 1B, 2B and the ballasting on 3B. Agents are believed to have been James Leishman (9L), William Radford jnr. (3L), and Hugh Ross (2B, 3B, 3B extension, 6B) assisted by William Henderson (3B, 3B ext., 1B and 2B) and Charles Nixon working on the bridges at Bath and Bristol, and 3B Ballasting.

107. Material relating to the GWR contracts is to be found at the PRO, Bristol University Library, and with British Rail.

108. *PRO/rail/530/4, /530/7*. Total value of tenders over £340,000. Hugh McIntosh was awarded the Altofts contract for £78,000 (*London Committee* 8 June 1837); the south of Belper contract for £20,050 (*London Committee* 24 April 1838); Belper station (*Herapeth's journal* 7 Dec 1839). David had the Milford contract (*London Committee* 12 September 1837) for £93,122, Duffield for £43,144 (*London Committee* 12 May 1838) the Derby contract for £19,000 with James Leishman as partner, (*London Committee* 19 June 1838) and Metley for £26,655 (*Leeds Committee* 11 September 1838).

109. On the North Midland Alexander Mackenzie Ross, former agent, was resident engineer. James Leishman, who signed Hugh's death certificate, was based at Duffield, as was John Barton.

110. On the Manchester and Leeds Hugh had the Kirkthorpe contract (*PRO/rail/34/2*). His tender was for £46,000 but claims of £20,000 were outstanding in 1842, and only settled by the arbitration of Robert Stephenson. Samuel Horne who completed the line after Hugh's death, was agent at Wakefield, along with McKennis Butler, and Joseph Thornton (?). Samuel later worked for Mackenzie and Brassey on the Trent Valley and other lines. His brother Joseph (d. 1854) was also a McIntosh agent. See 123 below.

111. *Wakefield journal*, 4 September 1840, reprinted in *Railway times*, 12 September 1840, p. 773.

112. *Gentleman's magazine*, 31 August 1840, pp. 441, 445.

113. This is referred to in the minutes of the canal company.

HUGH McINTOSH (1768–1840), NATIONAL CONTRACTOR

114. T. Telford (draft) Report on the Union Canal (Telford mss Edinburgh and Glasgow Union Canal, ICE Archives).

115. William Betts (1790–1867) acted as McIntosh's agent on a succession of contracts across southern England. In July and August 1840 he tendered on behalf of David McIntosh for eight contracts on the Manchester and Birmingham Railway. His second son, William (1820–60) was with him on the Midland Counties Railway at Leicester in 1841 (Census returns). After Hugh McIntosh's death the family operated for a short while as William Betts and Sons, contractors, taking contracts on the South Eastern railway. Edward took over existing contracts in 1845 (*London gazette*, p. 6725) and the firm was finally dissolved in 1847 (*London gazette*, p. 4485). William Betts (jnr.) went to Montreal, possibly to work on the Grand Trunk Railway, and died in North America. The father had a share in the profits of William Mackenzie's contract on the Manchester and Salford Junction Canal, and was involved in the promotion of the Chester–Holyhead Railway, and Brymbo Ironworks, all with heavy involvement of former McIntosh associates. In 1844 he purchased Bevois Mount near Southampton, but remained involved in engineering projects. He left c. £30,000. His own father was also called William.

116. Hugh U. Ross witnessed David McIntosh's signature on GWR contract 3B and continued to be involved with the McIntoshes affairs after Hugh's death, pursuing claims on the North Midland railway. It is unclear whether this Hugh Ross was A. M. Ross' brother, or his uncle, also named Hugh.

117. Alexander Mackenzie Ross (1805–1862) was viewed as a contractor's engineer by C. R. Moorsom according to William Mackenzie's diaries. He was a pupil of his uncle Hugh, working on McIntosh projects from 1823 until 1836 when he joined Robert Stephenson's staff to work on surveys of the London–Brighton line. After McIntosh's death he was involved with the Brymbo Ironworks scheme, and designed the structures on the North Wales Mineral Railway to Wrexham. He was engineer with Robert Stephenson on the Chester and Holyhead railway, was engineer for the Mold railway, and the Londonderry and Inniskillen railway, all of which had major ex-McIntosh staff involvement. His most famous work was the Grand Trunk Railway in Canada. His precise role vis-a-vis Robert Stephenson regarding the line and the Victoria Bridge, Montreal was the subject of controversy at the time, and may have contributed to his death. He had a brother Hugh (c. 1809–186?), who was at David's house for the 1851 census, and it is unclear at present which Hugh did what for McIntosh.

118. William Radford senior c. 1790–1856. ICE candidates circular 611, 1840.

119. William Radford junior, (1816–1854) ICE Minutes of proceedings. vol. 14, (1854–55), pp. 136–137. After McIntosh's death he became engineer to the Regents Canal in succession to William Anderson.

120. Henderson was probably related to the McIntosh's by marriage, and appears to have retired to Bangeston House, Pembroke. He was a contractor in the 1840s on the London–Brighton line (with Thomas Gregory), and extended Pembroke Dockyard wall.

121. *Alloa Advertiser* 1884.

James Leishman (1800–1884) joined Hugh McIntosh's staff on the Edinburgh and Glasgow Union Canal in 1817. His eldest son was born in Yorkshire in 1832 suggesting he may have worked on the Aire and Calder works. Following the completion of Hugh's contracts he purchased an estate at Brownrig, Dollar. In the 1840s he was a partner of Edward Price, taking contracts on the North Staffordshire Railway, and, between 1847–1848, on the Londonderry and Inniskillen Railway. The partnership was dissolved in 1849. That year he visited Turin on behalf of Mackenzie and Brassey. He then appears to have retired, although he was a director of the North British Railway, and involved in the Dollar Gas Company. He left a fortune of £120,154.

122. William Mackenzie began working for McIntosh on the Edinburgh and Glasgow Union Canal, where his father had originally taken a contract, and helped with the estimates for the Gloucester and Berkeley Canal, before becoming resident engineer on the Mythe Bridge Turnpike.

123. *Minutes of Proceedings of the Institution of Civil Engineers*, vol. 2, 1843, pp. 117–118. Horne died in 1854 in Canada, possibly working on the Grand Trunk Railway.

10
Some railway facts and fallacies

Charles E. Lee

In the selection of a subject for his Address, it is customary for a President to take some aspect of that portion of the history of engineering and technology which is within his own specialisation. In my own case that is clearly transport, and particularly rail and road transport. It was stated on an earlier occasion that "a President is usually expected to choose a subject on which he is an authority." Except for the saving word *usually* in this dictum, I should not be able to address you, as I should be unwilling to attach to myself the description of an authority. I ask you, therefore, to accept this as one of the unusual occasions, when the Address is not submitted as an authoritative pronouncement, but rather as guidance for future historians. The subject is one with which I may claim an intimate acquaintance, as a result of studies extending over a longer period than the life of the Society.

Within the limits of a single address, it is impracticable to deal with more than a few points in such a vast subject as the history of railway engineering. I have, therefore, confined my remarks to matters prior to the last 100 years, and to particular aspects upon which our precise knowledge is slight. The subject of railways, by reason of popular interest therein, has resulted in a vast outpouring of published material, of which only a small proportion represents the result of original research. Much of it includes sheer conjecture as statements of fact, and even more repeats erroneous particulars gleaned from secondary sources. It seems useful, therefore, to record some of the landmarks of early railway history of which we have reasonably satisfactory evidence; to direct attention to major outstanding queries; and to mention some of the fallacies that are widely accepted as facts.

Initially, it is desirable to define what is meant by a railway in my remarks. I submit that the one common characteristic of the many forms of track which may be regarded as a railway is that the combination of track and vehicle is self-steering. The word railway or railroad certainly had a wider connotation, particularly in the early years of last century,[1] as it included parallel tracks (usually of stone) in a gravel highway, later often also called a tramway. Outstanding examples in modern times were the Commercial Road Tramway in the East End of London, built by James Walker and opened on 27 March 1830,[2] which lasted until 1871; that between Towcester and Weedon on Thomas Telford's London–Holyhead road, laid in 1837 (and wrongly called "Roman" by numerous writers), which survived until 1951–52; and the tramways of flat iron plates opened across the new Westminster Bridge on 24 May 1862.[3] Such lines, as also their predecessors in antiquity, were not self-steering, and therefore are outside the scope of these notes. They are nevertheless of interest to the railway historian, particularly in regard to track gauge, as indicating an approximate standardisation of axle lengths.

---

[1] E.g. John Claudius Loudon, *Encyclopaedia of Agriculture*, 1825, p. 542.
[2] *Mechanics' Magazine*, 3 April 1830, pp. 73–77.
[3] *South London Chronicle*, 31 May 1862.

PLATE I

(*a*) REMAINS OF A WAGON RUTWAY IN A ROAD IN THE NEIGHBOURHOOD OF SYRACUSE, SICILY

(Reproduced from *La Nature* of 26 August 1905)

(*b*) A STREET IN POMPEII *c.* 1930

(*Photo: Martin H. Press*)

PLATE II

(*a*) RUTS IN MALTA NEAR MUSTA SHOWING THE FAMOUS DOME OF MUSTA CHURCH

(*b*) SECTION OF THE *Diolkos* ACROSS THE ISTHMUS OF CORINTH EXCAVATED IN 1957

SOME RAILWAY FACTS AND FALLACIES

Our first certain knowledge of a self-steering road is provided by the rutways of ancient civilisations, of which traces may be found in Greece, Malta, Italy, and in other parts of the Roman Empire (Plates I and II (*a*) and (*b*)). Our knowledge, as distinct from conjecture, extends very little further than what may be seen in the remains. In few cases can they be dated closely; their extent is the subject of widely-differing views supported by inadequate evidence; and it has been contended recently[1] that the most extensive group of rutways—those in Malta—were not cut deliberately (Plate II, (*a*)) Captain Gracie, who advanced this view, said that the Malta ruts had been discussed in print for 300 years without agreement having been reached on how, when, or why they were made, or what vehicles used them.[2] It seems probable that they are of Phoenician origin, of about 2000 B.C., but the whole subject is worthy of close examination. There are discrepancies in recorded measurements, and few writers have stated with precision the points between which they measured. The Knight Boisgelin, writing in 1805, said "these ruts are on an average 4 ft. 5 in. from middle of rut to middle of rut." Later writers, with a knowledge of the modern standard railway gauge, have endeavoured to see a closer approximation to 4 ft. 8½ in. than is justified. Captain Gracie took 178 reliable measurements between the centres of the curved bottoms of the ruts, nearly all of which were between 4 ft. 5 in. and 4 ft. 9 in.

There are numerous examples of grooved roads, or rutways, in Greece, and their similarity to modern railways has been mentioned by many writers[3] from 1842 onwards, but they still await the detailed attention of an engineering historian. Weber gave the gauge as 1.63 metres (5 ft. 4 in.), but here again there are divergences between one observer and another, and no published results of adequate field work are known to me.

Because of numerous Classical allusions, we know of the paved stone portage roadway, called the *Diolkos*, by means of which ships were conveyed across the Isthmus of Corinth. The form of road was not known, and it was often assumed that ships were hauled on rollers over a flat surface. More than 23 years ago, I had the temerity to suggest that the *Diolkos* was possibly a rutway, and that ships traversed this on wheeled cradles. As a result of excavations[4] conducted by the Archaeological Service of the Greek Government, some 1,640 yd. of this road were uncovered in 1957, from the terminus on the Gulf of Corinth. An undoubted rutway was revealed, with ruts of which the inner edges were some 4 ft. 6 in. apart. The published figures are at present insufficient to enable a dimensioned cross-section to be prepared. The *Diolkos* was apparently some four miles long. The uncovered section showed that it ascended the steep slope from sea level in great sweeping curves.

Until recently we had sound reason to believe that it existed in the time of Aristophanes (427 B.C.) and continued in use at least until the time of Strabo (24 B.C.), some 400 years. The recent researches have indicated a date of construction about 600 B.C., in the time of Periander, the tyrant of Corinth (625–585 B.C.), and an active life which lasted until the ninth century A.D., in all, some 1,400 years. This is by far the longest period of use which has been attributed to any form of railway. Although the *Diolkos* may not be representative in all respects of the ordinary rutway—it was built for the special purpose of a ship railway—further details of the Greek researches are awaited with keen interest (Plate II, (*b*)).

---

[1] Instructor Captain H. S. Gracie, R.N., "The ancient cart-tracks of Malta," *Antiquity*, No. 110, June 1954, pp. 91–98.

[2] See Captain E. G. Fenton in *Man*, May 1918, p. 40; Professor Sir Temistocle Zammit in *Antiquity*, No. 11, 1928, pp. 18–25.

[3] William Mure, *Journal of a Tour in Greece*, Vol. II, 1842, p. 251; Professor Ernst Curtius, *Wegebau bei den Griechen*, Berl. Acad., 1854; Freiheer Carl Philipp Max Maria von Weber, *Die Schule des Eisenbahnwesens*, Leipzig, 1857; *inter alia*.

[4] Nicholas M. Verdelis, "How the ancient Greeks transported ships over the Isthmus of Corinth: Uncovering the 2550-year-old Diolcos of Periander," *Illustrated London News*, 19 October 1957, pp. 649–651; Professor Oscar Broneer, "The Corinthian Isthmus and the Isthmian sanctuary," *Antiquity*, Vol. XXXII, No. 126, June 1958, pp. 80–88; "Portage roadway for ships traced," *The Times*, 20 August 1957.

SOME RAILWAY FACTS AND FALLACIES

The use of artificial wheel ruts on Roman roads is not common. They occur in mountainous districts, for instance in several parts of Gaul and the Maritime Alps, and in some towns. Artificial wheel-ruts are also found in the Balkans and the Danube provinces. Their apparent use by the Romans was to restrict traffic to a single channel, either to avoid building a wide road in difficult country, or to provide a safe path (on the railway principle) in mountainous districts. A good example of the latter has survived at Bons (Isère), on the Briancon–Grenoble–Vienne road, where the wheel-ruts are a protection against running off the road down the steep slope. There are remains of a 3-metre tunnel (9 ft. 10 in.). On the rocky surface of the road are two parallel grooves to take the wheels of the wagon and hold them far more efficiently than could any parapet. These ruts are 1·44 metres (4 ft. 8¾ in.) centre to centre. The figure given by a French author many years ago[1] was 1·38 metres (4 ft. 6½ in.), and this has often been quoted since. It represented the distance between the inner faces of the grooves. These rutways were examined with care by H. Ferrand[2] and he gave the following dimensions: average width of grooves 0·06 metres (2½ in.); maximum depth of grooves 0·07 metres (2¾ in.); gauge centre-to-centre 1·44 metres (4 ft. 8¾ in.).

Short sections of Roman ruts have been measured in many places, and by many persons, including myself. So far as they have been recorded with sufficient precision to admit of adequate comparison, the centre-to-centre figures range from 4 ft. 6 in. to 4 ft. 10 in., but the widths of rut vary considerably, although the distance between inner faces is usually about 4 ft. 6 in. There is thus a general similarity to George Stephenson's gauge, but certainly not anything which could be held to account for a precise 4 ft. 8½ in.

After the fall of the old Empires, there is a gap in our story. Road construction and maintenance were at a low ebb, and even the use of the wheeled vehicle seems to have been severely limited. There is no evidence of continuity. The beginnings of the "modern" history of railways are to be found in the mediaeval mining industry of central Europe. Here, again, it is desirable to take stock of what we know, as distinct from what has been deduced or conjectured from the limited early evidence at our disposal. German mining historians attribute to the twelfth century the use of wooden tracklines in the Harz mines and in those of the Tirol, but we have little definite evidence, and none as to the form of these lines and their vehicles until the sixteenth century. The lines may have been self-steering, but proof is lacking until 1556.

There are earlier illustrations of mining transport on these wooden tracklines, but none is sufficiently detailed to show any form of self-steering. Probably the earliest is on a link of a silver collar or chain of office, preserved at Gand. The collar is believed to be German work of the fifteenth century. It consists of 16 links, of which every link portrays one phase of work from the extraction of silver from the mine to the sale of the finished article in the silversmith's shop. The fifth link on the right (counting from the central medallion) shows ore transport on a trackway (Plate III, (*a*)).

About 1519 a book on mining law called *Der Ursprung gemeynner Berckrecht* included what is probably the earliest printed illustration of this subject (Fig. 1), and a Flemish "composite" painting, "Les Travaux de la Mine," of 1544, by Lucas Gassel, shows a miner pushing a truck on rails out of an adit (Plate III, (*b*)). Sebastian Munster's *Cosmographiae Universalis* of 1550 includes a picture of a similar truck on lines. Munster tells us that, at the Leberthal silver mine, the vehicles ran on four small iron wheels. This is the first known mention of iron wheels in mining transport (Fig. 2).

Not until the publication in 1556 of Agricola's classic *De Re Metallica*, however, do we have a clear description and illustration of the self-steering railway principle. The reference is well known, and it is sufficient here to mention that he describes a truck with wooden unflanged wheels running on narrow-gauge rails. The bottom of the truck, however, had a projecting blunt iron pin or prong

[1] Florian Vallentin, *The Roman Road of l'Oisans*, 1877.
[2] Observations published in *La Nature*, 6 May 1905.

PLATE III

(*a*) LINK IN SILVER COLLAR ATTRIBUTED TO THE FIFTEENTH CENTURY, WHICH IS PRESERVED AT GAND, SHOWING ORE TRANSPORT ON A TRACKWAY

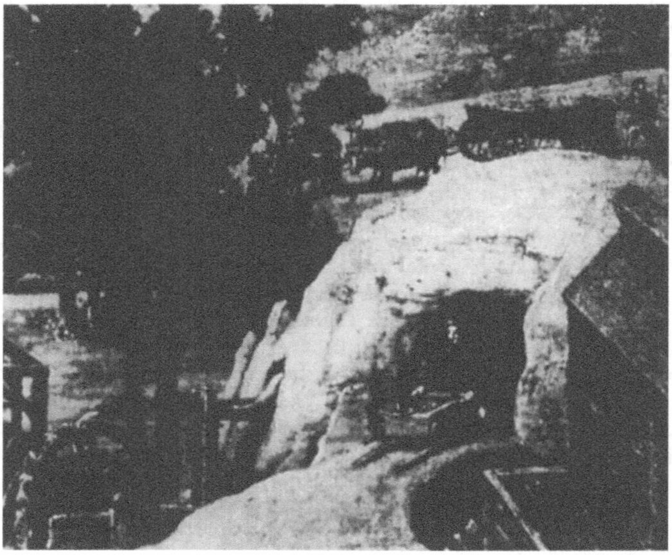

(*b*) MINING RAILWAY DEPICTED ON A PORTION OF THE FLEMISH PAINTING, "LES TRAVAUX DE LA MINE," OF 1544

PLATE IV

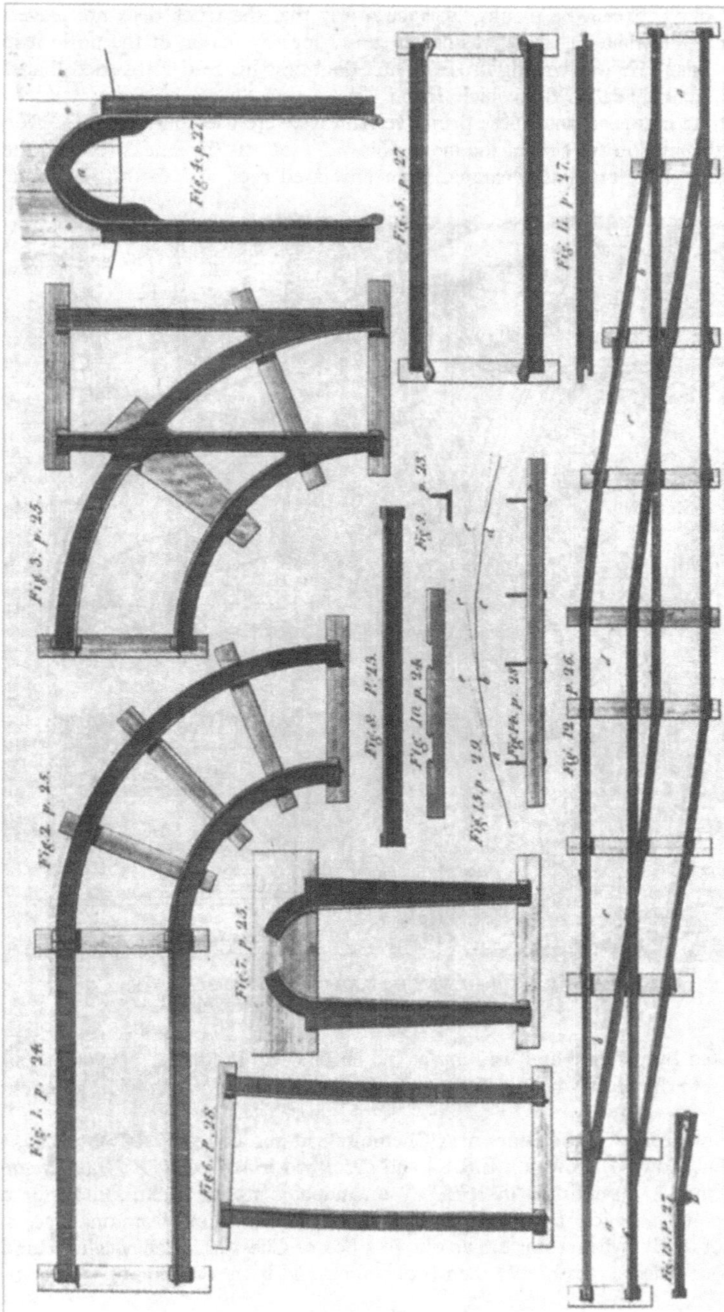

DETAILS OF CAST-IRON PLATE RAILS ILLUSTRATED IN PLATE 2 OF "THE COAL VIEWER, AND ENGINE BUILDER'S PRACTICAL COMPANION," BY JOHN CURR OF SHEFFIELD, PUBLISHED IN 1797

SOME RAILWAY FACTS AND FALLACIES

to run in the groove between the planks "in such a way that the truck does not leave the track" (Fig. 3). Agricola said that the truck was called *canis* (dog) by reason of the noise it made while passing over the rails. He was writing in Latin, and doubtless his readers realised that the Central European miners actually called the vehicle *Hund.*

In Agricola's time it appears that these primitive railways were used only from the working faces underground to the adits in the side of the mountains. As soon as the trucks reached the open air, they were emptied. The ore was separated from unwanted rock and earth. In some cases the

Fig. 1. MINING RAILWAY ILLUSTRATION OF ABOUT 1519 FROM "DER URSPRUNG GEMEYNNER BERCKRECHT"

metal was extracted by panning and washing at the entrance to the mine. In other cases the ore was brought down to the place where it was to be treated in a variety of ways but, apparently, not by railway.

Conditions below ground in the mines near Chemnitz had not altered greatly by 1669, when they were visited by Dr. Edward Brown(e), F.R.S., and described in his book, *A Brief Account of some Travels in Hungary . . .*, published in 1673. I am indebted to Sir Arthur Elton for calling my attention to this volume, which says "They carry their Ore underground from one place to another; or to the bottom of the Pit whence it is drawn out, in a Box or Chest which they call a Hundt or Dog; this runs upon four wheels, it is higher then (*sic*) before, and hath a tongue of Iron at the bottom, which being fitted into a channel of wood framed in the middle of the bottom of each passage, it can no ways deviate, but keep allwayes in the middle; and by this means a little Boy will run full speed with three or four hundred pound of weight of Ore or Earth before him, wherever you command him, without any light, through those dismal dark passages of the Mine: and it was very new to me to hear the rattling they make in the Mine, and the alteration of the sound as they are nearer or

Fig. 2. MINING TRUCK WITH FOUR IRON WHEELS RUNNING ON RAILS, FROM SEBASTIAN MÜNSTER'S "COSMOGRAPHIÆ UNIVERSALIS," 1550

(Note: Names of equipment in Latin, but names of workmen in German)

further from us; and to see them come with that swiftness out of the Rocks overturn their little Chariot, where they are to leave their Ore, then turn again and enter those dark Caves with such a force and swiftness."

As in the time of Agricola, it appears that such lines ran only from the working faces to the adits. It seems unlikely that, if the flanged wheel were in use by the middle of the sixteenth century, it

SOME RAILWAY FACTS AND FALLACIES

would have escaped Agricola's comprehensive survey. An exhibit in the Verkehrs-und Baumuseum in Berlin is a wooden mining wagon and section of wooden track, of 48 cm. (1 ft. 6⅞ in.) gauge, stated to date from the sixteenth century. It has four wooden one-piece flanged wheels on iron axles. These relics were taken from a Hungarian gold mine (Apostelgrube Brad, Transylvania) in 1889, and were formerly in the collection of A. Haarmann, General Manager of the Georg-Marien Mining and Blast Furnace Union at Osnabrück. It is an exhibit of great interest and importance,

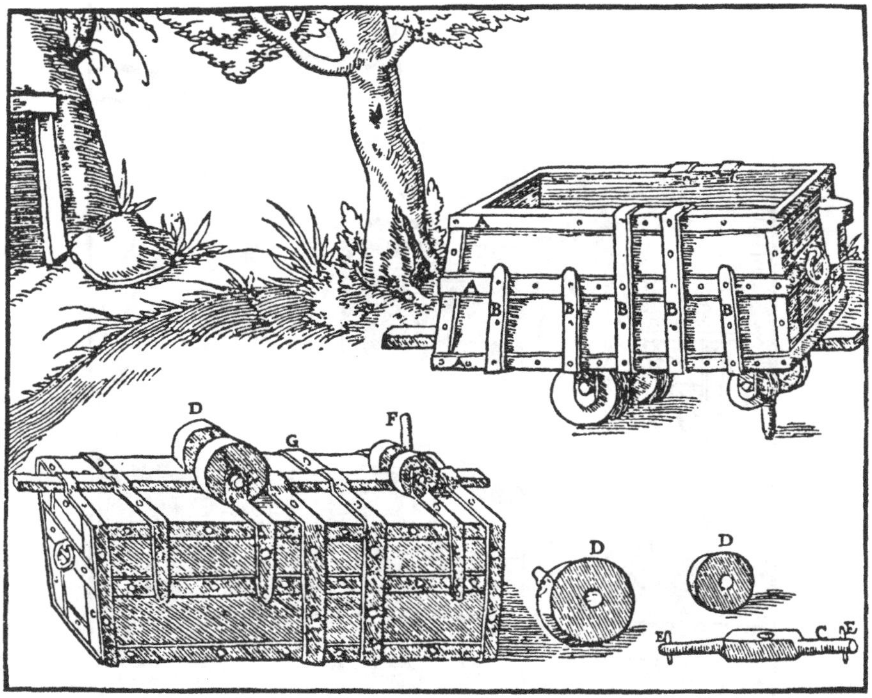

Fig. 3. Mining truck illustrated by Agricola in "De Re Metallica" (1556)

(A) *Capsae quadranguli ferrei* (iron-bound rectangular truck)
(B) *Eius bacilla ferrea* (its iron straps)
(C) *Axiculus ferreus* (iron axle)
(D) *Orbiculi lignei* (wooden wheels)
(E) *Parvi clavi ferrei* (small iron keys)
(F) *Magnus clavus ferreus obtusus* (large blunt iron pin)
(G) *Capsa eadem inversa* (same truck inverted)

but the dating lacks precise proof, although there are no known reasons for rejecting it. If correct, this is the oldest surviving rail vehicle in the world. I examined it shortly before the Second World War, and am pleased to be able to record that it survived the war, although the museum was bombed.

By 1669 the primitive one-piece flanged wheel had become a composite flanged wheel. Dr. Edward Brown again provides useful evidence. Railways were then in use to carry the ore from

SOME RAILWAY FACTS AND FALLACIES

the adits to the place where it was treated, and Brown describes them thus: "Not much unlike this [type of mine railway already described] is another instrument they have to bring the Ore from the Mouth of the Mine, or from the Hills down to the *Buchwerke* [*Pochwerk* or stamp mill] where they pound it and wash it; but instead of a tongue it hath eight wheels or four rowlers and four wheels, and the way is made with Firre in such manner and at such a distance that the rowlers rowle upon the wood of the Firre-trees. And these rowlers and wheels are so contrived, that these Chests can never overturn nor go out of the way, and a child draweth them, and sometimes a dog serves the turn. To one *Buchwerke* alone, they carry every week three or four hundred of these Chests full, and each Chest holdeth four hundred pound weight." It seems that Brown intended to describe, not a waggon on eight wheels, but one on four rollers, each backed by a wheel of larger diameter on the same shaft, acting as a flange.

At this stage the development of the railway on the mainland of Europe came to an end. There was no further fundamental change until the nineteenth century, although, at some date unknown, *Reibeisen*, or rubbing strips of wrought iron, were sometimes fixed on the wooden rails on curves and gradients at points of greatest friction. It will have been noticed that all the examples of continental mining railways were used in connection with precious metals; were of narrow gauges to meet the requirements of heavy concentrated loads of ore; and mainly had man or boy haulage or propulsion. This explains why the improved English railway, as developed principally in our coalfields, was referred to as an *englisher Kohlenweg* when it was introduced into the Ruhr district at the end of the eighteenth century.

Concerning the introduction of railways to this country, we have a considerable amount of contemporary or near-contemporary evidence, from which a plausible and consistent story may be compiled. Nevertheless, in its earlier stages the accuracy of this story is dependent on the correct interpretation of words which may have changed their meanings. The principal words are "tram," "rail," "wagonway," "tramway," and "railway," or, in the last two cases, the variants "tramroad" and "railroad." Some 23 years ago, I examined this matter of terminology in a book called *The Evolution of Railways*,[1] and it seems unnecessary here to repeat the evidence then adduced. I then advanced reasons for believing that the word "tram" in connection with transport was derived from the old Scandinavian *traam* or *trum*, meaning a beam of wood, and recorded that the word "tramway" or "tramroad" is not encountered until the late years of the eighteenth century. Our earliest references to what may have been self-steering railways use the words "railes" and "wagonway." There are the Middleton manuscripts, in which the earliest "rail" reference, heretofore interpreted as 1597, has been shown by Richard Smith[2] as being 1604. The first Shropshire reference to "tylting railes" is October 1605. "Reformed Cariages" for the conveyance of coal are mentioned in a petition to King Charles I on 10 July 1628, and a "wagon-way" in 1672. The form of rail and vehicle are unknown, and therefore whether these lines were self-steering is a matter of deduction from later practice.

We are on firmer ground in the 1676 description by Roger North of "rails of timber" with "bulky carts made with four rowlets fitting these rails." This, and Dr. Brown's description of 1669, are the earliest known mentions in print of a kind of flanged wheel, although, as we have noticed already, a mining wagon with flanged wooden wheels, on iron axles, in the Verkehrs-und Baumuseum, Berlin, is attributed to the sixteenth century. The first illustration showing beyond all doubt a flanged-wheel vehicle on rails is that of 1734 in Dr. John Theophilus Desaguliers's *Course in*

[1] Charles E. Lee, *The Evolution of Railways*, 1st edition July 1937; 2nd edition (revised and enlarged), November 1943.

[2] Richard S. Smith, *Huntingdon Beaumont: Adventurer in Coalmines.* Renaissance and Modern Studies, University of Nottingham, Vol. 1, 1957. *England's First Rails: A Reconsideration, ibid.*, Vol. IV, 1960.

SOME RAILWAY FACTS AND FALLACIES

Experimental Philosophy, depicting the Prior Park wagonway at Bath (Fig. 4). The equally clear drawings of a flanged-wheeled coal wagon given by Gabriel Jars[1] are of 1765 (Fig. 5).

The German practice of using *Reibeisen* on sharp curves and steep gradients is said to have been introduced into England "at least as early as 1716," but we have no contemporary evidence. The statement was made in 1841, and cannot be ignored, as reliable information may then have existed, which is no longer available, but it is inconclusive. There are also later stories of an unsuccessful attempt to *substitute* cast-iron for timber rails at Whitehaven in 1738. Abraham Rees states in his *Cyclopaedia* that the "first rail or waggon-way" at Whitehaven was opened on 4 August 1738. This

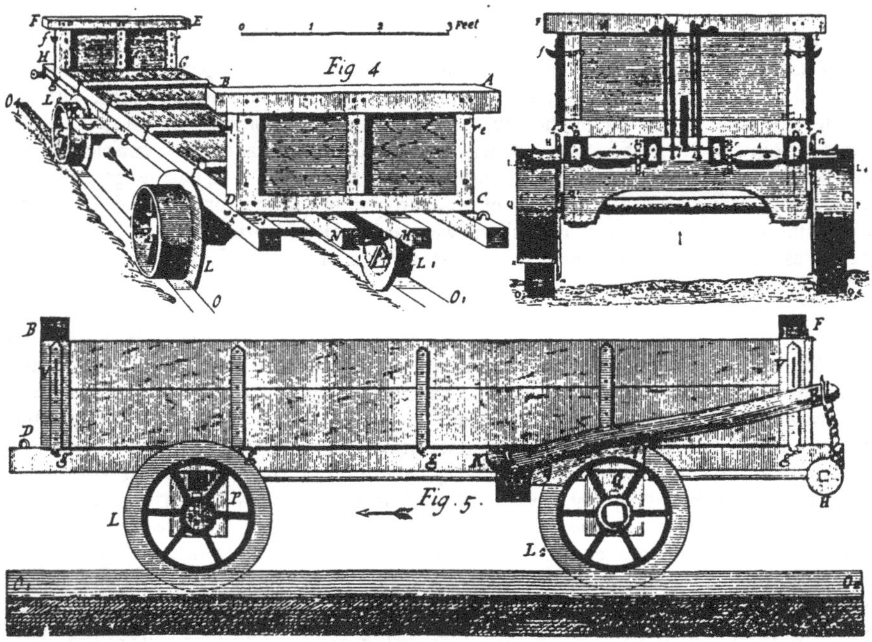

Fig. 4. FLANGED-WHEEL VEHICLE OF 1734 FOR THE PRIOR PARK WAGONWAY AT BATH
Reproduced from Plates 21 and 22 of "A Course in Experimental Philosophy"
by Dr. John Theophilus Desaguliers

would suggest that the experiment was made on a new line, and that the word substitute meant rather a change from normal practice than an alteration to an existing wagonway. The viewer at Whitehaven was Carlisle Spedding, one of the outstanding mining engineers of his period, and therefore there is nothing inherently improbable in the alleged attempt. Correspondence between Spedding and William Brown of Throckley (another leading mining engineer) has survived in the archives[2] of the North of England Institute of Mining and Mechanical Engineers. This correspondence contains a vast amount of information about the construction of wagonways, the erection

[1] Gabriel Jars, *Voyages Metallurgiques*, 1765.
[2] A selection was quoted by T. V. Simpson in "Old mining records and plans," *Trans, Inst. Mining Engineers* Vol. LXXXI, part 1 (March 1931), pp. 75–108.

of fire engines, and other details of mining practice, and has not been studied thoroughly, so far as I know. Should satisfactory evidence ever be forthcoming of the alleged Whitehaven experiment, an unsuccessful attempt will not occupy a place in the main stream of development, although lack of success by a leading man such as Spedding might well have deterred others for some years.

We do not know when iron was first *substituted* successfully for the familiar wooden rail. There is

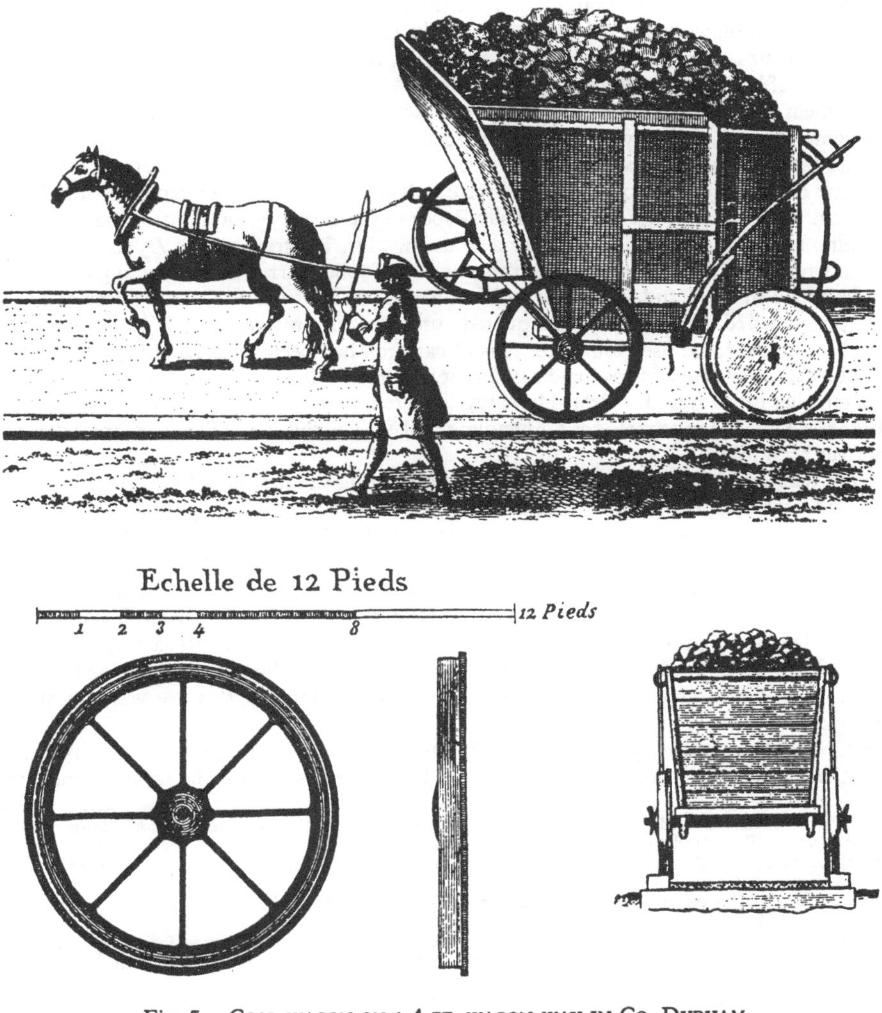

Fig. 5. COAL WAGON ON A 4-FT. WAGON-WAY IN CO. DURHAM

(with enlarged details of the flanged iron wheels) from "Voyages Metallurgiques" by Gabriel Jars (1765)

SOME RAILWAY FACTS AND FALLACIES

good evidence[1] for the first large-scale *use* of cast-iron rails at Coalbrookdale from 1767 onwards, but I submit that it has been widely misinterpreted. Ashton and Sykes[2] in 1929 said "Perhaps the most important step in the development of the railway was the substitution of metal for wooden rails. This, as is well known, was the achievement of Richard Reynolds, who, between 1758 and 1771, took up the wooden rails that had been laid in the 'fifties between Ketley and Coalbrookdale, and replaced them by plates of cast-iron bearing an inner flange to keep the wheels of the waggons in position."

In fact, it was not a substitution, at least initially, if at all, and the rails were not flanged plates. Jabez Carter Hornblower gave a detailed account in 1809[3] of the covering of existing oak rails with flat cast-iron plates, 5 ft. long, 4 in. broad, and 1¼ in. thick as a convenient means of stocking iron during a time of bad trade. When they were required "there was nothing to do but to take them up, and send them away as pigs." He added that the "scantlings of iron were not as those which are now laid in some places." Clearly this could not apply to flanged plates. If they were flanged plates laid on timber longitudinals (which, in any event, would not be a substitution), their removal would destroy the character of the road. If they were independent plates, their removal would eliminate the railway. Dr. Mott in his 1959 Horsehay Paper to the Society[4] saw some of the inconsistencies in the various accounts, but did not offer a solution. Dr. Raistrick, in his *Dynasty*, recorded the facts, but wrongly related them to the legitimate claim of John Curr. Before turning to Curr's work, it may be added that J. Randall's article on "Ironworks" in the *Victoria County History: Shropshire*, Vol. I, pp. 460–476, which Dr. Mott quoted, provided an answer. Randall said that the same wagons "continued to be used long after rails of cast iron had been substituted for the wooden rails, and until a narrower gauge of road and a different pattern of rail were introduced." The date of this change remains to be traced. Probably it was inspired by John Curr, who was certainly advising Coalbrookdale in 1793 as to methods for handling coal.[5] The full story of the Coalbrookdale railways has not yet been told.

Another date which is not known with certainty is that of the first use of the flanged plate rail for a line above ground. Here, again, part of the uncertainty is provided by terminology. John Curr, in *The Coal Viewer* (1797), claimed the invention of the flanged plate rail (Plate IV). He said "the making and use of rail-roads and corves were the first of my inventions, and were introduced at the Sheffield Colliery about twenty-one years ago." Throughout the book, he used the words "waggon ways" or "Newcastle roads" for wooden railways to accommodate flanged-wheeled vehicles, including a line at Froghill in Staffordshire, which had its wooden rail surfaced by cast-iron 1½ in. thick in 1777. Curr used the word "railroad" exclusively for his cast-iron flanged tram plates. His claim to have introduced them in 1776 is probably correct, but this use was almost certainly underground.

In *A General View of the Agriculture of Derbyshire* (1817), John Farey gives much railway information, and of the transition between wood and iron in the Derbyshire area he says: "Some few of the

[1] Dr. W. H. Chaloner and A. Birch, "The first cast-iron rails: New evidence," *The Railway Magazine*, September 1951. This evidence, from Rathbone manuscripts, and much other material previously available, was incorporated by Dr. Arthur Raistrick in *Dynasty of Iron Founders: The Darbys and Coalbrookdale*, 1953, Longmans, Green & Co.

[2] Thomas Southcliffe Ashton and Joseph Sykes, *The Coal Industry of the Eighteenth Century*, Manchester University Press, 1929.

[3] Observations by Jabez Carter Hornblower to the Committee on the Acts now in force, regarding the use of Broad Wheels and on the Preservation of the Turnpike Roads and Highways of the Kingdom. Appendix to the Third Report of the (House of Commons) Committee, 19 June 1809, p. 154.

[4] Dr. R. A. Mott, "The Coalbrookdale Group Horsehay Works," Part I, presented 11 March 1959, *Trans.*, Vol. XXXI, p. 284.

[5] Report by John Curr of Sheffield to Richard Dearman of Coalbrookdale dated 25 May 1793. Exhibited by John W. Hall at the Summer Meeting at Shrewsbury in 1924, and quoted by him in Notes read on 8 October 1924, *Trans.*, Vol. V, p. 1.

earlier Rail-ways laid in or near this county were of wood.... In the use of these wooden Rail-ways, the flanch or projecting rib for keeping the Waggon on the Railway was on the wheel; but now the flanches of iron Rail-ways are almost universally cast on the bars, and the wheels are plain, by which they are fitted for being occasionally drawn off the Rails on common roads. I have heard it said, that the earliest use of these flanched Rails above ground (for they were first introduced in the underground Gates of Mines, it is said) was on the South of Wingerworth Furnace; leading to the Ironstone Pits, by Mr. Joseph Butler, about the year 1788." Later Farey calls this the Wingerworth & Woodthorp Railway, and says it was of 20-in. gauge.

This date of 1788 for the first flanged plate rail above ground may be correct. We have no certain knowledge of an earlier one; and the caution in Farey's remark is in connection with the claim to primacy, not the fact nor the date of the Wingerworth line. Any earlier claim must be treated with great caution. We have noticed already the strong reasons for believing that the flanged plates at Coalbrookdale cannot be attributed to the 1767–71 period. The Catalogue of the York Railway Museum lists "point plate, run-on plate, and crossing of Outram Tramway (1775) from Pickering Lime Quarries, Yorkshire." This item was exhibited at the South Bank Exhibition in 1951, and I then queried the description. No satisfactory evidence was forthcoming. An enquiry to the then-proprietors of Pickering Lime Quarries produced the reply that, according to their records, the quarries were not being worked in 1775, nor until much later. There the matter of this particular exhibit must rest for the moment. I cannot give the answer to the mystery, but I quote it as evidence of the need to exercise caution.

The use of the words "Outram Tramway" in the catalogue entry may have been noticed. Samuel Smiles stated in his 1857 edition of *The Life of George Stephenson, Railway Engineer* that the word "tramway" was derived from Outram's name. He withdrew the statement from later editions. Subsequently this alleged origin was given wide publicity by Clement E. Stretton,[1] and, although both its inherent improbability and actual falsity have been shown on many occasions, the story will probably survive. Between the years 1871 and 1908, Clement Stretton wrote extensively on railway history, mainly during an uncritical period. The mixture of fact and nonsense which he produced has bedevilled the railway historian ever since, and has misled numerous museums, not excepting the Science Museum. We are indebted to him for saving many relics which otherwise would have been destroyed; we are not indebted to him for his excursions into historical guesswork.

The word tramroad made its entry into Parliamentary records in the phrase "Railways or Dram Roads" in 1799. The first use was apparently that in the *Journals of the House of Commons* on 18 February 1799, and thereafter it occurred several times, but by 1800 had become "tramroad." Undoubtedly, Benjamin Outram was the great proponent[2] of flanged plateways, and played an important part in the extension of their use, particularly in South Wales and Monmouthshire, but, in the many uses of the word tramroad in agreements and other legal documents from 1800 onwards, I have never found any indication of a link with Outram's name. Incidentally, the Stretton story is that the conjectural "outram-way" became contracted to "tramway." I have not encountered the use of the word "*tramway*" before 1825.

Stretton credited the invention of flanged plates to "Mr. Outram of the Ripley Ironworks," in some places, and to Benjamin Outram in others, apparently without realising that the former was Joseph, the father of Benjamin. Also, he named the father of Benjamin as "James," and Fred

[1] Clement Edwin Stretton was born on 3 October 1850 and died at Leicester on 20 February 1915.

[2] "Observations of Benjamin Outram on the Brecknock & Abergavenny Canal and Railways," 1 July 1799. Maybery Papers, National Library of Wales. Bundle 14. Benjamin Outram, "Report and estimate of the proposed rail-ways from the collieries in the Forest of Dean, to the Rivers Severn and Wye." Hereford, printed by D. Walker, High Town, 1801.

SOME RAILWAY FACTS AND FALLACIES

Bland repeated the error in a Paper[1] to the Society in 1931. Stretton also advanced some peculiar theories about the origin of the standard railway gauge of 4 ft. 8½ in., and supported them by stories of alleged events which are demonstrably false.

Joseph Outram, father of Benjamin, is reputed to have cast plates at his Ripley ironworks to John Curr's specifications, and therefore the relationship between father and son is not without interest. It seems that, for many years, there was a feud between them, a result of conflict between two determined characters, and Margaret Outram (Benjamin's wife) wrote[2]: "They had not spoken for seven years to each other, before our marriage [4 June 1800]; on learning this, I rested neither day nor night till I reconciled them, and for the five years I was a wife, they never had the smallest quarrel, though I must say, that had I not striven to keep peace betwixt them, war would have broken out, for they had both high and uncongenial spirits, and but for my softening and concealing everything that could displease, they would not have lived in harmony."

An interesting and little-known manuscript, undated, but probably about 1803 or 1804, preserved in the Outram family, is "Minutes to be observed in the Construction of Railways" by Benjamin Outram.[3] In this he uses the word railway for general purpose flanged-plate lines, but says "in Mines and other works underground where very small Carriages only can be used, very light rails are used, forming what are called Tram Roads, on a system introduced by Mr. Curr, and these sort of light Railways have been much used above ground in Shropshire, and other Counties where Coals and other Minerals are gotten." It will be seen that Outram makes no claim to invention, but acknowledges the prior rights of Curr. Also, it lends no support to any suggestion that the word "tramroads" was related to his own name. He was an outstanding engineer in his own right, and it is due to his reputation that false claims should not be made on his behalf. He was a founder in 1790 of the Butterley Company, at first called "Outram & Co." His partners were William Jessop, John Wright, and Francis Beresford. There is an officially-sponsored History of the Butterley Company which in many respects is unworthy of that great undertaking. Presumably in the absence of early books and records of the company, parts of the volume are a hotch-potch of uninspired borrowing from unreliable sources.

William Jessop is possibly the man who introduced the iron rail in fundamentally its modern form. The rail as we now know it is a beam, and its various sections are efforts to dispose the metal most suitably for that function. Flanged plate rails, of course, were also beams, but their early form of simple angle iron was not entirely satisfactory. A web under the tread was soon developed to reduce the risk of fracture, and the top of the flange was later made in the form of a segment of a large circle, instead of parallel to the tread. The flat plates of cast-iron laid on wooden rails were clearly not designed as beams; they were a protection for the timber.

William Jessop's claim seems to be based upon the statement[4] of Alexander Scott of Ormiston, who contributed an important essay to the *Transactions of the Highland Society* in 1824 on the "construction of railroads for the conveyance of ordinary commodities." He said: "The first Public Railway Company seems to have been established at Loughborough in the year 1789 under the direction of the late Mr. William Jessop. Here this eminent engineer introduced the Edge-rail, the upper part of which was of an elliptical figure, with flanges upon the wheels to guide them upon the tracks of the road." This is usually accepted as the introduction of the fish-bellied rail. Unfortunately the line cannot be identified. It has often been assumed to be the Loughborough–

[1] Councillor Fred Bland, "John Curr, originator of iron tram roads," 11 June 1931, *Trans.*, Vol. XI, pp. 121–129.

[2] Mary Frances Outram, *Margaret Outram*, 1778–1863, John Murray, London (1932). The author was great-grand-daughter of Benjamin and Margaret Outram.

[3] Printed as Appendix C to *Margaret Outram*, 1778–1863, *op. cit.*

[4] *Prize Essays and Transactions of the Highland Society of Scotland*, VI (1824), p. 131.

Nanpanton line of the Leicester Navigation Company, and, on this basis, Stretton produced a circumstantial story. He even celebrated the centenary of opening in June 1889. In recent years, research[1] into original records has blown this story sky-high. William Jessop was appointed engineer to the company on 5 July 1791, and the line was opened on 24 October 1794. The next early claimant was the Belvoir Castle Railway, also with William Jessop's fish-bellied rails. Once again we have a Stretton story, and a centenary celebration in 1893; also specimens of rail in museums. The short answer is that a scrutiny of the books in the Belvoir Castle Estate Office showed the line to have been built in 1815.[2] It was last used in May 1918.

We are thus left without a firm date or a certain inventor for the fish-bellied rail. The period of introduction would seem to be the 1790s, but the whole of the available evidence requires re-examination in the light of what we now know to be false.

As I have made various incidental references to the difficulties of terminology, and the early use of various words, mention should be made of the actual word "railway." At the beginning of the seventeenth century we found the word "railes" being used, and at least by 1676 these rails were for flanged-wheeled vehicles. Neither of the composite words "railroad" nor "railway" seems to have entered the language until very much later—"wagonway" was the normal term. Until recently, 1756 was the earliest known for "Railway" and 1774 for "rail-road." The former, which occurs in Abiah Darby's diary, under the date 1 mo. 31st. 1756, still stands, but Mr. Charles Hadfield's examination of canal minute books has now produced "Rail Road" and "Rail'd Road" in 1773 in the Chester Canal minutes. As indicating the period of change, two mentions of 1767 in the Birmingham Canal minutes may be recorded. The first says "Rail or Waggon Ways" and the second "Rail Way." The change seems to have come a little later on the North East Coast, with its long traditions of wagonways. A law report of 1787 uses "waggon-way" only, whereas one of 1798 uses "rail-ways or waggon-ways."

Incidental mention has also been made of the gauge of various lines, so far as it has been recorded. German mining railways, for heavy minerals and man propulsion, were of narrow gauge. The Berlin exhibit is of 48 cm. (1 ft. 6⅞ in.). At Prior Park, the 1734 drawings show a gauge of 3 ft. 9 in. Gabriel Jars gives us both drawings and textual evidence of 4 ft. in 1765. A letter dated 30 July 1754[3] from William Brown, senior, of Throckley, to Carlisle Spedding of Whitehaven concerning wagonways said "we Generally Lay our way of about 4½ Inch by 5 at about 4 foot 2 inches Inside and Inside." This was supported by the writer of the article "Coalery" in the second edition of the *Encyclopaedia Britannica* (Edinburgh), published in 1778, who said that 4 ft. was the common gauge. In the last three decades of the eighteenth century we find some slightly larger gauges being adopted, but without any uniformity. The Beamish Wagonway of 1763 was 4 ft. 4 in.[4]; the Wylam Wagonway (formed previous to 1763) was 5 ft.; and the Killingworth Wagonway (of which the first portion was built in 1762) 4 ft. 8 in.

With flanged tramplates there is a similar lack of uniformity. A difficulty here is that the method of measurement was not consistent. Some measured the clearance between the inner faces of the flanges, and others between the outer faces, as the latter constituted the minimum width of tread. John Curr advocated 2 ft. or less for use in mines, but surface lines were normally about 4 ft. or 4 ft. 2 in. A letter from J. T. Price to Edward Pease, dated 22 December 1818, which is preserved in the Railway Collection at the Newcastle Public Library, states that most of the Welsh tramroads

[1] A. Temple Patterson, "The making of the Leicestershire canals, 1766–1814," *Trans. Leicestershire Archaeological Society*, Vol. XXVII (1951); Robert Abbott, "The railways of the Leicester Navigation Company," *Trans. Leicestershire Archaeological and Historical Society*, Vol. XXXI (1955), pp. 51–61.

[2] Charles E. Lee, "The Belvoir Castle Railway," *The Railway Magazine*, June 1938.

[3] Quoted by T. V. Simpson in "Old mining records and plans," *Trans. Inst. Mining Engineers*, March 1931, p. 96.

[4] Letter from William Harrison to Thomas E. Harrison dated 2 August 1832.

SOME RAILWAY FACTS AND FALLACIES

were laid to a gauge of 4 ft. 2 in. By far the most extensive use of flanged plates was that of the South Wales area. Between 1800 and 1812 some 150 miles of line were built, and the total is stated in 1824 by Cumming[1] to have exceeded 400 miles. Benjamin Outram's Minutes (already quoted) say "the best width of road for general purposes is 4 ft. 2 in. between the Flanches of the rails, the wheels of the Carriages running in tracks about 4 ft. 6 in. as under."

In connection with early suggestions that railways of flanged plates might be used by ordinary road vehicles, it is of some interest to note that the Highway Act 1773 (13 Geo. III, cap. 84, sec. 12) imposes limitations on vehicles "unless the same shall be made and constructed in such manner that no pair of such wheels shall be wider than 4 ft. 6 in. from inside to inside to be measured on the ground." It would seem that only limited use of this road-rail facility was made, and, moreover, the South Wales tramroads speedily produced a special form of narrow-tyred wheel particularly suitable for use on flanged plates, but destructive to ordinary roads. An Act of 19 July 1823 (4 Geo. IV, cap. 95) provided penalties against both owner and driver if "any waggon or cart, built or constructed to be, and usually used on any railway or tramroad, shall be drawn or pass loaded on any turnpike road, out of and away from such railway or tramroad, for the distance of more than one hundred yards."

These are the facts about early railway gauges, as revealed by contemporary evidence. The variations are too great to admit of any theory of common origin. George Stephenson found 4 ft. 8 in. on the Killingworth Wagonway, and used it on the early railways he built, including the Hetton Colliery Railway in 1822, and the Stockton & Darlington Railway in 1825. The last-named retained this figure until 1840, and then it became 4 ft. 8½ in. Other railways of the early steam period varied between 4 ft. 8 in. and 4 ft. 9 in. There was no precise standardisation on the mystic 4 ft. 8½ in. for many years. Moreover, even excluding Brunel's famous 7 ft. 0¼ in., there were wider variations, such as 4 ft. 6½ in. in Scotland from 1831 to 1849; 5 ft. in East Anglia from 1839 to 1849; and 5 ft. 6 in. in Scotland from 1838 to 1847.

Parliament did not interest itself in specifying gauge until 1828. By that time it had incorporated 41 railway companies, beginning with the Surrey Iron Railway Company on 21 May 1801 (41 Geo. III, cap. 33), and had authorised railways (including those of canal companies and private persons) in 104 Acts. Then, in a Liverpool & Manchester Railway Act of 26 March 1828 (9 Geo. IV, cap. 7) authorising diversions, it enacted that "the distance between the inside edges of the rails of the said railway or tramroad shall not be less than 4 ft. 8 in. and that the distance between the outside edges of the rails of the said railway or tramroad shall not be more than 5 ft. 1 in." A similar requirement was included in the Stockton & Darlington (Middlesbrough branch) Act of 23 May 1828 (9 Geo. IV, cap. 60); in the Leeds & Selby Railway Act of 29 May 1830; in the Leicester & Swannington Railway Act of 29 May 1830; and others. It is an interesting speculation as to how such a requirement could be observed with modern rails, particularly on curves. The British Standard flat-bottom rail has a head width of 2¾ in. and a base width of 5½ in.

As a final point, I wish to direct attention to a much more modern feature of railway history, but just within our limit of 100 years ago, that is the subject of some rather surprisingly contradictory statements. This is the introduction of the water trough between the rails to enable steam locomotives, or their tenders, to pick up water while in motion. There are four dates given in reputable places, and one of the false dates will probably continue to be quoted indefinitely. The invention was that of John Ramsbottom, Chief Mechanical Engineer of the London & North Western Railway, and the patent specification was No. 1,527 of 23 June 1860. However, Sir George Findlay, General Manager of the L.N.W.R., stated in his authoritative book[2] that "these troughs were first introduced by Mr.

[1] T. G. Cumming, *Origin and Progress of Rail and Tram Roads*, Denbigh. The preface is dated November 1824.

[2] Sir George Findlay, *The Working and Management of an English Railway*, 1889.

Ramsbottom in 1857." This date was also given in *The Railway Gazette*[1] in 1913, and was attributed to a statement of the late Francis William Webb. In a paper[2] given to the Institution of Locomotive Engineers in 1931 by H. Chambers, the same year was quoted. The year 1859 has been given in various reference books[3] and also in some official publications of British Railways,[4] although others have given 1861.[5] The fourth—and correct—date of 1860 has been used by Wilfred L. Steel,[6] E. L. Ahrons,[7] Lt.-Colonel Harold Rudgard,[8] and myself,[9] among others, but the reader had to make his own choice of which source to accept, in the absence of any quoted contemporary evidence. Statements about the location of the first installation were also contradictory.

The matter has engaged the attention of Mr. Michael Robbins for some years, and the evidence he has adduced[10] establishes clearly the facts that the first trough was laid at Mochdre on the Chester-Holyhead main line (and not at Aber as often stated) in the autumn of 1860; was inspected by the L.N.W.R. directors on 23 October 1860; and was in constant use throughout the severe winter of 1860–61 and thereafter. This installation was removed frcm Mochdre in September 1871 and laid at Aber on 14 November 1871. With the aid of this evidence, the grass-covered earthen embankments which once enclosed the reservoir feeding the Mochdre trough were traced. These earthworks, however, were destroyed by ploughing about March 1958.

Although the water trough is used widely in this country, the adoption of the device elsewhere has been limited. The U.S.A. had a few "track pans" from 1870 until 1956, and two survive on the Western Region of the French National Railways. With the decline of steam traction, the days of the water pick-up trough must be drawing to a close, but at least this useful invention has survived to reach its centenary. It seems probable that the cast-iron troughs now in use at Aber are indeed the original ones, the first in the world, and I have no doubt that they will engage the attention of the astute Curator of Historical Relics of the British Transport Commission when they become redundant.

Throughout the printed version of this Address, you will find numerous annotations indicating the sources of information, to assist those who wish to pursue their researches further, or perhaps to controvert something I have said. You will require no detailed proof from me, however, of the statement that the development of railways in this country has given us unexampled opportunities of assembling both records and relics exemplifying an important feature of our heritage. The efforts to this end of some of the individual railway companies were deserving of high praise, but others showed less enthusiasm. With nationalisation, the British Transport Commission took the enlightened step of appointing an Archivist of Historical Records and a Curator of Historical Relics. The latter is aided (I hope) by a Consultative Panel, with which our Society is associated. It would be difficult to praise too highly the work which these officers have done, and are doing, to aid the study of an important section of the history of engineering. It is our obvious duty to assist them where we are able. In expressing our thanks also to the British Transport Commission, we must not forget that that body is primarily a trading organisation which has been charged by Parliament with the task

[1] *The Railway Gazette*, 23 May 1913, p. 641.
[2] *Journal of the Institution of Locomotive Engineers*, Vol. 21 (1931), p. 450.
[3] E.g. *The Railway Year Book* in all its editions.
[4] E.g. *The Track of the Irish Mail.*
[5] V. S. Haram, *The Irish Mail Centenary*, 1948, pp. 16 and 22.
[6] Wilfred L. Steel, *History of the L.N.W.R.*, 1914, p. 334.
[7] E. L. Ahrons, *The British Steam Railway Locomotive*, 1825–1925, p. 369.
[8] Presidential Address to the Institution of Locomotive Engineers in 1948.
[9] *The First Main-Line Railway*, L.M.S. Centenary supplement to *The Railway Gazette*, 16 September 1938.
[10] *The Railway Magazine*, July–August 1949 and April 1952. The principal evidence quoted is Ramsbottom's paper of 31 January 1861, *Proceedings of the Institution of Mechanical Engineers*, 1861, p. 43; *The Engineer*, Vol. 12 (30 August 1861), p. 121; *The Practical Mechanic's Journal*, Vol. V, second series (1 March 1861), p. 311; entries in the minutes of L.N.W.R. committees; and a letter from Ramsbottom to Lord Chandos, Chairman of the L.N.W.R., dated 17 October 1860.

of balancing its budget. It has undertaken all that could be expected in the way of preserving the evidences of history, and more than many persons thought was likely. Is this properly the unaided task of the B.T.C. however? Could not a National Trust be established, and aided by a substantial Parliamentary grant, as well as donations from industry? This is no new policy. It is well recognised in other spheres, and its adoption in this connection would be but recognition that transport is no mean part of our industrial—and indeed cultural—heritage.

11

The influence of landowners on route selection

Frank A. Sharman

The proposition

Little detailed consideration seems to have been given to railway and canal route selection[1] and when the matter has been considered the chief concern has been with general economic factors, engineering questions, the location of other transport undertakings and the influence of water supply. One possible factor influencing the selection of routes, which has received only scant attention, is the attitudes of the landowners along the proposed line. One writer who has considered the topic is Dr J. H. Farrington, whose conclusion – and it is fair to say that this was not a major finding in his work – was that "the effect of vested interests was to produce minor alterations in the physiographically and economically optimum route. These alterations might be small deviations, pehaps with heavy cutting or embankment"[2]. It is the accuracy of this proposition which will be considered in this article.

General principles

How can one test such a proposition as Dr Farrington's? One way of doing it is to start from the premise that, if a route is planned from A to B then there are factors which define the ideal route; eventually there is the route actually used. The ideal and the actual route may vary and, if they do, one can start to analyse which factors influenced that variation.

A canal or railway can have no primary purpose other than the carrying on of trade between two termini[3]. There is no theoretical reason why the ideal route should not be dead straight and dead level[4]. The factors which cause variations from this in practice appear to be as follows. Firstly, there are technological and geomorphological considerations which indicate the limits of engineering possibility. For example, both a canal and a railway have, in general, a maximum practical gradient of 1:100; on canals, sharp turns are possible, on railways they are not; in the case of canals there is also a water supply factor which dictates the need to find a line with a single summit if that is at all possible[5]. Secondly, there are factors such as the potential competition of existing transport undertakings, the opposition of vested trade interests (such as the town councils who fear a breach of a local trade monopoly) and also the potential benefits of complementary undertakings. Thirdly, there may also be general environmental factors though, except in those cases where railway stations were kept outside cities such as Cambridge, this does not seem to have weighed heavily in canal and railway route design.

Clearly these factors cause variations to be made from the ideal route but it is submitted that, even when they are all taken into account, they by no means leave only one possible line:– a wide variation is still, in most cases and over much of a route, possible. The last remaining factor which can affect the final result is the influence of the landowners.

A proper assessment of the extent of the landowners' influence would need to take into account several matters. In the first place, could landowners affect the route at all; what devices had they by which they could exert influence to vary the line? That question must be dealt with here only briefly[6]. The important point is that if any landowners on a proposed line disapproved of the scheme the promoters would either have to go round him or get compulsory purchase powers by Act of Parliament to force the sale of his land. Parliamentary Standing Orders required prior publicity for the Bill; Parliamentary procedure provided an opportunity for aggrieved landowners to be heard. Therefore, there was a possibility of the Bill's being defeated, drastically amended or seriously delayed by the opposition of the landowners – and the promoters had to allow for them.

The practical effect of the exercise of the landowners' influence can be tested in a number of ways. Some evidence is collected under each of the following four headings.

Landowners in favour

If the affected landowners were all supporters of the railway or canal then, arguably, the actual route taken is likely to correspond closely with the economically, technologically and geomorphologically optimum route. But evidence on this is hard to come by – the easy cases leave little mark on the historical record, and even a willing vendor of land might yet seek some deviation in the route. A detailed examination of the routes of, for example, the Pocklington Canal and the Manchester, Bolton & Bury Canal might prove to be revealing[7] as indeed might an examination of the route of the Thames & Severn Canal where, at the outset, the landowners were in agreement even though land acquisition problems were to arise later[8].

Causing the line to be abandoned

Another test of the influence of the landed interest in whether or not they could prevent a line being built at all. On this there is little evidence, doubtless because if a scheme were abandoned at its earliest stages on account of the opposition of landowners little or no trace of the scheme will now remain in the records. Similarly if a scheme failed because the Bill for it were rejected by Parliament the effect of the landed opposition cannot always be disentangled from a variety of other possible reasons for rejection: Committees of Parliament had an awkward habit of not recording their own debates but merely giving a 'yes' or 'no' answer. Landowners' objections were only one of the matters to be weighed in the balance, and how the balancing was done can only be deduced from secondary evidence which is hard to come by.

But some evidence may be collected. The first Bill of the London & Birmingham Railway Company was rejected by Parliament, and in this case it is quite clear that it was because of landowners' objections. A year later the Bill was passed. The Berkshire & Hampshire Junction Canal Bill suffered exactly the same fate – although in this case the canal was never built, for other reasons. It also seems that a scheme to extend the Kennet & Avon Canal down to Bristol was prevented by landowners' opposition, but a later Act for the same scheme was passed despite the objection that it would be 'in every respect ruinous to private property'[9]; this canal, too, was never built, for other reasons. The landowners staved off legislation for the improvement of the River Don for many years[10]. The Dorset & Somerset Canal Company abandoned the whole of its proposed southern extension to Poole because of landowners' opposition and thereby they lost much of the commercial purpose of the canal which was to link Bristol with the English Channel[11]. At least two of the counsel for landowners opposing the 1825 Bill for the Liverpool & Manchester Railway emphasized to the House of Commons Committee that the rights of private property should only be interfered with in the face of clear evidence of public necessity and practical convenience, and the Bill was defeated, at least in part because of the landowners' objections, though mainly on account of inaccuracies in the deposited Plans and Sections. A new Bill was passed the following year when a new route had been devised which was less objectionable to the landowners[12].

This limited evidence may indicate that landowners could not, in the long term, completely defeat a project that was pushed forward by promoters, was not economically hare-brained and was not along a route which directly effected the property of a large number of influential men.

Varying the line

So to what extent did the promoters have to vary their line in order to get their Bill through Parliament? A number of examples may be examined to give some indication of the extent to which deviations were made.

The following comment, made by the engineer John Smeaton, in his report to the promoters of the Trent & Mersey Canal on the route designed by James Brindley, is quoted by Farrington:

> '[the route] is as little injurious to private property as the nature of the work will admit of, Mr Brindley having judicially designed the course thereof to pass through a great number of level commons and waste grounds and, in general, through the most barren lands that could be found in any wise to agree with the route thereof.'[13]

Of this observation, Dr. Farrington remarks that 'this reference to such an influence on canal routing appears to be unique.' However, it is submitted that the considerations which influenced Brindley's choice of route were such obvious ones as usually to be not worth remarking on; no engineer could afford to pay no attention at all to the expense and opposition his choice of route might generate.

In 1819 in Rees's *Cyclopaedia*[14] an article on canals, thought to have been written by Thomas Telford, describes how the line is surveyed initially. Once a level has been found it is followed 'till some obstacle as a gentleman's park, houses, gardens, orchards, mills, roads, etc. present themselves at a distance': the engineer then sets about finding a way round the obstacle. This clearly reflects two points: firstly the likelihood of the country gentleman's being an effective opponent of the Bill and, secondly, the tenderness Parliament showed for houses, parks and gardens.

In 1831 George and Robert Stephenson advised the London & Birmingham Railway Company of three principles to be observed in planning a route: the need for the lowest relief, the need to avoid expense and the need to avoid parks and pleasure grounds in every practicable case.

In 1831, when the dispute between the landowners and the railway company was at its height, an advertisement inserted in a local newspaper on behalf of the railway company announced that 'The utmost endeavours will be made to avoid all molestation to noblemen and gentlemen's seats near which the proposed line is intended to pass. The railway company has given directions to the surveyors to spare no expense for this purpose. It is always ready and happy to communicate with all parties who wish an alteration in the direction of the railway.'[15] In the end the company had deviations in its line to avoid the parks of both Lord Essex and Lord Clarendon, which deviation necessitated the building of a tunnel 1 mile 57 yards long; further deviations were made to avoid the parks of a Mr Thornton at Brockhall, a Mr Lidingstone of Elmdon, the Countess of Bridgwater, and probably others too[16]. The Earls of Essex and Clarendon had earlier caused similar trouble to the Grand Junction Canal Company. In March 1795 the company paid £5,000 to the Earl of Clarendon for permission to pass through his park, Grove Park, and that sum did not include the price of the land. They also paid the sum of £15,000 to the Earl of Essex to pass through Cassiobury Park and most of that amount was a *douceur*[17]. The Act[18] authorizing this deviation from the previously authorized line was duly passed in the same month as the payment was made.

Other specific instances of deviations can be given. Of the course of the summit level of the Grand Junction Canal at Braunston, Farrington says [19]:— 'At the eastern end it seems that the limiting factor was the large estate of Norton Park. Although reference to such a specific influence has not been found, it may reasonably be assumed that the level might have been continued further south if Norton Park had not necessitated its descent to a level lower down the valley, which enabled it to skirt the park.' The Basingstoke Canal Company went to considerable expense in changing their line to avoid Tylney Hall, taking it through the very long Greywell Tunnel in order to do so[20]. The Oxford Canal contains a remarkable length which comes almost full circle upon itself round Wormleighton Hill; the shorter, direct route would have involved a cutting but, as the engineer Samual Simcock pointed out, it would also be 'much more injurious to the estate of Earl Spence': further, the longer route would need no lock on it so that 'the navigators will have no business to stop . . . so that the apprehended danger from the inroads of bargemen will be the less.'[21]

In 1793 Robert Whitworth, the engineer of the Dorset & Somerset Canal, recommended one of a number of possible routes as the best 'provided the consents of gentlemen, near whose seats the same must necessarily pass, can be obtained', and the line eventually decided upon was amended in various places to allow for opposition[22]. When the Hampshire & Berkshire Junction Canal Company had lost their first Bill, because there was insufficient time for the Commons Committee to hear all the landowners' objections, the company started to prepare their new submissions by considering alterations in the line to reduce the number of objectors[23]. In 1826 the Liverpool & Manchester Railway having lost one Bill the previous year, had a new route surveyed by Charles Vignoles who 'by so skilfully routing his line, was successful in cutting the heat out of much of the opposition's case against the railway. The majority of landed proprietors were appeased . . . '[24]. But this new line, 'because of the route adopted in order to avoid irritation of the landed and waterways interests, . . . had certain decided disadvantages[25]. The Bill, however, was duly passed[26]. In 1857 the Oswestry & Newtown Railway instructed their engineer, in making the survey, 'to take the line south of the Severn between Welshpool and Newtown so as to avoid opposition from a certain Mrs Owen who seems to have been a force to be reckoned with in Mid-Wales railways promotion.'[27] Even when the enabling Act was passed, and the line defined to a greater or lesser extent, variations were still possible. The Warwick & Napton Canal varied their line, not by passing an amending Act, but simply by entering into a deed with the landowner concerned[28].

These examples at least throw some doubt on Dr Farrington's conclusion, but the cases referred to so far mainly concern the choice of line in rural areas. In urban areas there seems to be little doubt that the large landowners did affect the route but in the opposite way. The question has been considered in great detail by John Kellett, who says in his book that:

'it is one of the recurring themes of this book that the underlying pattern formed by units of land ownership is one of the critical factors in explaining the routes cut through Victorian cities by the railway builders, and the choice of site for the termini and yards'.[29]

This conclusion is backed by much detailed evidence which cannot, and need not, be repeated here. Two quotations will suffice. On the siting of stations Mr Kellett says: 'The main consideration in their cities was to achieve the cheapest and simplest approach and terminus, with the minimum of disturbance to property . . . .'[30] The way this was achieved was as follows:–

'For the railways in particular, the prospect of dealing with a few owners offered a simplicity and speed which alone made it possible to carry through large schemes within reasonable margins of time and cost. It was well worth while paying through the nose to conclude such a series of deals privately, and to disarm the opposition. . . . Reaching a settlement with the larger urban landowners had the further advantage of avoiding the endless reduplication of legal and survey charges connected with small cases, the endless delays which arbitration procedure presented and the high awards sympathetic juries tended to give.[31]'

In other words, in the urban areas the railway companies sought out the large estates for their routes, whereas it appears, in the rural areas they tried to avoid them.

Other benefits accruing to the landowner

There were also numerous minor effects of the need to make an accommodation with the landowners. The general point is expressed by the engineer Robert Whitworth when reporting to the Thames & Severn Canal Company on the merits of two lines they were considering. Of the one he says[32] that it can be made 'with very little detriment if any at all to owners or occupiers of the mills or to any other persons whatever' but 'as it goes through a populous vale, and very near several gentlemen's houses, and through some of their orchards and gardens, it will require more than a common neatness in the execution to make it agreeable – and consequently the expense will be more.'

The form of increased expense Mr Whitworth was referring to here was landscaping the canal. This would either be in the form of putting the towpath on the opposite side of the canal from the uphill side preferred by the engineers[33] to lessen the risk of intrusion (as for instance happened on the Kennet & Avon Canal in a number of places), or of ornamental bridges instead of the usual plain ones. This occurs in many places on many canals, some of them, for instance, occurring on the Grand Junction Canal as a result of the agreements with the Earls of Essex and Clarendon for taking the canal through their parks, one of the conditions of which was that the canal be made as ornamental as possible. But expenditure on this form of landscaping could be heavy, especially when it went so far as to involve – as appears to have been more frequently the case with railways than with canals – hiding the line in a cutting or tunnel.

Finally the landowner could often get the benefit of making the promoters buy him off – either by direct bribery, or by the insertion of clauses in the Bill giving privileges to the landowners, or by bribery disguised as greatly increased compensation for land acquired.

Conclusion

It is only possible to draw an interim conclusion from all this but it seems clear enough that the final line of a canal or railway might be quite considerably affected by the landowners along the optimum route. In the final analysis all the constraints which operated on the engineer when he was fixing the line became economic ones. There were very few physical limits on where a canal or railway might go: the real question was whether the costs of the works would make the undertaking unprofitable. The same is true of the landowners' opposition: the real question was whether the deviation was economically worthwhile bearing in mind the increased cost of construction and maintenance and whether the longer route would result in a significantly less efficient transport undertaking. Usually most landowners had their price and the economic question was whether it was worth paying it to get the line through.

Whether or not one agrees with Dr Farrington's conclusion that 'the effect of vested interests was to produce minor alterations' depends very much on what one thinks is 'minor'. One way or another the effects of allowing for landed opposition could lead to the success or failure of the undertaking and in that sense no deviation would be 'minor' to the company concerned: everything contributed to the viability of the scheme. For the engineer the landed opposition was just as important as the other constraints upon him and he had to achieve a balance between them and attempt to minimize the effect of each. The effects of vested interests on the route were, in fact, no greater and no less than the market would allow, and no greater and no less than other economic factors. In physiographical terms, it is submitted this article has at least thrown doubt on whether the effect of landed opposition was no more than 'small deviations' – the effect pervaded the whole route design from the start of planning the route to the finish and from one end of the route to the other. It may be possible to devise a precise quantification of each of the elements influencing route selection but that lies in the remoter realms of historiography.

REFERENCES

1. The chief concern has usually been with general economic factors, engineering matters, the effect of rival and complementary undertakings, water supplies and so forth. A good example of a detailed analysis can be found in J. H. Farrington 'The Leeds and Liverpool Canal: A Study in Route Selection', *Transport History* Vol 3 No 1 March 1970.
2. J. H. Farrington, *Morphological Studies of English Canals* PhD thesis, University of Hull, 1970; published in part under the same title, University of Hull Occasional Papers in Geography No 20 1972.
3. Canals whose primary or secondary function is irrigation or water supply are not of concern here. Canals deliberately built to a meandering line in order to allow of greater industrial access or to serve more agricultural units are not relevant to the argument: each industrial access point or agricultural unit is, for the purposes of this analysis, the desired economic terminus.
4. In highway design there are safety reasons for a dead straight line not being the ideal.
5. J. H. Appleton *The Geography of Communication in Great Britain,* Oxford University Press 1961, pp 1–19.
6. The question of landowners' influence on legislation has been much canvassed in respect of Inclosure Acts but widely disregarded elsewhere. The best published general work is probably Sheila Lambert *Bills and Acts* Cambridge University Press 1971. The present writer has produced a thesis ('The

History of the Law and Procedure for the Compulsory Acquisition of Land' Nottingham University) which deals largely with the question; it awaits publication.

7. See: Baron F. Duckham *The Inland Waterways of East Yorkshire*, East Yorkshire Local History Society 1972, p.60; and V. L. Tomlinson 'The Manchester, Bolton and Bury Canal Navigation', Transactions of the Lancashire and Cheshire Antiquarian Society Vols 75 and 76, 1965 and 1966.
8. Humphrey Household *The Thames & Severn Canal* David & Charles 1969 pp 29–41.
9. Kenneth R. Clew *The Kennet & Avon Canal* David & Charles 1968 pp 54 and 76.
10. T. S. Willan *The Early History of the Don Navigation* Manchester University Press 1965.
11. Kenneth R. Clew *The Dorset and Somerset Canal* David & Charles 1971 p.32.
12. Robert L. Carlson *The Liverpool & Manchester Railway Project 1821-1831* David & Charles, 1969 pp 125, 129.
13. Quoted from the Reports of John Smeaton in the Library of the Institute of Civil Engineers.
14. Rees's *Cyclopaedia* London, 1819 Vol 6.
15. Aris's *Birmingham Gazette* 24 January 1831.
16. Peter S. Richards 'Some Geographical Aspects of the Construction of the London & Birmingham Railway and its influence on the growth of Towns along the Route', MA thesis, London University, 1957.
17. Alan H. Faulkner *The Grand Junction Canal* David & Charles 1972 pp 31-32.
18. 35 Geo 3 c.8.
19. Farrington, Thesis p.267.
20. P. A. L. Vine *London's Lost Route to Basingstoke* David & Charles 1968 p.34.
21. Anthony Burton *The Canal Builders*, Eyre Methuen 1972, p.103.
22. Clew, *The Dorset & Somerset Canal* p. 24.
23. Vine, op cit p.94.
24. Carlson, op cit p.145.
25. Carlson, op cit p.145.
26. 7 Geo 4, c 49.
27. G. Martin Smith, in a private communication to the author.
28. Warwickshire Record Office CR770, Box 5.
29. John R. Kellett *The Impact of Railways on Victorian Cities*, Routledge & Kegan Paul 1969 p.125.
30. Kellett, op cit p.4.
31. Kellett, op cit p.126
32. Gloucester Record Office TS 182/2.
33. Farrington, Occasional Paper p.40.

12

England's first rails: a reconsideration

Richard S. Smith

Few of the really epoch-making developments in technology can be said to be the exclusive product of any one nation, but it is probably true to say that railway locomotion in all its essentials was a uniquely English contribution. The making of special ways of one kind or another for wheeled traffic had been known to antiquity, and Agricola's famous treatise shows that such devices were in use in German mining practice by the sixteenth century.[1] But from the first appearance of the wooden railway in England at the beginning of the seventeenth century until the triumph of Stephenson's Rocket, a straight line of descent can be traced within this country.

As has often been pointed out, all the prerequisites for railway locomotion were worked out in connection with collieries, and it was at collieries that the earliest wooden rails known to us were laid down. These were at Wollaton and Strelley, near Nottingham, Broseley on the Severn, and Bedlington, Cowpen and Bebside on the North-East coast. It is a curious fact that the rails in these three areas, which are the only ones certainly known to us before the Restoration, were all laid down in the first decade of the seventeenth century. Much of what is known about them is to be found in that important source for early technological history, J. U. Nef's *Rise of the British Coal Industry*,[2] and his account has been followed by others in this field.[3] Since he wrote, however, new material has become available on the Wollaton rails, and a re-examination of all the known documentary evidence available for these and the other early rails suggests certain revisions in the accepted story.

Wollaton and Strelley

Since the publication of the Historical Manuscripts Commission's *Report on the Manuscripts of Lord Middleton* in 1911, the rails referred to there in connection with the famous Wollaton colliery have often been quoted as the first to appear in this country, and the date assigned by the *Report* to the reference, 1597-8[4], has been taken as the founding date of the British railway system. In a previous paper[5], I pointed out that this date was incorrect, though the correction is not such as to affect the priority of the Wollaton rails. Another assumption that the rails ran from Wollaton to the river Trent[6] has never been questioned, but this too is not borne out by the evidence.

Unfortunately, detailed records for the Wollaton pits have not survived for the last two decades of the sixteenth century, or the opening years of the next, after which the pits were ' laid down ' for some years. But a few ' Sinking Books ' containing maintenance and capital expenditure accounts for certain years between 1549 and 1580 have survived. As might be expected, they show that the problem of maintaining the routes inside and around the coalfield was an old one. The following extracts are typical of many others:

7 November 1551.	' Paid to men of Radforde for caryeng of ix lodes of pipill stones for the heye waye at the yate. iiis.'
14 November 1551.	'Paid for removying of the yate and settying it in a newe place at the neder feilde it beyng very foule. xiid.'[7]

Curiously, there are a few references in these early books to the mending of the ' rail ' or ' rails ' but in such contexts that it seems virtually certain that the reference is to a rail fence enclosing the coalfield, and not to a wagon-way. Two such entries are the following:

12 June 1551.	' Paid for mendyng the rayle and the yate at the hey (High) field ivd.'[8]
30 November 1554.	' Paid for hangyng of the yate and mendyng of the reyles in to the deye howse close. ivd.'[9]

Such entries are infrequent, and are mostly associated with repairs to the gate into the coalfield. Moreover, during the years 1578 to 1580, when the pits were nearly at the peak of their activity, and very full accounts are available, there are no entries which could be construed as implying a wagon-way. The following entries, however, are of interest in this connection:

8 February 1578.	' Paid to Wyllson and his fellow for dytching of xxiv rode in crosse dytches in the parke for carriage and for a waynewaye. iiis. iiid.'[10]
11 April 1579.	' Paid to Moke and his fellow for vi daies a pece sawing sleepers in Shortwood. ixs.'[11]

As C. E. Lee points out,[12] the new conveyances for carrying coal on rails seem to have been known from the beginning as 'wagons' to distinguish them from the 'waynes' or carts in which the coal had previously been conveyed, and a 'wayneway' implies no more than a specially prepared road for carts. It is tempting to think that the sleepers were being prepared for some such special carriageway, thus anticipating the much later use of the word, and providing a ready-made bed for a railway when the idea should emerge, but it is also possible that the reference is to joists to be used in a colliery building.

The surviving Sinking Books to 1580, then, provide no evidence for a wagon-way at Wollaton, and we may now consider the document[13] to which the date 1598 was assigned by the *Report on the MSS of Lord Middleton*. This account provides a statement of the charge, receipt and delivery of coals 'to all persones, railes and bridges' [i.e. Trent Bridge] for a twelve-month period from October to October, and the year, which occurs three times at the head of the document, is wrongly written '1698'. Francis Willoughby, the naturalist, who worked through the collection with John Ray after the Restoration, spotted the obvious error, and endorsed the document in his unmistakable hand '40 Eliz' (1598). It is this correction that has been followed in the *Report*, but there are several reasons for rejecting it, and preferring 1608 as the date intended. The document is endorsed 'Fosbrooke', the name of a coal factor who was closely associated with Sir Percival Willoughby. There are many references to him in the Middleton MSS., none of which is earlier than 1608. Again, the figure of 13264 rooks, given as the 'Wholl gettes' for the year covered in the document, coincides very closely with an independent figure, 13353 rooks, submitted by the Beaumonts for the year ended October 1608.[14]. No figure has survived for the output of the Wollaton pits in the year 1597-98, but an account for the eleven months ending August 1597 gives output as 6631 rooks,[15] and it is hardly likely that output doubled in the succeeding year. Finally, the document gives details of a delivery of coal to 'Sir Percivall', who was not knighted until 1603.

So much for the negative evidence. What positive evidence is available for fixing the date of the Wollaton rails? I have pointed out elsewhere that a charge of £86 'for the half-cost of the rails' is included in the accounts submitted by Huntingdon Beaumont to

Sir Percival Willoughby for the year ending October 1605.[16] But a careful examination of all the documents fixes the date of construction rather more precisely. Beaumont first appeared in the Nottingham area round 1600, and soon began to make offers to Percival Willoughby for the Wollaton pits. Willoughby's first lease to him in the Middleton MSS. is dated October 1601,[17] leasing the pits in Crow Wood, Wollaton, for three years. A similar lease followed in April 1602 of the coal in 'Bretlawne' (Bretland), Wollaton. Both leases were supplanted by another dated July 1602,[19] which confirmed the previous leases, though on different terms. By this agreement, Beaumont undertook at the expiry of his lease to leave as many horses and engines as were then occupied at the works, together with other houses and implements, but in none of these leases is there any reference to rails.

The indenture of July 1602 also states that Beaumont has taken over the coalmines in Strelley, the parish adjoining Wollaton. This seems to have been a preliminary agreement, for Beaumont's main lease of the coal in the manors of Strelley and Bilborough was dated 2 October 1603.[20] Although the pits in Wollaton were not 'laid down' immediately, the intention from the beginning was obviously to concentrate production at Strelley, and we may assume that as soon as the Strelley lease was safely in his pocket he proceeded to construct his rails across Wollaton. On 1 October 1604 he concluded a new agreement with Sir Percival and his heir,[21] undertaking to pay them half the clear gains from the Strelley pits, whilst they in return confirmed his permission to convey water from Strelley and Bilborough through the sough in Wollaton 'as now it doothe', and also to carry coals through Wollaton 'alonge the passage now laide with Railes, and with suche or the lyke Carriages as are now in use for that purpose.' This indenture, then, becomes the first document in England to make reference to a wagon-way, and proves that the rails were in existence by 1 October 1604.

When the geography of the Wollaton and Strelley area is considered, the reasons for Beaumont's decision to construct his rails becomes apparent. Wollaton no doubt owed its long sixteenth century prosperity to the fact that its section of the exposed coalfield was the nearest to Nottingham and the river Trent. The early Wollaton pits were situated not far from the boundary with Strelley and Bilborough, and the local coal factors would have made their way along the Wollaton Lane, whence it would be only a short distance to the workings. The road to Strelley, on the other hand, involved a considerably longer distance, much of it uphill. When Beaumont transferred his operations to Strelley, therefore, he constructed his rails from the new workings through Wollaton to the Wollaton Lane, where the factors had been accustomed for over a century to collect their coal.

Conclusive evidence that the rails ended here, and not at the Trent, is provided by the agreement drawn up by Fosbrooke and dated 15 June 1609,[22] by which coal was to be stacked ' at Wollerton lane end at the new rayles end.' The advantage of extending the rails to the Trent did not escape Beaumont,[23] but whilst it was not difficult to get permission from his partner, Sir Percival Willoughby, to lay rails through his manor of Wollaton, it was a different and more expensive proposition to obtain wayleave through other people's land, and there is no evidence to show that the attempt was ever made.

A valuable reference to the rails, hitherto unnoticed, occurs in a complaint to Chancery in January 1615 by Francis, brother of Sir Phillip Strelley, who had succeeded to the title to the Strelley estates.[24] Francis maintained that Beaumont was exploiting his lease so vigorously that the coal would be all wrought out or lost. ' And the said Huntingdon Beaumont hath used new and extraordinary invencions and practises for the spedy and easy conveyance of the said coales, and especially by breaking the soyle for layinge of rayles to carry the same uppon with great ease and expedicion the length of two myles or thereaboutes, and by drawinge of certen carrylagges laden with coales uppon the same rayles whereby a hundred and twenty rooke of coales in a weeke att the least being worth twenty poundes are conveyed away uttered and solde more than otherwise there could bee by the ordinarie and customarie course of the gettinge and carryinge away of the said coales.' In his reply, Beaumont made no attempt to deny the novelty of his 'extraordinary invencions and practises,' but claimed that in a confirmation of his lease in 1608, Nicholas Strelley, son and heir of Sir Phillip Strelley, had granted him, for great sums of money, free liberty to lay rails in any part of Strelley and Bilborough, to use rails from pit to pit, and to carry away coals upon the rails ' by horses, wagons, or other carryages.'[25]

Nicholas Strelley does not claim that the two miles of railway has been laid within Strelley and Bilborough, and he is almost certainly referring to the rails laid across Wollaton. The distance from the nearest part of Strelley to the Wollaton Lane is about two miles, and it is interesting to note that this area in Strelley contains clear evidence of bell-pit workings, though they cannot, of course, be assigned to this period with any certainty. There is a gentle slope nearly all the way down to Wollaton Lane, so that the terrain was quite favourable to the introduction of rails.

Although there is no evidence that the Willoughbys had ever handled the marketing of their coal in the sixteenth century, both Sir Percival Willoughby and the Beaumonts, urged on no doubt by their desperate financial needs, attempted to obtain a share of the profits of distribution in the early seventeenth century. The competition this implied was bound to lead to hostility, as is shown in the letters from Robert Fosbrooke, Sir Percival Willoughby's agent in the carrying trade.[26] But for the time being at least, the hostility was kept within bounds, and Fosbrooke could ask his patron, in a passage which clearly shows the advantages and significance of the wagon-way, 'I beseeche you to take order with Sir Thomas [Beaumont] that we maie have libertie to bringe coales downe the rayles by wagen, for our caridages onely, and we will bringe them downe by raile ourselves, for Strelley cartway is so fowle as few cariadges can passe.'[27]

In November 1611, another agreement between Sir Percival Willoughby and Huntingdon Beaumont confirmed Beaumont's right to have passage for his coals ' down the rails through Wollaton ' for the remaining six-and-a-half years of his original Strelley lease, and Sir Percival undertook not to get coal in any of his Nottinghamshire manors for the next six years.[28] It is not possible to say whether the rails continued in use throughout this period since the only reference to the rails after this date, apart from the Chancery petition mentioned above, is in a summary account from Fosbrooke dated 1613-14.[29] Relations between Sir Percival and the Beaumonts became increasingly difficult as his hopes of profit from the Strelley venture faded, and promptly in November 1617 the Wollaton pits were brought back into action. The letter announcing this news to Sir Percival also informs him that ' Our sale for coles is yett but smalle by reason of the fairness of the waye to Stirley [Strelley],'[30] which may indicate that the use of the rails was already being denied. In any event, Sir Percival's agreement to permit the use of the rails through Wollaton expired soon after, and it was now in his interest to forbid the passage of Strelley coal through his manor. The attempt to re-establish the Wollaton pits, however, was not successful, and after several shutdowns, the works were finally abandoned in 1635.[31] Very little is known of the Strelley coalfield after the death of Huntingdon Beaumont in 1624, but it seems that here too disputes over the title to the manor and the ownership of the coal led to intermittent working.[32] It is difficult to believe that the use of rails was ever given up once their advantages had been demonstrated, but there is no further reference to their use in Nottinghamshire in the seventeenth century.

BROSELEY

J. U. Nef was the first to point out that rails were also laid down at Broseley, Shropshire, in the opening years of the seventeenth century. ' At Broseley in 1606,' he writes, ' the colliery owner, James Clifford, had set down ' tylting railes ', ' which were made fast to certen of the grownde.' Along these a ' very artificiall engine ' of timber conveyed coal downhill to the river.'[33] Actually, Nef has conflated the details from two Star Chamber Proceedings, and in the process has given the credit for the introduction of rails at Broseley to the wrong person. In order to correct the picture, it will be useful to summarise each case in turn.

The interrogatories in the first[34] of these actions, in which reference is made to a 'very artificiall engine', were taken in May 1606, which has no doubt supplied Nef's date, but in fact the riotous behaviour to which they refer took place the previous November, 1605. The complainants, Richard Willcox and William Wells, both of whom are described as gentlemen of Broseley, had an estate in two pastures named Callcotts, together with a passage thereto fourteen feet broad. They also had part interest in a coalmine ' in a parcel of ground neere adioyneing unto Calcottes called Birchleasowe . . . And your saide subiectes mindeinge and intendinge not to reape to themselves onelie a privatt commoditie and benefitt by the said delphe and coalemyne, but rather purposeinge the generall good of the whole contrey thereaboutes did about October now last paste [i.e. 1605] cause and procuer to be made and sett up a very artificiall

Engine or Instruments of Timber which stood your subiectes in above twoe hundreth markes, therewith to convey coles from the said ground called Birchleasowes by and throughe other certain growndes parcell of the saide demeasnes of the said Mannor of Broseley now in the possession of your saide subiectes into and through the said Calcottes and soe unto the banck of the River of Severn, from thence to be conveyed by water . . . to the great ease and common good of all the countrie thereabouts which is verie much distressed with wante of fewel.'

But James Clifford who was lord of the Manor of Broseley, and was working pits near Calcotts, gathered together ' sundrie riotous routers and unrulelie persons ' in the night of 14 November and made an attack upon the engine ' as that xx *li.* did not suffice to reduce and bringe the same into such repair and serviceable use as it was before.' No sooner had the damage been repaired than Clifford again rallied his rioters and made another attack upon ' the Engin aforesaide and the railes thereof,' this time causing damage, it was claimed, to the extent of £30, though the attackers had intended, the complainants maintained, " utterlie to distroye the Engin aforesaide which your Subiectes were licenced and authoriside to sett up and make by youer highnes moste honorable privie counsell by speciall direction from them sente unto your highnes Councell established in the Marches of Wales, unto whom the purpose and use of the said Engin beinge made knowne was exceedinglie well liked and allowed of longe before it was finished, in regard it was for the common and generall goode of the whole countrey thereaboutes as before is sett forth.'

Amongst the replies to the complainants' charges, that of the defendants Richard Bassett and John Davis had the audacity to suggest that the damage might have been inflicted by the complainants themselves: ' But it is thought and spoken in the said parish of Broseley . . . that the Complainants or some of his Copartneres of the said Coaledelphe, understandinge that the cariadge of the said coales to the saide River of Severne would have undone and beggered the Complainants and their Coparteners by reason theire charges in digginge pittes, landinge the coales, bringinge them to the said River of Severne . . . with other things verie chargeable belonginge thereunto, would have bine more in value than the money they should have receaved for the same coales at the river syde, did of purpose to prevente that inconvenience and mischiefe cutt and hacke the same Engine and the railes thereof. And this defendant is the rather perswaded to believe the same to be true, for that the railes of the same Engine were not so cut or hackt but that the same served againe by turninge the said railes and layinge the upper syde downewarde. And that xs. or thereabouts (as this defendant hath heard) did suffice to repaire the same Engine, and the railes thereof againe, which if they had bine cutt or hewed by any Adversaries they would by all likelyhoode have so spoiled the same railes as that they should not have served againe.'

One of the reasons for the hostility between the competing mine-owners is given in the reply of Jeffrey Hall and Thomas Haddon, who claimed that so far from the operations of Willcox and Wells being in the public interest, they had led to there being " yearelie gotten more coales than can be transported by water.' These defendants also provide some idea of the length of the rails, for they state that Birchleasowe was a mile from the river bank. It seems that a mile of track at Broseley cost £133. 6s. 8d. (200 marks), whilst two miles at Wollaton, according to the account for 1604-5 cost only £172, but it must be remembered that Willcox had an interest in court in exaggerating the expense he had been put to.

The other action, in which reference is made to James Clifford's ' tylting rails,' was brought by Clifford against Willcox, Wells and others,[35] and interrogatories were taken in the early part of 1608. Clifford described himself as ' seised in his demeasne of and in the manor of Broseley . . . of which mannor there are divers lease-holders and coppyholders for life,' amongst whom Richard Willcox held an estate including the pasture of Calcotts. For some years Clifford had an agreement with Willcox which permitted him to work the coal in Calcotts, to erect houses there, and to have access from Calcotts over other lands in the possession of Willcox. But about two years ago Willcox had leased part of Calcotts to William Wells to prevent Clifford getting coals there. He had then begun an action at common law against Clifford's men working in Calcotts, and, falsely claiming a writ of possession, in ' October last past ' [i.e. October 1607] he had taken forcible possession of Clifford's cottages in Calcotts, and committed other riotous acts including carrying off ' your subjects tylting railes in Calcotts aforesaid being your said subjects moveable property.' Clifford thereupon prepared ' a new paire of tilting railes,' which in turn were carried off by another riotous assembly on 6 November.

The sequence of events is established by the interrogatory administered to the men who had removed the ' tilting rails ' at Wells' orders. They were asked (Item 4) whether on 26 October last Wells had commanded them ' to goe with him unto the Rayles that leadeth from Thomas Haddon's pit to Severne side in Calcotts aforesaid where Robert Prescott with others was then carrying coales to Seaverne side aforesaid, and did you by the appoyntement of the said William Wells there offer to cut the Roope that coales were lett downe withall to Seaverne side aforesaid . . . ? ' Item 5 continues ' weare you or eyther of you the third day of November last in the nighte at the pulling downe and taking away the tilting Rayles placed upon the foresaid Railes ? ' Item 6 asks whether they went upon 6 November last at the command of William Wells ' with divers others unto the Railes aforesaid, and there tooke away other tilting Rayles that were newly placed.' Wells himself admitted that he had ordered his men ' to cutt the ropes wherewith Prescott and his companie did lett the coles downe to Seaverne ', and further that ' he this defendant himself did pull owte the pynnes of the Tilting Railes ' . . . and then gave orders ' to carry the said tilting Railes into the said cottage.' Another defendant, Thomas Wakelam, admitted that he ' did help to carry certen Rayles (which were made fast to certen of the grownde in Calcotts aforesaid) into one of the said cottages . . "

The explanation of the special term ' tilting rails ' is no doubt to be sought in the local circumstances. Clifford's pits in Calcotts were not far from the Severn; in another action,[36] Clifford refers to his ' great charges in wynninge of the coalemynes in Calcottes under the great banckes of Severne in the said parish of Broseley.' The banks of the Severn are quite steep in this area, and Clifford was obviously using steeply inclined rails, down which the coal was wound by a rope to the riverside. For this reason, his rails were distinguished from the ordinary rails along which a wagon was drawn by a horse, (as with Beaumont's rails in the North),[37] or by oxen (as with Willcox's rails previously mentioned).[38]

Bedlington, Cowpen and Bebside

It has long been recognised that it was Huntingdon Beaumont, in his ill-fated venture into coal-mining on the North-East coast, who first introduced rails to the area where they were later to undergo such a significant development.[39] As we shall see, the documents indicate that he was responsible for the construction of three sets of rails in this district, but unfortunately, it is impossible to pinpoint the date when his first railway was laid down. Beaumont's earliest lease in the locality appears to have been an assignment dated September 1605 of the pits in Bedlington from a local gentleman, Roger Fenwick of Bichfield, Northumberland, who in turn held from the Bishop of Durham. This assignment is merely quoted in another document,[40] and the original has not survived, but it is quite likely that it gave Beaumont permission to lay down rails, and that he proceeded at once to the construction of a wagon-way. At all events, in September 1608, when Fenwick had obviously been bought out, the Bishop of Durham granted to Beaumont and his partners a lease for twenty-one years of all the coalmines within the lordship of Bedlington 'with sufficient way leave to digge sincke wynne gett and drawe coales and the same to carrye and recarrye either upon rayles or other carryages.'[41] We may assume that the specific mention of rails means that a wagon way was already in use or was shortly to be laid down in Bedlington.

But it was not only at Bedlington that Beaumont introduced his new device. He also took over pits and saltpans in Cowpen and Bebside, which were Crown property and had been leased to Peter Delavale and Ambrose Dudley in 1594 for 31 years.[42] Presumably Beaumont took over an assignment of their lease soon after his first appearance in the area in 1605, and here he proceeded to lay down two sets of rails, as is made clear in the report of the final overthrow of the works at Cowpen and Bebside. This account, which is dated 1 July 1636,[43] records that Richard Paramore of London and Matthew Sanders of Shankton, Leics., who had taken over from Beaumont and the merchants in July 1614, carried on for two years without any success. At the end of the two years they deserted the works, leaving behind them three boats 'and many and diverse Rayles, Utensells and implements . . . And that after their desertion and leavinge as aforesaid, one Edward Delavale, gent., tooke and carryed away all the said Rayles sett upon the land and ground of Bebside for five hundred paces in the Waggon way on both sides of the way. And that the said John White [and others] as farmers of the said Works and saltpans entred upon the said two keeles and boate and upon the saide Rayles in Cowpen aforesaid sett upon the wagon way of each side of the way for five hundred paces together and upon all the Utensills and implements of what kinde soever by the said Mathew Saunders and Richard Parramore sett up and made about the said works and by them there left as aforesaid, and took them and carryed them away and converted them to their own use.'

The Court ordered that a process should be served on Delavale and White and the others so that they should answer to the Court for the value of the property they had carried off, but I have been unable to find any further reference in the Land Revenue Enrolments. The Court's assumption that the rails left behind by Saunders and Parramore had been laid down by them is a natural one, but there is every reason to believe, as stated above, that they had in fact been constructed by Huntingdon Beaumont soon after his arrival in the district.

The Question of Priority

Perhaps the most interesting aspect of this re-examination is that the gap between the appearance of the first two sets of rails at Wollaton and Broseley is considerably narrowed. As we have seen, the Wollaton rails are probably to be assigned to the year October 1603 — October 1604, whilst the first Broseley rails can be firmly dated to October 1605. Wollaton, then, still has the priority, but only by a short margin, and the claim that before the Broseley rails were laid, the project was ' well liked and allowed of longe before it was finished ' by the Council for the Marches of Wales reduces the gap still further. Unfortunately, neither the Privy Council registers nor those for the Council in the Marches of Wales have survived for this period,[44] so that it is not possible to date Broseley's application for a licence from these sources. There is nothing to indicate that Huntingdon Beaumont sought permission before laying his rails, and it may well have been the uncertain nature of the wayleave from Birchleasowe that was responsible for the Broseley application.

Now that the Broseley rails are seen to follow those at Wollaton so quickly, the question arises whether Broseley copied Wollaton's example, or whether the idea was taken over by both collieries from an outside source. We know that there was an active interest at

Wollaton in Continental pumping devices,[45] and there is evidence in the Middleton MSS. that Sir Francis Willoughby was acquainted with Agricola's great treatise. A recent German writer on the history of technology states categorically ' For centuries past fixed rails on wooden blocks had been common in German mines. When Queen Elizabeth brought skilled miners from Germany to help develop the British mines, they introduced the system into Britain.'[46] It is interesting to note that the German miners who came to Keswick under the auspices of the Company of Mines Royal had in their possession a copy of Sebastian Münster's *Cosmographia Universalis,* 1550, which includes an illustration of a truck being pushed along a narrow-gauge railway.[47] But contrary to Gartmann's confident statement, there is no evidence at all that they laid down rails in connection with their workings in this country.

The reference to Beaumont's ' new and extraordinary invencions ' and the application from Broseley to the Council in the Marches of Wales for a licence certainly suggest that the rails were looked upon as a quite new device at the time. C. E. Lee has further argued that a primitive precursor of the flanged wagon-wheel was an essential part of the invention from its first appearance in this country.[48] He points out that the fact that the carriages used upon the rails were called ' wagons ' to distinguish them from ' waynes ' or carts suggests a special type of vehicle which could not be used away from the track. He admits, however, that the differentiation is not conclusive, and indeed there is a reference in the Keswick accounts already mentioned to ' Wägen ' as well as ' Karren ',[49] although as we have seen there is no evidence to support the existence of rails there. More conclusive is the reference in the Broseley action quoted above to the rails which although hacked about on the upper surface could be turned over and made to serve again. This certainly implies that the wheels were running upon the rail itself, and hence flanged in some way. The early wheels would probably have been made of wood, and it is interesting to note that flanged wheels cut from solid blocks of timber have been found in old workings in the Broseley area.[50]

All the evidence, then, strongly supports Wollaton's claim to the cradling of this important discovery. Wollaton's operation during the sixteenth century had been on a scale not exceeded by any other inland colliery known to us, and there is ample evidence that a considerable body of technical skill had been built up there. It was not only a wagon-way but also boring rods and new pumps that Huntingdon Beaumont took with him to astonish the North. We have seen that rails were not in use before Beaumont's arrival at Wollaton, but the ability was certainly there to solve difficult technical problems, and Beaumont's initiative and energy could be relied on to take advantage of any new ideas. At the same time, with his eyes on the profits to be obtained from large-scale production, he would be unlikely to bother to patent the invention, as one might expect an independent inventor to have done. Once in being, the significance of the wagon-way would have been appreciated in the coal-mining districts of the Midlands, and it need cause no surprise that the example was quickly followed at Broseley, where the local situation particularly favoured the introduction of rails.

For two centuries the wagon-ways were virtually confined to the collieries whose needs had called them into existence, until the advent of steam locomotion enabled them to demonstrate what potentialities they possessed. But the appearance of the device at the beginning of the seventeenth century is significant for more than its future possibilities. During most of the sixteenth century, England had been absorbing what she needed for her rapid economic growth from the more advanced technology of the Continent. In the later decades of the century, native inventors began to show that in many fields Britain was no longer in tutelage. It is surely not fanciful to see in the early spate of projects and inventions, of which the wagon-way is an outstanding example, the background to Sir Francis Bacon's advocacy of the power of applied science to improve the human condition, an advocacy which, as is well known, was to provide the programme for the later Royal Society.

REFERENCES

(1) G. Agricola, *De Re Metallica*, 1556, *trans.* H. C. and L. H. Hoover, London, 1912, pp. 156, 170, 343.

(2) J. U. Nef, *Rise of the British Coal Industry*, 2 vols., London, 1932.

(3) Notably by C. E. Lee, *The Evolution of Railways*, 2nd. ed., London, 1943.

(4) *H. M. C. Middleton MSS.*, p. 169.

(5) R. S. Smith, 'Huntingdon Beaumont: Adventurer in Coalmines', *Renaissance and Modern Studies*, (1958), I, pp. 115-153.

( ) See, e.g., J. U. Nef, in C. Singer *et al. eds.*, *History of Technology*, London, 1957, vol. 3, p. 81.

(7) Middleton Collection, University of Nottingham (hereinafter referred to as 'Midd. Coll.') MiAc 40.

(8) *Ibid.*

(9) Midd. Coll. MiAc. 42.

(10) ,, ,, MiAc. 46.

(11) ,, ,, MiAc. 47.

(12) C. E. Lee, *op. cit.*, p. 24.

(13) Midd. Coll. MiX 4/2.

(14) ,, ,, MiX 3/75.

(15) ,, ,, MiAc 49.

(16) ,, ,, MiX 3/65.

(17) ,, ,, MiX 3/3.

(18) ,, ,, MiX 3/5.

(19) ,, ,, MiX 3/6.

(20) ,, ,, MiX 3/10.

(21) ,, ,, MiX 3/15.

(22) *H. M. C. Middleton MSS.*, p. 172.

(23) R. S. Smith, *loc. cit.*, p. 125.

(24) *Chanc. Proc. Jas. I*, S 27/4.

(25) The confirmation is contained in a document dated 30 April 1608, in which Nicholas Strelley, who had recently married Sir Percival Willoughby's daughter Bridget, makes a settlement of his estate. It apparently granted permission to lay rails in Strelley for the first time. Midd. Coll. 6/170/157.

(26) *H. M. C. Middleton MSS.*, pp. 175-7.

(27) *Ibid.*, p.177.

(28) Midd. Coll. MiX 3/26.

(29) ,, ,, MiX 4/15.

(30) ,, ,, MiX 5/1.

(31) ,, ,, MiX 5/2.

(32) See *Edge MSS.*, Notts. County Record Office.

(33) J. U. Nef, *op. cit.*, vol. 1, p. 244.

(34) *Star Chamber Proc. Jas. I*, 310/16.

(35) *Ibid.*, 109/8.

(36) *Ibid.*, 86/18.

(37) [W. Gray] *Chorographia: or a Survey of Newcastle-upon-Tyne*, Newcastle, 1649.

(38) *Star Chamber Proc. Jas. I*, 310/16.

(39) W. Gray, *op. cit.*; N. Wood, *A Practical Treatise on Rail-Roads . . .*, 2nd. ed., 1832.

(40) Midd. Coll. MiX 3/21.

(41) „ „ MiX 3/23.

(42) *Land Revenue Enrolments*, LR1/203. This and the succeeding reference are from the account in the *History of Northumberland*, Newcastle-upon-Tyne, 1909, vol. 9, pp. 227-8, where the references are incorrectly to the *Land Revenue Miscellaneous Books.*

(43) *Ibid.*, LR1/202.

(44) M. S. Giuseppi, *Guide to the Manuscripts preserved in the Public Record Office*, London, 1924, vol. 2, p. 128; P. Williams, *The Council in the Marches of Wales under Elizabeth I*, Cardiff, 1958, p. 362.

(45) *H. M. C. Middleton MSS.*, pp. 172-5.

(46) H. Gartmann, *Science as History* (*Sonst stünde die Welt still*), London, 1960, p. 36.

(47) W. G. Collingwood, *Elizabethan Keswick*, Kendal, 1912, p. 65.

(48) C. E. Lee, *op. cit.*, p. 24.

(49) W. G. Collingwood, *op. cit.*, p. 87.

(50) *V. C. H. Shropshire*, 1908, p. 454.

13
The Butterley Company and railway construction

P.J. Riden

I

THE FIRM OF BENJAMIN OUTRAM & CO was founded in Derbyshire in 1790 by two men, Benjamin Outram, a local surveyor then engaged in the building of the Cromford canal, and Francis Beresford, a member of a Derbyshire landed family and a promoter of the same canal.[2] The canal was intended to make the rich deposits of coal and ironstone in the Erewash valley between Langley Mill and Pinxton more easily accessible, and to enable coal and other goods to be carried to Cromford, which lies off the coalfield. At Langley Mill it joined the Erewash canal, which in turn joined the Trent near Long Eaton, thus providing an outlet to the midland and southern canal system and to the North Sea. Outram and Beresford were quick to appreciate the opportunity offered by the new waterway and in June 1790 Beresford bought the Butterley Hall estate on the line of the canal near Ripley. He and Outram established a company to mine coal and ironstone on the estate and an ironworks was built to produce pig iron and various types of casting. In 1791 two other men were brought into the partnership: William Jessop, then nearing the height of his career as a civil engineer and at the time principal engineer of the Cromford canal, and John Wright, a wealthy Nottingham banker who in 1791 married Beresford's daughter Elizabeth, and whose ancestors, before the bank was founded in 1760, had been iron merchants.

Stimulated by considerable capital investment, the growth of the canal system and war demand, the ironworks and mines at Butterley expanded, with Outram as manager and sole acting partner. In 1797 the company began working pits on the nearby Codnor Park estate and a few years later leased minerals under Wright's estate at Butterley

Park, between Butterley Hall and Codnor Park. By 1806 another furnace had been built and in 1810 the construction of a second ironworks began at Codnor Park, so that by 1813 there were three furnaces at Butterley and two at Codnor Park. The company also built a puddling forge at Codnor Park, their first incursion into the production of wrought iron on a large scale, which, probably because of the post-war slump, did not go into commission until 1818. A pause in the company's growth is detectable in the years after 1815 until expansion began again in the early 1820s. Large areas of minerals were leased in both Derbyshire and Nottinghamshire and new collieries were sunk some distance from the ironworks. In 1828 a third furnace was built at Codnor Park. By the end of the period under discussion here the Butterley Company, as the firm had become after Outram's death in 1805, now managed by Jessop's son William, were the largest coal-owners in the East Midlands and at the same time among the more important members of the ironmaking community centred on Sheffield.

In their first forty years Butterley produced a remarkable variety of cast iron goods, ranging from 'a heater for a tea urn' to the ironwork for Vauxhall Bridge. Broadly speaking, they specialised in structural ironwork for docks, bridges and canals; pipes for gas and waterworks; and rails and other goods for railways. Of their early products they are probably best remembered today as makers of cast iron rails. The type of rail with which Butterley were most closely identified during this period was the L-section plate rail, first used in England in 1787 by John Curr in a colliery at Sheffield. The following year it appeared at two places on the surface, one of them Joseph Butler's ironworks at Wingerworth, not far from Butterley, and before long it became widespread in many parts of the country, with Benjamin Outram as its leading exponent. After his death others carried on the work until by the 1820s, except in the North East and a few other areas, the plate rail had largely superseded the older edge rail.[3] Outram never claimed to have invented plate rails but he did, as we shall see, regard the plateways he built as an improvement on those of Curr and Butler. The main differences between his plateways and earlier lines were the use of stone blocks to support the rails instead of wooden sleepers,[4] the choice of a much wider gauge than either Curr or Butler, and a corresponding use of larger wagons. William Jessop the elder also built several railways, using both plate and edge rails. Not unnaturally both men turned to the ironworks in which they had an interest for rails for lines they built and, as a result, Butterley acquired a reputation for the production of cast iron rails which outlived both Outram and Jessop (who died in 1814). Thus in 1821, when the Stockton & Darlington promoters were considering what type of rail would be best suited for their line, they sought the advice of Jessop's son William, who despite the ubiquity locally and nationally of plate rail was not enthusiastic:

> we have seen, indeed made the Rails for many in the Neighbourhood of Newcastle, & we can pronounce without any hesitation or doubt, that the edge Rail Ways are very superior to the Tram Roads of this County & Wales, and we should very gladly alter all our own roads to that plan [if it] were not attend'd with so great an expence. The expences of Repairs are less, & they a[llow] of greater Weights being carried on them than Rails [of] the same Weight on the Tram plate & Trams will also do considerably more work. The advantages are so many as not to leave a doubt on the subject and all those who have experienced both will agree in that opinion.[5]

The precise nature and extent of the Butterley Company's predominance, if any, in the market for the supply of cast iron railway goods in the early nineteenth century has until now been the subject of comment rather than investigation.[6] This article attempts to describe and evaluate Butterley's activities as suppliers of rails and other castings to private customers and to canal and railway companies between 1790 and 1830, using, for the first time, the company's own records, the bulk of which are now in public hands.[7] It is not, among other things, a study of Outram or Jessop as railway engineers. The year 1830 is a convenient terminal date, marking the beginning of the modern railway era, which led to changes in the market for railway goods as orders for rails of unprecedented size went not to the smaller inland districts like Derbyshire but to the larger iron-producing areas, notably South Wales.[8]

II

The first railway with which Outram was connected, apart from the earliest lines at Butterley works, was probably that built by the company from their limestone quarry at Crich to the Cromford canal near Ambergate, a distance of a little over a mile, which was opened in 1793 as soon as the canal was finished thus far.[9] In the same year he began to supervise the construction of the Derby canal's plateway from Little Eaton to Smithy Houses, and in 1794 was appointed engineer of the Peak Forest canal, whose line included a railway from Bugsworth to limestone quarries about seven miles away. The Peak Forest's decision in September 1794 to buy rails from Butterley may well have been the first order of any size placed with the firm. The original rails for the Derby line came from Joseph Butler, a point to which we shall return.[10]

In 1796 comes the first indication of the company supplying a railway to a private customer, George Morewood, who owned a colliery near Swanwick.[11] Other lines connected with the growing canal network in the East Midland coalfield followed: at Brinsley for Joseph Wilkes in 1797 and possibly at Measham shortly afterwards; at Beggarlee for Barber Walker & Co (1799); at Pinxton for D'Ewes Coke (1802); and at Belvoir in 1815 for the Duke of Rutland.[12]

Further afield, in 1798 John Lees of Fairfield near Manchester bought a railway for his mill, and in 1808 George Shiffner built the first railway in Sussex at a chalkpit at Offham Hill with rails from Butterley.[13] In the North East at least two waggonways were relaid with Butterley rails: at Benwell in 1809 and at Backworth the following year.[14] The Peak Forest canal ordered further rails from the company in 1798 for its temporary plateway at Marple and in the next few years Butterley executed a number of other contracts for public railways. The Ashby canal began buying rails for its plateway branches in 1799 and a year later the Grand Junction canal, who a few years earlier had bought rails from Butterley for contractors' lines during the building of the canal, went to the company for its Blisworth

Hill railway. In 1801 the London Docks were equipped with railways from Butterley, and in 1802 the Trent & Mersey canal relaid the Caldon Low line as a plateway with Butterley rails.[15] Butterley also built two complete lines for railway companies before 1830: the Croydon Merstham & Godstone Iron railway, authorised in 1803, and the Cromford & High Peak railway of 1825.[16] They quite possibly built other railways between 1790 and 1830 but no record of their construction has survived in those of the company's papers which are at present accessible. In addition to the lines for which they were the sole supplier of rails and other goods, and sometimes the providers of labour as well, Butterley also sold varying quantities of rails to owners of many other railways, public and private.

Outram preferred to have personal supervision of railways of any length which were laid with Butterley rails: 'In works of considerable extent it is most agreeable to me to undertake the whole road to make by Contract or to have the management of the Construction as much depends on the manner in which the work is executed.'[17] On occasion he would only supply rails if the line was built to a satisfactory standard. When the Morewoods were building their railway from near Swanwick to the Cromford canal, instead of placing the work in Butterley's hands from the start, they employed someone else to prepare the track-bed before asking the company to supply and lay the rails. This the company were reluctant to do, as Outram's younger brother Joseph wrote in September 1796:

> [I] have taken the level of your intended Gang road & am sorry to find so much time & expence has been applied to so bad a purpose in forming the ground. The situation admits a gradual descent of nearly 1¼ Inch in a yard which would enable you to bring down all your Coals upon a good Iron gang road at a very easy expence, but instead of finding it to descend gradually some parts of it prove to be near 3 inches at a yard & others in different gradations to a dead Level & some absolutely ascending the wrong way. There the Horse would be overloaded & in the other extream would be overpowered by the push of the Waggons & the Present Road being cut thro' Hills forms a ditch for the reception of the Water which in rainy seasons would render the road wet & dirty. These are great disadvantages & under such a good Road can't be. To remedy these inconveniences the best & cheapest way appears to be to form an entire new road which may be done

for much less than this has cost you. Shall gladly serve you in the best manner we can but should think ourselves highly inexcusable to lay a good road on so ill formed a basis.[18]

Three weeks later Benjamin Outram himself complained that someone had interfered with his brother's surveying and that:

I have been still more surprized this day to receive a Message from Mrs Morewood . . . by which I learn she is persuaded to adopt the Line at present Cut & be at considerable expence to regulate it & wishes us to send the Rails directly to be laid down upon it. I inclose my Brothers sketches of the Line & Sections by which it appears the proposed line will be about 12 yards longer than the present one but will be made complete in less time and for less expence than the latter will cost to alter it so as to answer the purpose to any tolerable degree and after all the present line would be a bungled piece of work neither profitable to Mrs Morewood or creditable to us and if Mrs Morewood is determined to have it done in that imperfect manner I hope she will excuse us for declining to have any thing to do with it & accept out thanks for her kindness in offering us the order for the Castings. Many of the Rails and Waggon Wheels are compleated but we can dispose of them elsewhere for I would rather advise her to use common Carts than be at a great expence to make an imperfect Road . . . I cannot keep considering our Credit as outballancing the profit in a business so near home.[19]

Two days later he wrote again, pointing out certain faults in the road but expressing willingness to do the job on certain conditions. Butterley would provide the plate-layers: 'We have some men laying rails for us at Crich that may come & do yours as soon as ever the ground is ready.' The railway was finally built by Outram and was in use by the beginning of 1797.[20]

Ten years later, when George Shiffner was building his line in Sussex the younger Jessop adopted a similarly intransigent attitude. Jessop supplied Shiffner with detailed plans for the line, which included a tunnel and an inclined plane, on which Shiffner tactlessly sought a second opinion, as well as approaching another ironworks for the rails. Jessop reacted angrily:

It was not without due consideration that the Plan recommended for your inclined Plane was constructed and the experience of my Father & myself induced us to adopt this mode as liable to the fewest objections. But if you be induced to prefer the mode pointed out by Mr. Rand, I can only say that I think I may venture to predict that you will be disappointed in your expectations; and I must beg leave to decline the furnishing any Plans thay may involve me in the discredit which I think will attach to it, but we shall be very ready to execute any Orders for any part of it from Plans which may be furnished by any persons that you may chuse to entrust with it . . . I think it unnecessary, at present, to reply farther to your letter.[21]

Butterley's attitude towards the new plateways for which they were responsible is also apparent in a draft proposal to build the Surrey Iron railway, in which the company undertook that 'the said Railway . . . shall be executed in as compleat and perfect a manner as any of Benn. Outrams improved Railways have here to fore been done.'[22] Most Butterley railway contracts of this period, however, like those of other foundries, included a more concrete guarantee to replace rails broken by fair wear and tear within a limited period of laying down. Thus in 1796 the company sent Joseph Wilkes details of rails for his line from Langley Mill to Brinsley Colliery:

Prices proposed for Cast Iron Rails to be delivered at Brinsley Whf.

Those weighing 18lb to the Yard each Yard 13s 6d per Cwt, 24lb 12s, 30lb 11s, 36lb 10s 6d, 40lb 10s. Any of these rails that break from a defecit in the casting in 3 months to be replaced . . . The 18lb Rails are for Carriages of ½ Ton loading, 24lb for 1 Ton, 30lb for 30 cwt, 36lb for 2 tons to 50 cwt, 40lb for 50 to 70 cwt. Will make good all broken rails of the 3 stoutest sorts for £5 pr Cent addition to the above prices, provided they are properly laid down on a bed of Stone and fairly used with Carriages not exceeding the above weights and not broken by any other means than the passage of such Carriages over them.[23]

The following year John Lees bought rails with the guarantee that 'Any Rails that break within the Year to be exchanged for sound ones at Manchester.' Similarly in the contract for the Marple line rails broken 'by fair use in the first 12 months' would be exchanged. Rails sent from Butterley for the Ashby canal's railway, like those at Blisworth, were guaranteed for three years.[24] On the other hand the company declined to guarantee the rails sent to John Buddle in 1809 for the Benwell waggonway for twelve months, as Buddle had requested, on the ground that the durability of rails depended more on the care with which they were laid than the quality of the iron and Butterley were not to be the plate-layers. They would, however, replace badly cast rails.[25] As the Brinsley tender shows, the company insisted that the conditions be strictly adhered to if the guarantee was to be honoured and the railway be a credit to the makers. When Thomas Walker asked for changes in the pattern of the rails for the Beggarless railway Joseph Outram replied that a:

> great objection is your adding & paying for Metal without strength or service and, when we guarantee for three Years this is material, for by diminishing the strength of a 36lb Rail 4lbs or more, which this would do, would be subjecting us to guarantee the same wt for a 32lb Rail as a 36 ought to carry when a diff[erenc]e of nearly 4cwt ought to be made according to our Calculation, therefore if you insist on your plan being adopted we must either make the Rail 40lbs or reduce the wt we guarantee to carry to 36cwt of 4032lbs avoirdupoise.[26]

With the same letter Outram sent a draft agreement for the construction of the line, which he hoped 'will be noways objectionable. I thought it better to express it as fully as possible to avoid misunderstandings.' These agreements were of a fairly standard form, setting out exactly how the line was to be built and on what terms. The specification was usually closely similar to Outram's recommendations in his 'Minutes to be Observed in the Construction of Railways'.[27] The contract for the Marple railway stipulated that the trackbed was to be 18ft wide, 'effectually drained' and 16ft covered with small broken stone to a depth of 6in. At Blisworth there was to be a bed 24ft wide composed of small stones, gravel and brickbats 12in deep. On the Surrey railway the 18ft wide bed of 'gravel or other durable material' was to be at least 1ft thick. In this case Outram & Co would be responsible for all earthworks while the railway would build the bridges, culverts and soughs, in such a way as not to interrupt Outram's work of laying the track. If there was interruption Butterley would be compensated.[28] The preparation of a solid bed for sleepers Outram regarded as very important: 'I am sorry time would not allow of a Bed of Broken Stones under the Sleepers,' he wrote to Wilkes shortly after the Brinsley line was finished, 'it is of great consequence to the stability of a Road & I hope at Measham you will prepare a bed of small stone about 6 Inches thick.'[29]

On this bed were laid the sleepers, for which Outram wherever possible used stone blocks, apparently his own innovation, instead of transverse wooden baulks. The weights specified, such as 112lb at Blisworth and 150lb on the Ashby line, correspond with those suggested in the 'Minutes'. For the Grand Junction line the blocks were to come from local quarries, while at Marple the blocks and as much of the other stone as convenient was to be got near the canal, 'doing no unnecessary damage thereby', and boated toll-free to the works. The blocks for the Offham Hill railway were also obtained locally, whereas those for the Croydon line came from quarries in Derbyshire at Little Eaton and Cromford.[30] Joseph Wass and Samuel Sims, both of Lea, near Cromford, supplied large quantities of 'rail blocks' to Butterley at 2½d each. The blocks for the Belvoir Castle railway were charged to the Duke of Rutland at 1s 2d each and those for Croydon at 2s, showing how much long-distance water carriage could add to the cost of a railway, since the blocks almost certainly originated in Derbyshire.[31] Wooden sleepers were supplied occasionally, for example to d'Ewes Coke at Pinxton. Here nearly a mile of railway was at first laid on stone blocks but Coke was later credited 3d a block when about 430yd were relaid on wooden sleepers.[32] Once the blocks were in place the space between had to be filled with more stone. A hole 1½in in diameter and 6in deep was drilled in the sleepers into which a plug of 'heart of oak' 5in long was placed. Into this was driven a wrought iron spike which held down the rails. Finally the path between the rails on which the horse would walk was stoned and the railway was complete.[33]

We have already seen that Outram preferred to supervise the construction of railways built with Butterley rails personally and for a number of lines the company supplied at least some of the more specialist labour. In 1805 the Croydon company were charged £4 10s 'Cash paid Block layers expences going to Croydon' and the proposal for the Surrey line specified that Butterley would do the plate-laying. Sometimes the company appear to have provided all the labour, as for example at Belvoir, where the Duke of Rutland was simply charged for 'forming, making & completing the Railway . . . per Contract'.[34] Often part of the contract was let to a sub-contractor, and, like the heavy purchases of stone blocks from local quarries, these sub-contracts show

how the activities of a large company such as Butterley created employment outside its own payroll which would not otherwise have existed. Of the contract to build the London Docks railway in 1801–2 (worth a total of £3,124 10s 2d) over a fifth, £669 18s 5d, was paid to one Bough for 'forming Railways laying do & for Sleepers & Nails'. At Pinxton £85 4s 6d of a contract totalling £1,003 14s was 'paid Edwd. Banks for boating Blocks finding plugs & Nails & laying 1664½ yds of road'.[35] Edward Banks in particular, who did a good deal of subcontracting of various kinds for the company, including the construction of a turnpike road for which Butterley were the main contractor, and who for a few years after 1805 took over the company's entire quarrying and limeworking operation at an annual rent approaching £2,000, must owe at least part of his remarkable rise to wealth and fame to work for Butterley.[36]

On other occasions, Butterley simply delivered rails to a wharf and the customer arranged for them to be laid, as happened at Brinsley and Fairfield. In the case of the Backworth waggonway on Tyneside, far distant from Butterley, the company delivered the rails to John Buddle, who had already made a separate agreement with the colliery to lay the railway. Likewise at Offham, Butterley supplied plans and rails but both the stone blocks and labour were obtained locally.[37] For the Cromford & High Peak Butterley supplied the company with every imaginable item, beginning with surveying equipment and ranging through bricks, cast iron window frames, machinery for the inclined planes (including eight steam engines), to bridgework, as well as rails and wagons, but there is no suggestion of their providing any labour. On the contrary, several accounts appear in the ledger in the years immediately following 1825, for example with Joshua Hodgkinson, Job Jowett, John Upton and James Leaf, all of whose addresses were either Cromford & High Peak railway or Buxton and who bought rails and wheels, which were probably with contractors working on the line.[38] When Butterley were the main contractors, it was generally the subcontractor who provided the nails and various other items for the railway. The company did not supply wooden plugs, wrought iron nails or stone blocks except sometimes for lines built under contract.

Butterley were essentially producers of iron, mainly during this period cast iron, and were thus suppliers of iron rails. For other goods they were content to give work to sub-contractors.

The maintenance of a line appears to have been the owner's responsibility and only in the case of James Fletcher's railway from Ripley Colliery to the Cromford canal, which the company themselves used, are Butterley found repairing lines.[39]

A few accounts give examples of the cost of complete railways. The Pinxton line, computed to be 1,664yd 18in long, with 435yd laid on cheaper wooden sleepers, cost £1,003 14s in 1802, including £178 0s 11½d, about 30 per cent, clear profit. In 1815 Rutland paid £3,272 for the 2 mile railway at Belvoir. In 1801 the company offered to build 16,237yd of the Surrey Iron railway for £29,000; while the extension of 8½ miles authorised two years later cost £41,579. The Blisworth Hill railway, just under 3½ miles long, cost £9,750 for a double road in 1799.[40] In a class of its own was Butterley's account with the High Peak, on which the railway owed £40,000 in 1843, when a private act enabled the line to be mortgaged so that the debt could be settled.[41] Where rails alone were supplied the price was dependent on weight and was ultimately based on the price of pig iron in Derbyshire and Yorkshire, laid down at quarterly meetings of ironmasters held throughout this period. They abandoned any attempt to regulate the price of castings directly in 1809.[42] The cost of rails thus fluctuated between about £8 and £13 a ton at the works or nearby, to which the cost of transport had usually to be added.

During most of the period under discussion here the book-keepers at Butterley described all rails as *gang rails*, a term restricted to the East Midlands but there used almost invariably, until at least 1820, for both underground and surface rails. *Pit rails* then begin to appear for underground track, gang rails continuing on the surface until the 1840s, when *bridge rails* were sent to locomotive-worked railways. The majority of rails made at Butterley before the 1830s were plate rails. The major exceptions locally were the Cromford & High Peak and Belvoir Castle railways, but large quantities of edge rails also went to the North East to John Buddle and John Goodchild. The other local railway company, the Mansfield & Pinxton, also bought edge rails, as did a few lead mines.[43] Chairs, always called *pedestals*, are scarce before 1830, the largest orders coming from Buddle for Benwell and Backworth.[44] The chairs and almost all the rails supplied by Butterley during this period were cast iron; wrought iron rails are not found until the later 1820s and then only infrequently.

John Curr recommended rails 4 or 6ft long for his underground railway and at Wingerworth Joseph Butler's rails were 4ft long.[45] Butterley rails were usually 3ft in length, sometimes 4ft, and very occasionally longer. The trustees of the Bridgewater estate bought some 6ft rails in 1806 and some of the edge rails sent to Goodchild in 1812 were 9ft long.[46] Ledger entries do not generally give any indication of pattern but the sketch (reproduced on p 97) probably shows a typically Butterley plate rail and wheel of the early nineteenth century, as evidenced by several examples recovered from lines for which Butterley made rails. Lighter rails for underground use, such as one found in Codnor Park, may have lacked the base at each end of the rail and have been generally lighter in section; heavier rails, like those for the Croydon line and a contractor's railway at Southwark Bridge, besides being thicker all round, would have had a rib cast on the outer edge of the tread.[47] The edge rails supplied to Rutland, the Mansfield & Pinxton and the High Peak were heavy fish-bellied rails with integral chairs.[48] Sometimes, as in the Southwark Bridge order, Butterley and not the customer were responsible for the exact design of the rail: here Rennie simply asked for rails of a given weight and left Jessop or his foundry manager to devise a particular pattern. They were in fact almost identical to the Croydon rails but rather heavier.[49] John Buddle, on the other hand, was very careful to say exactly what he wanted.[50] Customers sometimes had their initials inscribed on their rails. In 1805 Thomas Pearson and Humphrey Goodwin, who were working Pentrich Colliery, ordered rails marked P&G and, more typically, the Grand Junction, Croydon and High Peak companies all had their rails appropriately initialled. A rail probably originally supplied to William Jolliffe has been found with a partially legible inscription including the word OUTRAM.[51]

We have already seen Outram offering Wilkes rails for his line at Brinsley at weights ranging from 18lb to 40lb a yard, although in his 'Minutes' he recommended a minimum weight of 30lb a yard for surface use. Eighteen-pound rails (or 4ft rails weighing 24lb) are the lightest found in the ledgers and seem to have been sold mostly to mines, suggesting underground use. The next lightest rails weighed about 21lb or 28lb for a 4ft rail. When John Radford wrote in 1797 asking about 4ft rails Joseph Outram told him that 'the weight of them what we generally use weigh from 26 to 28lb pr Rail'. Buddle's edge rails were slightly heavier at 29lb a yard for 3ft rails supported on 7lb chairs.[52] The most popular weights for Butterley rails during this period, presumably mainly for surface use, were between 33lb and 39lb a yard. Wilkes eventually used 33lb rails for his Brinsley line, while those at Caldon Low weighed 34lb. Rails of about 37lb were used by both the Ashby and Grand Junction companies and by Coke at Pinxton, as well as on a contractor's line in Scotland during the building of the Caledonian canal. Also in Scotland the Kilmarnock & Troon railway used 39lb rails.[53]

Later lines, such as that at Belvoir with 40lb edge rail, or those carrying heavier traffic, bought heavier rails. At Marple the rails were to contain ½cwt of iron and be capable of bearing wagons loaded with 5 tons 'but to be commonly used with Loads from 30 to 45cwt each'. The rails at the London docks weighed 42lb, as did those on the Croydon line.[54] The Peak Forest bought rails in the early 1800s weighing 47lb, and in 1812 Goodchild's 9ft rails weighed 200lb. The heaviest Butterley rails of the period were the 72lb plate rails sent to Southwark Bridge in 1816 and the 74lb edge rails on the Cromford & High Peak, supplied in the later 1820s.[55]

Besides ordinary gang rails Butterley made three other patterns. Where a plateway crossed a road it was laid either with rails with unusually low flanges or with trough section rails, over which road vehicles could pass with reasonable facility. These were known at Butterley as *road rails* and generally weighed between 47lb and 50lb on lines where the other rails were in the 33–9lb range. Turn-outs were laid with *cross* and *bevelled* rails; typically cross rails weighed about

75lb, bevelled 55lb. Butterley did not make cross and bevelled rails in which the turn-out diverged from the main line at an angle of more than 30°; for sharper junctions a turnplate, which Butterley could also supply, had to be used.[56] The number of rails supplied to the Grand Junction canal between 1800 and 1804 gives some idea of the quantity and proportions of different types of rail needed for a complete line. The Blisworth Hill railway was a double plateway 3 miles and 720yd long, for which Butterley sent 39,482 gang rails and 2,338 road rails each 3ft long, together with enough cross and bevelled rails to make 125 turn-outs.[57] These extremely generous quantities suggest frequent breakages as well as extensive sidings at each terminal wharf.

Butterley did not at first make complete railway wagons but only the cast iron wheels and sometimes the axles, leaving the underframe and body to be added by someone else chosen by the customer. In their early days the company did not even bore their own wheels or fix the axles: 'we have several sets of Waggon Wheels cast that will be suitable for your road,' wrote Outram to Mrs Morewood during a temporary reconciliation, 'and if you chuse I will order the man who bores them and makes the Axles for us (J. Fox of Derby) to make Axles and fit them up for you'. James Fox of City Road, Derby, established his machine tool business in 1785 and later became famous for quality of workmanship and advanced design.[58] The company later had their own boring and turning mill. In 1796 Outram offered Wilkes 'Waggon wheels on an improved construction 14s Pr Cwt. Fluted Axles to work without Brasses or Bushes of any sort at 8d pr lb.' Four years later the price had risen to 16s a hundredweight for a wheel 18in in diameter, the size most commonly supplied by Butterley at this time, and 21s a hundredweight for wheels bored from solid metal, with wrought iron axles a further 7d a pound.[59] In 1801 Joseph Outram described a typical Butterley wagon: 'Those we usually make are 4ft 2in wide, the beds for the Axles 3 feet asunder from Centre to Centre and the length of the Bodies or Boxes 5ft 6in, if for carrying blocks of Stone only they may be too large and unwieldy but if for a variety of materials you cannot have a better size than the above mentioned.'[60] A few years earlier, in May 1795, as the Derby canal's railway was nearing completion and the local colliery owners were thinking about the type of wagon they would use on the line, Benjamin Outram wrote a most interesting letter to John Radford, William Drury Lowe's agent at Denby Colliery, mentioning a visit to Coalbrookdale made at the instigation of the Peak Forest canal committee:

> I have examined the Model for the Wheels on the light construction and will cost them of the best soft Metal, pay all carriage, stand the risque of Boring & deliver them when bored at Smithy Houses for 13s pr Cwt, you paying the expence of boring. I nevertheless am fully persuaded the other will be found the best. I have seen the Railways near Coalbrookdale last week where they have every sort of Waggon Road that has been put in use and I am convinced ours is preferable to any of them, the best they have adopted are small waggons carrying abt 15 Cwt, on Cast Metal Roads formed with light Rails, the Wheels are low (abt 12 or 15 In diameter) have each a separate Axes which turned and runs in Brasses. The Wheel has also liberty to turn round on the Axes, to avoid Friction as much as possible, but an Horse seldom draws more than 3 Tons, whereas I believe if you adopt Waggons of the best construction to carry 2 or 3 Tons each one Horse will take 6 Tons down and bring the waggons back, and Waggons necessary to contain 100 tons will be made for less expence on the large than on the small construction. The strong Waggons will take up much less room, on the Pit bank, the Road, & the Canal Wharf, indeed our passing places will be inadequate to pass a great trade on small waggons without hindrance to the numbers that would be employed, and as the Road is capable of carrying the large ones I cannot see what advantages can possibly arise from the use of the small ones, the small Waggons are usually made with low wheels which as above mentioned cannot readily be strained so much as to get off the Rails, the model now made is so high that if the Wood Axis gives way at all (and that must be expected after a little use) they will be liable to get off the Road. Waggons made without a Bolster to turn will be found very Inconvenient, at the Turns of the Road and at the Wharfs, this applys to both large and small, the large if made without Bolster would be made for considerably less expence, on the whole if a fair tyral is made of a dozen Waggons of the Large size, and they are not found to do more business and for less expence than any now in use I will take them to myself. I think after what I had before said of taking the risque on myself they should be fairly tryed & not given up on the opinion of any individual who never saw such in work. The comparisons hitherto made have between the large old illconstructed Waggons working on Wood, and small ones better constructed working on Iron, small ones are much more applicable to short private Roads than long public ones on acct on the Rails, but I believe they are not the best for Mr Lowe.[61]

On their own line at Crich the company used wagons made wholly of wrought iron to withstand the violent loading and unloading of limestone by mechanical tipplers, and these were made entirely at Butterley. They were, on occasion, supplied to outside customers, for example John Goodchild of Sunderland in 1812 at £27 each.[62] After about 1820 wagons feature more prominently in the ledgers, always as *gangwaggons* and usually at prices between £12 and £25, which suggests the company were now making complete wagons themselves. The wholly wrought iron model, which the company also used for carrying coal and ironstone, seems to have been the most popular.

Finally, Butterley's railway customers before 1830 bought various special items. In 1802 Anderson Eades & Co bought a 'tippling frame' for 12 guineas, presumably some kind of shute for unloading wagons such as Farey saw at Crich, and in 1810 the Gloucester & Cheltenham railway bought three 'machines' of an unspecified nature for £33 12s each.[63] Several customers bought equipment for either self-acting or steam-operated inclined planes, the best known being the series of nine for the High Peak. Besides stationary steam engines, Butterley also made two early locomotives, those designed by William Brunton and William Chapman, both in 1813. Brunton's, which was tried on one of the company's own lines since he was at the time their engineer, appears in the ledger as 'Horse made for Crich Railway' £240 and another of the same type which went to Newbottle Colliery as 'do for Nesham & Co'. Chapman's 'locomotive steam Engine fitted complete' cost £480 11s 2¾d.[64]

III

From the Butterley Company's surviving and accessible records it is possible to identify nearly a hundred different customers for railway goods between 1790 and 1830, besides the company's own pits and works. Of these about a dozen were buying rails or other equipment for public lines. The rest were mostly mine owners, or in one or two cases the owners of quarries or a mill. A few were contractors using temporary lines during the construction of some public works.

All but a handful of the rails for private lines, apart from contractors' railways, went to Derbyshire, Nottinghamshire or Leicestershire. However famous Butterley may have been they could not compete with local ironworks for castings for private railways. Areas with a large mileage of iron railways, such as east Shropshire or South Wales and with a plentiful supply of ironworks had no need to go to Derbyshire for their rails. An interesting contrast is Shiffner's line in Sussex, where a London firm, Bailey Ward & Co, offered rails at 10 guineas a ton against Butterley's £13 but in the end Shiffner went to Butterley, presumably placing some value on their reputation.[65] Within the East Midlands, although Butterley sold rails to a large number of collieries and other railway owners, as well as building several of the best-known private lines, their importance can be exaggerated. As Farey found in 1805, most collieries of any size had a railway to the nearest canal and almost every pit used rails underground.[66] The mileage was so great that no single firm could have had any kind of monopoly. Furthermore, an important part of the East Midland coalfield, the Rother valley in north-east Derbyshire, was entirely untouched by Butterley. This is partly explained by the presence of several ironworks around Chesterfield; partly also by the absence of water communication between the Erewash and Rother valleys, a difficulty which despite several attempts at least one of which the company supported financially was never overcome.[67] A district twenty miles from Butterley was as much *terra incognita* as South Wales; their market for private railways was not merely almost entirely local, it only faced south from Butterley down the canals and along the Trent. It also diminished noticeably after about 1815. As the East Midland canal network was being built the local colliery owners laid railways from their pits to the nearest canal. This ended with the Belvoir Castle line. After 1815 we know of no other private railway being built by Butterley and the sharply reduced volume of sales of rails between then and 1830 may be accounted for in part by a dearth of new lines in the coalfield. Local collieries continued to buy rails but only in small amounts, presumably for replacements and minor extensions. The fall in sales may also reflect an increase for other ironworks in the district, such as

Alfreton or Morley Park. Alternatively, Butterley, chiefly a producer of cast iron, may have lost ground towards the end of the period to other local companies better able to produce the wrought iron rails then coming into fashion.[68]

About sixty railway companies were incorporated between 1801 and 1830, although several of the lines authorised were never built. They were mostly promoted in two periods either side of the decade 1813–23, during which only a handful of acts were passed.[69] Of the two incorporated before about 1813 which bought rails from Butterley (the Croydon and the Kilmarnock companies) the elder Jessop was engineer to both, just as he was also engineer to the Caledonian canal, their only other Scottish customer.[70] Almost all the other railways were in or around South Wales and the border. As with private lines it is clear that although Butterley could not compete with the South Wales ironmasters on their home ground, in the neutral territory of Surrey or southern Scotland they evidently had a sufficient reputation to persuade the new companies, as they persuaded George Shiffner in Sussex, that their rails were best. It seems reasonable to suppose that Jessop may have helped them to decide to buy from Butterley, and it is interesting to note by comparison that whereas the South Wales railway owners willingly followed Outram's advice and converted their lines to plateways he could not, since they were virtually all iron masters themselves, persuade them to relay with Butterley rails.[71]

Those English canal companies which also owned railways were more susceptible to Outram's recommendations. As engineer to the Peak Forest canal, he laid two plateways with Butterley rails. Called in to advise the Ashby company on railways, he secured the contract for the rails for his own ironworks. Another engineer, John Rennie, also placed several large orders for rails, and other ironwork with Butterley, including those for the London Docks railway, the Caldon Low plateway and a temporary line at Southwark Bridge.[72] Most canal-owned railways were built in the 1790s or 1800s and during these two decades Butterley can be seen to be suppliers of rails of national rather than local importance, apparently with few challengers outside districts with their own iron industry. However, certain qualifications are necessary.

It may be unwise to view the period in question as a unity. Butterley were expanding throughout these years and were at first a comparatively small concern. In 1799 Outram told the Ashby canal that an order for the rails for their line would 'almost wholly occupy' the works for fifteen months, and indeed it appears to have overstrained their capacity, since at one stage Butterley bought 927 rails from Joseph Butler for the Ashby contract and passed them on at cost. Butler also, as we have seen, supplied the original rails for the Derby canal's plateway, as well as sharing the contract for the Leicester Navigation's railway in 1793, engineered by Jessop, with another Rother valley ironmaster, Ebenezer Smith of Chesterfield.[73] It may be that Joseph Butler, whose line at Wingerworth predates the one at Crich by five years, and about whom very little is known, was briefly as important a supplier of rails to canal companies as Butterley were later when they had grown bigger and more famous, and Outram had been recognised as the foremost engineer of his day. The complete absence of Butterley ironwork from the Rother valley may conceal an influential position for Butler, who owned three railways and widespread integrated coal and iron interests in the district. Butler sold no further rails to the Derby canal after 1793, which instead bought rails or sought tenders from Butterley, Claytons of Nottingham, Mold of Morley Park, Wheeldon's foundry in Derby, Saxlebye's furnace at Alfreton, and Hurt's at Alderwasley, or in other words, from nearly every ironworks in and near the Erewash valley.[74] The fortunate survival of early Butterley ledgers and the complete disappearance of similar records from other local ironworks inevitably tends to exaggerate the position of Butterley until evidence from the company's records is tested against outside evidence such as the Derby canal minutes.

In the railway speculation of the middle and late 1820s, Butterley played little part. The successful schemes of these years were mostly in South Wales, Scotland or the North West, and not even in Lancashire could Butterley make any impression. Their only railway contract between 1820 and 1830 was admittedly for the line with what was then the largest authorised capital of any British railway and was potentially very profitable but, however spectacular it may be today as a monument to Butterley and to Josias Jessop, the Cromford & High Peak was at the time a massive loss. It was technologically backward in its use of inclined planes and cast iron rails and was atypical of the railways being promoted around 1830, none of which bought rails from Butterley. It is tempting to suggest that the contract for the High Peak went where it did largely because the engineer was the brother of one of the partners at Butterley and the other partner, Wright, was the largest shareholder in the railway.[75] It is a little odd that Butterley were not the contractors for the other local railway company, the Mansfield & Pinxton, authorised in 1817, of which again Josias Jessop was engineer and Wright a leading promoter. Here the contract for the rails was split between Butterley, Oakes and several small ironworks, and for no clear reason Butterley refused to execute their share of the job, although they had previously supplied rail patterns to the other ironworks concerned. Only in 1823 did Butterley first sell rails to the company.[76]

In conclusion, it cannot be denied that Butterley occupied an important, possibly uniquely important, position in the railway scene between the introduction of plate rail and about 1815. Their founder, Benjamin Outram, was the leading exponent of plate rail. He and the elder William Jessop were ideally placed to secure work for their ironworks and appear to have taken every opportunity. Locally, the company was probably the largest single supplier of rails in the East Midlands outside the Rother valley, if only because by 1815 they were the largest ironworks in the area. Nationally, even though they could not compete in parts of the country with their own local iron industry (and although until the activities of some of the larger South Wales iron companies have been more fully investigated it would be unwise to be more dogmatic) it appears that Butterley secured more important contracts for public railways than any other ironworks in the country. After about 1815, sales to private railway owners and canals fell as fewer new lines were built, and when railway promotion recovered in the 1820s Butterley were no longer the nationally famous firm they had been when Outram and the elder Jessop had been alive.

St Edmund Hall, Oxford

References

1 I am much indebted to V. A. Hatley Esq, M. J. T. Lewis Esq, P. Stevenson Esq, Messrs Butterley Building Materials Ltd, Messrs The Derby Canal Co, Derby Borough Library (DBL), East Sussex Record Office (ESRO), Derbyshire Record Office (DRO), the North of England Institute of Mining and Mechanical Engineers (NEIMME), the Public Record Office (PRO, BTHR), and Sheffield City Library (SCL) for help with this work.

2 The early history of the company is more fully described and documented in P. J. Riden, 'The Butterley Company 1790–1830' (Oxford BA dissertation, 1973).

3 M. J. T. Lewis, *Early Wooden Railways* (1970), 293, 318.

4 The use of stone blocks has not been traced earlier than 1793 at Crich. I owe this point to Dr Lewis.

5 M. Elsas (ed), *Iron in the Making. Dowlais Iron Company Letters 1782–1860*, Cardiff (1960), 172–3.

6 Eg, B. Baxter, *Stone Blocks and Iron Rails* (*Tramroads*), Newton Abbot (1966), 44; M. J. T. Lewis, op cit, 292.

7 DRO, D503. The pieces used here and cited without additional reference are Furnace Ledgers B, C and D and a file of loose papers, D/103. In addition Mr Stevenson has kindly made available a transcript of a company letterbook (here cited as Letterbook) covering the years 1794–1801, of which the original is not at present accessible.

8 G. R. Hawke, *Railways and Economic Growth in England and Wales 1840–70*, Oxford (1970), 239.

9 J. Farey, *A General View of the Agriculture and Minerals of Derbyshire*, ii (1813), 422.

10 C. R. Hadfield & G. J. Biddle, *The Canals of North West England*, Newton Abbot (1970), 307; DBL, Derby Canal Co Minutes, 23 December 1793.

11 Letterbook, 1796 passim; Farey, op cit, iii (1817), 338.

12 Letterbook, 9 May 1797, when Outram offers to supply rails for Measham; ibid, November 1798; Ledger B, 101, 733.

13 Letterbook, 10 March 1798; R. M. Robbins, 'The First Sussex Railway', *Ry Mag*, July 1971, 355–7.

14 NEIMME, Buddle Coll, xxiii, 55–85 passim; iv, passim (Benwell); xxiii, 106–15 (Backworth).

15 Hadfield & Biddle, op cit, 308 and Ledger B, 25; Letterbook, 24 July 1799 and Ledger B, 12; Ledger B, 4ff; ibid, 49; J. Lindsay, 'The Butterley Coal and Iron Works 1792–1816', *Derbys Arch J*, lxxxv (1965), 34, and Ledger B, 111.

16 Ledger B, 155ff; Ledger D, 95ff. A draft proposal to make the Surrey Iron Rail-

The Butterley Company and Railway Construction, 1790–1830 51

way, 1 June 1801, survives in D/103 but since there is no account in Ledger B for sales of rails to the company and the line is not among those listed by Jessop in Elsas, op cit, 172, it seems unlikely that the proposal was accepted.

17 Letterbook, 8 March 1798.
18 Ibid, 3 October 1796.
19 Ibid, 22 October 1796.
20 Ibid, 14 October 1796 and 11 January 1797.
21 ESRO, Shiffner MS, 1967.
22 D/103, draft proposal to make Surrey Iron Railway, 1 June 1801.
23 Letterbook, between April and September 1796.
24 Ibid, 1 January 1798; ibid, January or February 1798; C. R. Clinker & C. R. Hadfield, 'The Ashby-de-la-Zouch Canal and its Railways', *Trans Leics Arch & Hist Soc*, xxiv (1958), 63; V. A. Hatley, *Rails Over Blisworth Hill*, Northampton (1970), 4.
25 NEIMME, loc cit, xxiii, 25 March 1809.
26 Letterbook, 30 November 1798.
27 B. Outram, 'Minutes to be Observed in the Construction of Rail-ways', *Recreations in Agriculture*, iv (1801), 473ff.
28 Letterbook, January or February 1798; Hatley, op cit, 4; D/103, loc cit.
29 Letterbook, 9 May 1797.
30 Hatley, op cit, 4; Clinker & Hadfield, loc cit, 63; Letterbook, January or February 1798; ESRO, Shiffner MS, 1972; Ledger B, 155.
31 Ledger B, 78, 79, 133 and 155.
32 Ibid, 101.
33 Outram, loc cit.
34 Ledger B, 235; D/103, loc cit; Ledger B, 109.
35 Ledger B, 80 and 101.
36 Riden, op cit, passim.
37 Letterbook, 9 May 1797; ibid, 1 January 1798; NEIMME, loc cit, xxiii, 106; ESRO, Shiffner MS, 1961–73, passim.
38 Ledger D, 95ff, 247, 225, 151 and 280.
39 Ledger B, 538; Farey, op cit, iii, 338.
40 Ledger B, 101; ibid, 109; D/103, loc cit; Ledger B, 259; Hatley, op cit, 4.
41 7 & 8 Vict, c xviii.
42 SCL, Jackson Coll, 1297, 19 July 1809, et passim.
43 A. Rimmer, *The Cromford & High Peak Railway*, Lingfield (1967), 11; C. E. Lee, 'The Belvoir Castle Railway', *Ry Mag*, June 1938, 392; Ledger B, 573 and 687; J. A. Birks and P. Coxon, 'The Manfield & Pinxton Railway', *Ry Mag*, July 1949 and Ledger D, 279. When Farey (op cit, iii, 288) described Wilkes's line at Measham, for which Butterley may or may not have supplied the rails (n12 above), it too had edge rails. When a deputation from the Grand Junction Canal committee visited both Measham and Brinsley in August 1799, however (*The Repertory of Arts and Manufactures*, xiii (1800), 167–9), Measham was in no way distinguished from the plateway at Brinsley, which, had it been laid with edge rails, it surely would have been.
44 Ledger B, 573 and 717.
45 J. Curr, *The Coal Viewer and Engine Builder's Practical Companion*, Sheffield (1797), 23; Farey, op cit, iii, 289.
46 Ledger B, 290 and 687.
47 P. J. Riden, 'Plate Rails at Codnor Park', *Ind Arch*, X no 1 (1973); C. E. Lee,

'Early Railways in Surrey', *Trans Newc Soc*, xix (1940), 72; D/103, W. Jessop to G. Goodwin, 16 June 1816.
48 Lee (1938), 393; Birks & Coxon, loc cit; Rimmer, op cit, 11.
49 D/103, W. Jessop to G. Goodwin, 16 June 1816.
50 Information from Dr Lewis, quoting NEIMME, Buddle Coll.
51 Ledger B, 268; Hatley, op cit, 9; Lee (1940), 75; Rimmer, op cit, 11; K. W. Bean, 'Plate Rails at Godstone', *Ind Arch*, vii (1970), 188.
52 Letterbook, 3 June 1797; NEIMME, loc cit, xxiii, 106.
53 *The Repertory of Arts and Manufactures*, xiii (1800), 168; Lindsay, loc cit, 34; Ledger B, 12, 4, 101, 204 and 526.
54 Lee (1938), 393; Letterbook, January or February 1798; Ledger B, 49 and 155.
55 Ledger B, 22 and 687; D/103, Jessop to Goodwin, 16 June 1816; Ledger D, 95.
56 Letterbook, 30 April 1801.
57 Hatley, op cit, 5; Ledger B, 4, 13 and 18.
58 Letterbook, 24 October 1796; F. Nixon, *The Industrial Archaeology of Derbyshire*, Newton Abbot (1969), 128–30.
59 Letterbook, between April and September 1796; Ledger B, 7.
60 Letterbook, 30 April 1801.
61 Ibid, 1 May 1795; Hadfield & Biddle, op cit, 307.
62 Farey, op cit, iii, 338; Ledger B, 687.
63 Ledger B, 233 and Farey, ibid; Ledger B, 546.
64 Ledger C, 64; Ledger B, 715; C. F. D. Marshall, *A History of Railway Locomotives down to the end of the Year 1831* (1953), 61ff.
65 ESRO, Shiffner MS, 1964, 1967, 1968.
66 Farey, op cit, iii, passim.
67 Ledger B, 454.
68 I owe this suggestion to Dr Lewis. Although some sales of wrought iron appear in the Furnace Ledgers there was almost certainly a series of subsidiary Forge Ledgers in which most such sales would have been posted. In the absence of these it is impossible to draw definite conclusions on this point.
69 J. Priestley, *Historical Account of the Navigable Rivers, Canals, and Railways of Great Britain* (1831).
70 C. F. D. Marshall, *A History of British Railways down to the Year 1830*, Oxford (1971), 77, 133; Priestley, op cit, 129.
71 Lewis, op cit, 293.
72 S. Smiles, *Lives of the Engineers*, ii (1861), 198; J. R. Hollick, 'The Caldon Low Tramways', *Ry Mag*, June 1937, 435; D/103, Jessop to Goodwin, 16 June 1816.
73 Clinker and Hadfield, loc cit, 63; Ledger B, 12; R. Abbott, 'The Railways of the Leicester Navigation', *Trans Leics Arch & Hist Soc*, xxxi (1955), 51ff.
74 DBL, Derby Canal Co Minutes, 14 February 1803, 1 May 1804, 10 December 1804, 19 August 1805, 11 August 1807, 8 September 1807, 8 August 1809 9 August 1814, 13 September 1814, 6 July 1824.
75 Hodkins, loc cit, 42, 46.
76 G. G. Hopkinson, 'The Inland Navigations of the Derbyshire and Nottinghamshire Coalfield, 1777–1856', *Derbys Arch J*, lxxix (1959), 29; PRO, BTHR, MPX 1/1, 22 August 1817, 22 September 1817; Ledger C, 220, 279.

14

Cast iron edge-rails at Walker Colliery, 1798

A.W. Skempton and A. Andrews

Introduction. In his *History of British Railways down to the year 1830* (Oxford, 1938) Dendy Marshall draws attention to undated drawings in the Boulton & Watt Collection showing cast iron rails and stone sleepers for Lawson Main waggonway, and correlates the drawings with a statement by Nicholas Wood in *A Practical Treatise on Rail-Roads* (London, 1825 p.46) that 'the late Mr Barns' employed stone supports for the rails 'in forming the first iron Rail-road which was laid down in the neighbourhood of Newcastle-upon-Tyne, viz. from Lawson main colliery to the river, in 1797'.

The drawings are of exceptional interest as they show details of well designed 'fish belly' edge-rails with chairs and cross sleepers. It is therefore desirable to place on record a summary of the considerable amount of information concerning these rails which exists in the Boulton & Watt correspondence files.

Thomas Barnes. 'The late Mr Barns' to whom Wood refers is Thomas Barnes (c.1764-1801) viewer of Walker Colliery, and one of the most able mining engineers of his day. At least sixty of his letters to Boulton & Watt have survived, chiefly dating from 1794 to 1798. They show him actively concerned with the installation of pumping and winding engines at Tyneside collieries, opening up new mine shafts, testing ropes and, in 1797, ordering, from Soho Foundry, cast iron segments for lining part of the King Pit at Walker; a very early example of this technique in mining or tunnelling.

Location. Walker Colliery, about two miles east of Newcastle, had a central group of pits with a waggonway to the Tyne, and an outlying pit known as Lawson Main at Walker Hill. According to an engraved *Plan of the Collieries on the Rivers Tyne and Wear....by John Gibson, 1788* a separate waggonway ran from Walker Hill, joining another from Byker colliery, and thence to the river bank about a mile upstream from the Walker staiths. Before 1798-99 these waggonways were of the traditional wooden rail construction (M.J.T. Lewis *Early Wooden Railways.* London, 1970). From the Barnes correspondence with Boulton & Watt we learn that one of their winding engines was installed at Lawson Main in December 1797 and the Low Main coal seam there, at a depth of 133 fathoms, was won in October 1798. No doubt the new iron railway formed part of this development of the pit.

The Drawings. The drawings (Fig. 1 and 2) cover both sides of a single sheet of paper, watermarked 1797, in Portfolio 169 of the Boulton & Watt Collection at Birmingham Library. They are reproduced here by permission of the City Librarian. One side, entitled 'Profile of the Cast Iron Rail Way for Lawson Main', shows an elevation and a plan of rails and sleepers on a scale of one inch to a foot; on the reverse are shown side and end views of a single rail together with full-size sections of a rail and chair.

CAST IRON EDGE-RAILS AT WALKER COLLIERY 1798

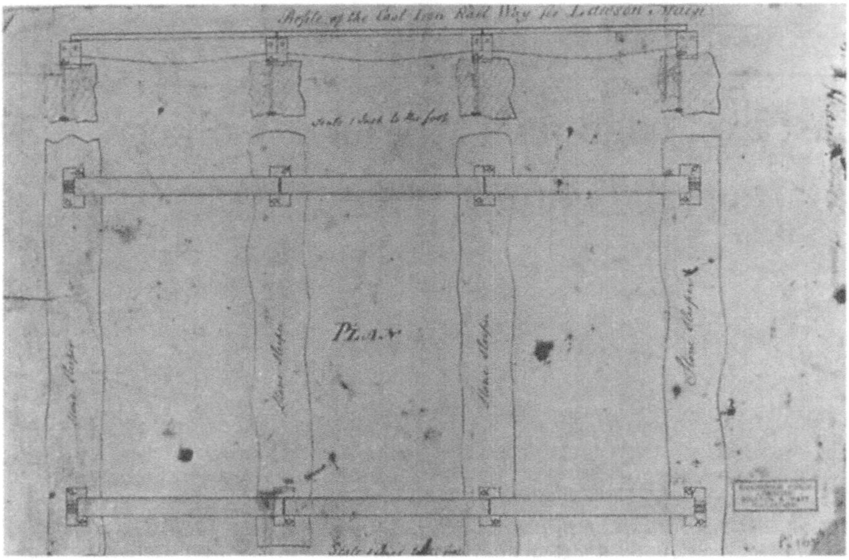

Fig. 1. Profile of the Cast Iron Rail Way for Lawson Main.
By courtesy of the City Librarian, Birmingham.

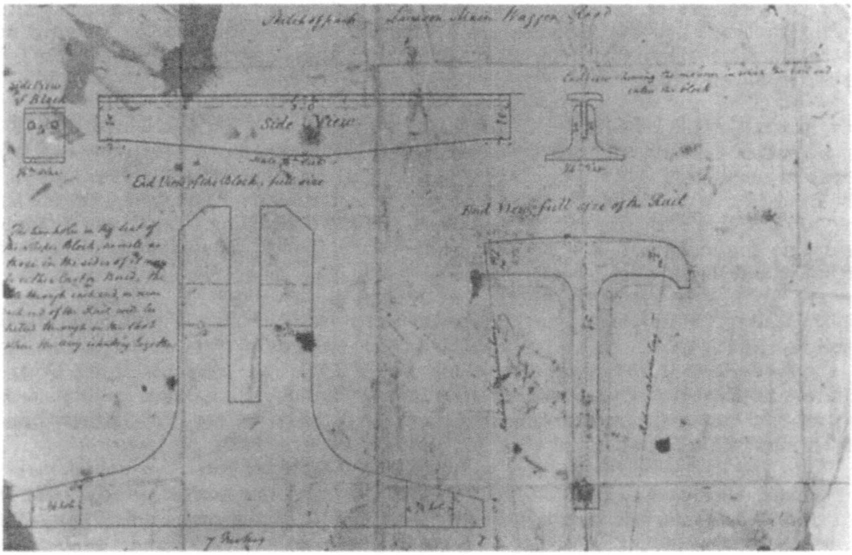

Fig. 2. Details of Rail and Chairs.
By courtesy of the City Librarian, Birmingham.

CAST IRON EDGE-RAILS AT WALKER COLLIERY 1798

The rails, 3 ft long, are of T section varying in depth from 3¾ ins at the ends to 5 ins at mid-span. The horizontal flange is 3 ins wide overall and 7/16 ins thick, and the vertical web has a thickness of 3/8 ins. The bearing surface for the wheels is rounded on a 12 ins radius. The gauge, as scaled from the drawing, is 4ft 3 ins.

Documentation. The following extracts from correspondence in the Boulton & Watt Collection are relevant:

(1) T. Barnes, Walker Colliery 25 Sept 1798, to John Southern at Soho (Box 6):
Barnes mentions in passing 'I must write to Mr W[att] about the Cast Iron Rail Road.'

(2) T. Barnes, Walker Colliery 27 October 1798, to John Southern at Soho (Box 6):
'I had the honour of a visit from Mr Watt...(who) did us the favour to look at Lawson Main & the King Pit'.

(3) J. Watt jnr, Soho 5 Nov 1798, to Mr Barnes at Walker Colliery (No.21: Office Letter Book):
'Upon shewing the drawing of the Rail to our founder, he expressed an opinion that it could not be made of these reduced dimensions for the price stated (£13 p.ton)... [and] they could not be strong enough to support the weight mentioned. A consultation being held of all the Engineers here, it was agreed to avail ourselves of the latitude given by you of altering the pattern as we thought best. We ordered it to be strengthened, so as to be about ½ Inch thickness.. [in] the fin or lower part. The Rail cast from it weighs 1qr 8 lbs .... Some rails [will be] cast the strength of which will be ascertained experimentally & yourself made aquainted with the result'.

(4) Thos Barnes, Walker Colliery 16 Nov 1798, to Boulton & Watt (Box 6):
'I shall be very glad to know the result of the trial as soon as possible. Meanwhile though I chearfully acquiesce in any alteration you may think proper to make in the pattern I assure you that the experimental Rails I got cast here & which Mr Watt Senr. saw upon the Waggon Road near my House do not exceed the dimensions I sent to Soho & they appear to stand their ground with great firmness under the present Work which is 90 Loaded waggons each 4 Tons weight passing over them every day'. He adds that the chairs will be cast locally to save cost 'as they are not so very essential to the goodness of the Work as the Rails'.

(5) John Southern, Soho 4 Dec 1798, to Mr Barnes (No. 68:Foundry Letter Book):
'We have made some castings from the rail-way pattern both from the Air furnace and cupola. One from the latter broke with 7 ton applied in the middle, another with something less than 5½ tons and one from the Air furnace with rather less than 7 tons. .. The pattern 5 inch deep ½ inch mean thickness, the casting weighing about 36 lbs'.

(6) T. Barnes, Walker Colliery 18 Dec 1798, to John Southern at Soho (Box 6):
'From the experiments you have made , .. I think the pattern will be strong enough at 36 lbs to the Rail viz. 5 Inches deep & ½ In mean thickness of the feather, as the working load will never be more than 1 Ton or ¼ of the weight of the Loaded Waggon. Having 3 Rails by me (exclusive of those placed in the Waggon Road near my house over which all the Waggons of the Colliery have now passed about 3 Months without doing them any injury) when your favour of the 4th came to hand I resolved to try their strength which turned out as follows: 6t 17c, 6t 6c, 5t 16c. These Rails were each 5 Inches deep & the feather only 3/8 In thick, weight 32 lbs each which is the size & the weight of those in the Waggon Road which Mr Watt Senr. saw when he was at Walker.... I trust you will be able to cast these rails free from the defect of those I got cast here, which are all more or less warped'. Barnes then gives details of his test rig; a simple lever system applying the load at mid-span.

(7) John Southern, Soho 22 Dec 1798, memorandum to the Foundry (Foundry Order Book 1797-1805):
'For Mr Barnes. 1400 yards of the rail to be cast, say for 700 yards of road, from the pattern that gives a rail of 35 or 36 lbs weight. Please to take the greatest care to prevent both the pattern and the castings from warping'.

(8) John Southern, Soho 7 Feb 1799, to Mr Barnes (No. 68: Letter Book):
'The Rail Way is ½ finished & on the fairest way of being finished before the frost leaves us. We have had very tempestuous weather, and driving snows, which have stopped or delayed very much, all the mails & coaches to this place'.

(9) James Watt jnr. Soho 9 May 1799, to Mr Barnes at Newcastle (No. 21: Office Letter Book):
'When I was in London some time ago, I met with a Copy of Smeaton's Edystone Lighthouse, which I hoped might prove an acceptable addition to his reports & therefore left word with our Agent to forward it.... Your Rails I hope will give satisfaction. They seemed to me pretty well done'.

(10) T. Barnes, Walker Colliery 11 Feb 1800 to James Watt jnr. at Fenchurch Street (Muirhead Box IV Bundle B
'I have sent by yesterday's Post addressed to Messrs B & W [the final balance of payment for]... the Lawson Main Waggon Road'.

CAST IRON EDGE-RAILS AT WALKER COLLIERY 1798

Summary. By September 1798 Barnes had laid down a short length of iron rails on one of the Walker waggonways. These rails, cast locally, were as shown on the drawings, i.e. 5 ins deep with a 3/8 ins web, weighing 32 lbs exclusive of the chair. He sent the drawings to Boulton & Watt to have a large number of rails cast to this pattern for the Lawson Main waggonway. This was in October 1798. The Soho firm advised a somewhat sturdier section weighing 36 lbs with a ½ ins web, and tested three castings. Barnes pointed out that his original 32 lb rails were sufficiently strong to carry the 1 ton wheel loads imposed by the loaded waggons, and proved this by tests. Nevertheless he accepted the heavier section and an order for 1400 of the 36 lb rails went to the foundry in December 1798. The rails were delivered by May 1799. There is no reason to suppose they were not laid soon afterwards. The iron railway of 1799 almost certainly ran from Lawson Main at Walker Hill down to the junction with Byker waggonway. The chairs, and no doubt the stone sleepers, were obtained locally.

Comments. The date of 1797, hitherto universally accepted, has to be revised. But this does not alter the facts that we have here (i) the earliest known drawings of a cast iron edge-rail, (ii) the earliest known rail chairs and (iii) the earliest certain example of the 'fish belly' profile applied to rails.

15

Embankments and cuttings on the early railways

A.W. Skempton

Introduction

Between 1834 and 1841 nine main lines of railway were built in England totalling 660 miles in length complete with tunnels, bridges, viaducts and stations and involving the excavation, in cuttings, of some 70 million cu.yd (54 million cu.m) of material, most of which was used in making embankments. Construction on this scale had no precedent. The earthwork quantities, in particular, are astonishing. Apart from the vast total amount, about 2 to 3 million cu.yd of excavation (and banking) was achieved on each line per year averaged over the whole period of its construction (Table 1); figures not much less than the rate of work on the Great Central Railway in the 1890s, using 40 steam navvies and upwards of 100 locomotives, and substantially exceeded, on works of a comparable nature, only with the introduction of modern earthmoving plant on the first 55 mile length of the M1 Motorway in 1958, where 11 million cu.yd were excavated and placed in (well-compacted) embankments at an average rate of 7 million cu.yd per year[1]. Yet one looks almost in vain in modern histories for details of the methods employed, the results obtained or the difficulties encountered in making embankments and cuttings on the early railways.

Table 1: Excavation Quantities

RAILWAY	DATE		LENGTH	EXCAVATION		
	WORK BEGAN	LINE OPENED	MILES	TOTAL Million cu. yd.	VOLUME PER MILE cu. yd	VOLUME PER YEAR Million cu. yd
Liverpool & Manchester	Mar. '27	Sep. '30	31	2.7	90,000	0.8
Grand Junction	Jun '34	Jul. '37	78	5.5	70,000	1.8
London & Birmingham	Jun '34	Sep. '38	112	13.0	115,000	3.1
Great Western	Feb. '36	Jun. '41	118	9.8	85,000	1.8
Midland Counties	May '37	Jun. '40	57	5.4	95,000	1.7
London & Brighton	Sep. '38	Sep. '41	48	6.9	145,000	2.3
Great Central	Nov. '94	Jul. '89	94	13.3	140,000	3.5

By contrast, as these were subjects of great concern to the engineers and contractors involved, there is plenty of information available chiefly in reports and parliamentary papers, together with articles in the *Civil Engineer & Architect's Journal* and an important paper in *Proceedings of the Institution of Civil Engineers* as well as several books, all published before 1846. This material forms the basis of the present paper.

Management

The high rates of construction were achieved by good organisation and a large work force. Data are given in Table 2, where the numbers of men employed include bricklayers and masons as well as navvies, but not the brickmakers and quarrymen[2]. Locomotives and staionary engines were used on some of the earthwork operations, and in addition there would be several small steam engines for pumping excavations.

Table 2: Workforce

RAILWAY	DATES	NUMBERS EMPLOYED			
		MEN	HORSES	LOCOS	STATIONARY ENGINES
Great Western	Jul. 1837	6250	450	4	2
Midland Counties	Dec. 1838	4030	460	2	1
Midland Counties	May 1839	5140	510	3	1
London & Brighton	Jan. 1840	4370	690	5	0
London & Brighton	Jul. 1840	6210	960	5	0

As for organisation, the plan adopted on the London & Brighton Railway is typical: under the engineer-in-chief John Rastrick there were three resident engineers in charge of three divisions of the line and another responsible for the tunnels, each with three or four assistants. The 48 mile length was divided into 14 contracts, plus separate contracts for three long tunnels, and on every contract there was a superintendent or agent directing the gangs, many of which worked as sub-contractors.

Excavation and End-Tipping

All 'soft ground' including clay, gravel and the softer variety of Chalk was excavated by pick and shovel. Hard rock had to be blasted. For example, in Blisworth cutting on the London & Birmingham Railway up to a ton of gunpowder was used per week in blasting the limestone (270,000 cu.yd out of a total 870,000, mostly excavated between December 1836 and May 1838)[3]. In an intermediate category of soft rock a (horse-drawn) plough proved to be most effective. This seems to have been introduced by the contractor Joseph Thornton in 1835 while excavating Keuper Marl in the deep cutting at Yardley, also on the London & Birmingham, where, as usual, the excavated material was taken into an adjacent embankment. According to an account written four years later:

> *after a few trials, and by increasing the strength and altering the form of the plough, the plan was crowned with a success far beyond what was originally contemplated; for it was found that, in addition to dispensing with a number of men employed in undermining, wedging and breaking up, it reduced the material to such small pieces that the labour of several men, who used to break it up at the foot of the embankment, was saved; and many excavations are now entirely worked with the plough.*[4]

Clearly it was desirable so far as possible to balance 'cut' and 'fill', both in total and over relatively short distances. Admittedly it was not difficult to excavate a cutting and simply run the material to spoil banks nearby, or to build an embankment from borrow pits along each side of its base (the process of 'side-cutting'). But the total cost would be roughly double that involved in building embankments of material from cuttings; while extra land was required outside the normal boundaries, and useless spoil banks or pits would be left alongside the line. Locally such methods had to be adopted to overcome a deficiency of banking material or an excess of excavation, and sometimes they were used to expedite construction. But as a general rule embankments were formed from cuttings, not too far distant, the excavated material being carried in earth waggons drawn by horses or, for distances of more than about two miles, by locomotives. An integrated system of cutting and embanking was therefore achieved, capable of handling large quantities in a reasonably short time.

A W Skempton

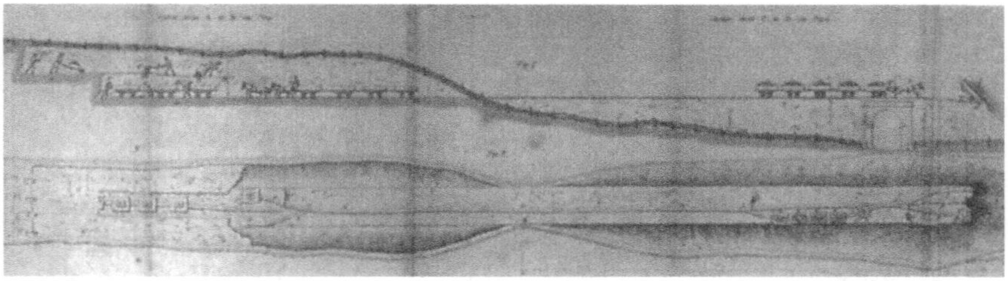

Fig 1: Cutting and embanking, from 'Resident Assistant Engineer' (J.D. Hughes), *A Practical Inquiry into the Laws of Excavation and Embankments upon Railways* (1840).

Now the best way of building an embankment, especially in clay, would have been to form the bank in shallow layers, say 2 to 4 ft thick, running out each of them to the full length, and following with the upper layers after each of the lower ones was laid and compacted with 'beetles' or punners. The bank is thus progressively raised and there is time for some consolidation in the bank and foundation. This was the method (with 4 ft layers) specified by John Macneill in 1836 for embankments on the Slamannan Railway in Scotland (12 miles long, built 1837-40) where the strata chiefly consists of Boulder Clay overlying Coal Measures and the total excavation amounted only to 1 million cu.yd, the largest embankment being 40 ft high with 2:1 slopes and 180,000 cu.yd. of fill[5].

It was also quite usual to compact in layers that portion of an embankment immediately adjacent to the abutment of a viaduct[6]. But construction of the whole of an embankment in shallow layers was too slow a process, and one not well suited to the combined operations of cutting and embanking on a large scale. Instead, the method most commonly used on the railways by the mid 1830s was to run out the bank to its full height at once, by end-tipping from the advancing head of the bank on a level with the bottom of the cutting.

The procedure is illustrated in Fig. 1. The cutting has been started by opening a 'gullet' 8 ft deep. Above this the 'second lift' is being worked on the face and sides with barrows. These, when filled, are wheeled over stages and emptied into earth waggons. The loaded waggons in a 'set' of six are led towards the bankhead and then drawn out singly at about 6 m.p.h. by a horse until 20 yd from the end, when the horse is suddenly detached and the waggon (Fig. 2) is precipitated against a wooden bumper and thus tilted over[7]. Meanwhile another waggon is 'teamed', and so on until the set is ready to be drawn back to the cutting.

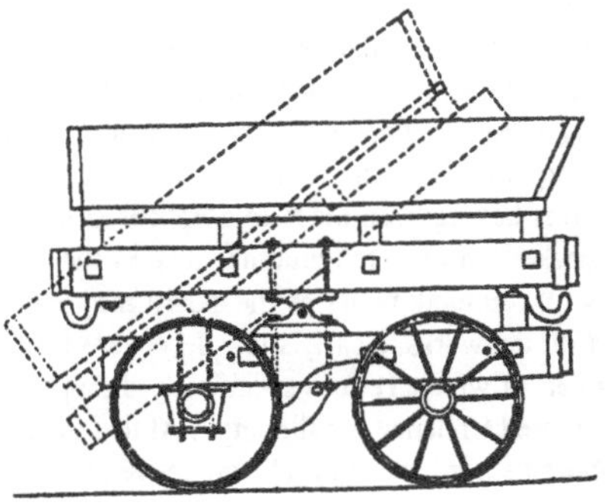

Fig 2: Earth waggon, from *Civil Engineer and Architect's Journal* 6 (1843).

The embankment in Fig. 1 is being tipped from three 'teaming roads' on a bank little wider than its finished dimension and with side slopes, at the angle of repose, which merely require

Embankments and Cuttings on the early Railways

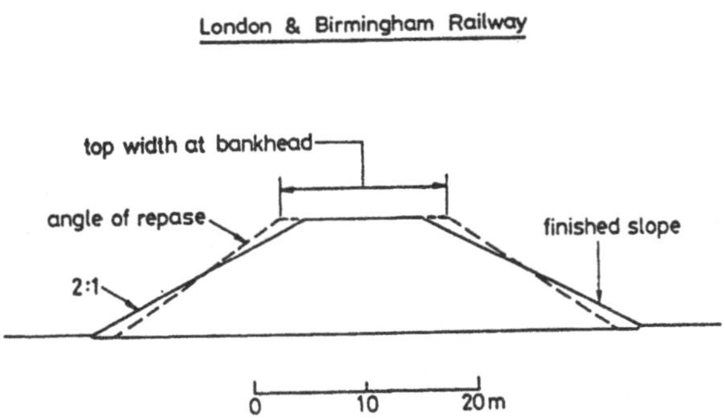

Fig 3: Cross section of Watford embankment, July 1835.

trimming to the finished inclination of 1.5:1 in this example. More often the finished slope was 2:1. The working width at the head would then be greater (Fig. 3), allowing four or even six roads, the slopes afterwards being dressed off to the final angle and turfed and/or seeded with grass (as were the cutting slopes)[8].

Earth waggons held about 2 cu.yd. A typical workforce would consist of 40 to 80 men with 6 to 12 horses. Two shifts were often employed in summer, working either from daylight to dark or throughout 24 hours (with rest periods of course) the number on nights generally being less than in daytime. Only in exceptional circumstances were two shifts employed on earthworks in winter. Experience showed that work could proceed, on average, for 250 days in a year or 5 days a week, allowing for Sundays and wet or frosty weather; perhaps rather less with clays[9].

The greater rate of construction achieved by high end-tipping, as opposed to building in shallow layers, was gained at the expense of increased settlements; but on railways (unlike canals and structured roads) settlement could readily be corrected by packing more ballast under the sleepers and, when necessary, imposing a speed restriction until the bank had become consolidated.

The method may have originated in making the 'batteries' (embankments) on colliery railways in the north-eastern coalfield, and there is a strong likelyhood that George Stephenson adopted end-tipping on the Stockton & Darlington Railway (25 miles in length, built 1822-25). Although two ridges of high ground at the western end, near Bishop Auckland, were negotiated by inclines worked by stationary engines, most of the line had fairly flat gradients suitable for locomotives, requiring cuttings and embankments up to 24 or 30 ft in height; and in his design report of 18 January 1822 Stephenson makes the point that, in contrast to a more 'undulating' line proposed by another engineer in an earlier scheme, he had arranged the levels so that the banks could be built of material excavated from cuttings and carried on not too long a 'lead', in tipping waggons on temporary rails[10].

Certainly he used high end-tipping on the Liverpool & Manchester line. Earthwork operations began there in March 1827, one hundred waggons having been made to Stephenson's design[11] Every fortnight the three resident engineers marked up on a longitudinal section the progress achieved. Sections of Broad Green and Newton embankments at the positions reached in February 1829, published with Telford's report on the works, clearly show they were formed by end-tipping at the full height; a process, moreover, of which Telford specifically disapproved on the grounds of delayed consolidation and an increased tendency for slipping[12]. At that time Broad Green embankment had advanced to a point where its head was 42 ft (13 m) above ground level, and at Newton the height was 24 ft. However, another section shows the 55 ft (17 m) bank across Sankey valley being built in 15 ft lifts, though each was formed by end-tipping. Joseph Locke, who had been resident engineer on the western district of the line, records that on Broad Green embankment, with 1.5:1 slopes and a 25 ft top width (allowing space for three roads), the utmost quantity that could be teamed in a day, employing 'as many men as we could place there', on two shifts from daylight to dark, was 800 cu.yd[13]. The material was chiefly broken-up (Triassic) sandstone brought on a mile-long lead from Olive Mount cutting.

In principle the procedure of building in two or more lifts increased the rate of construction, as more 'roads' could be used on the lower (and wider) portion which obviously was kept in advance of those above. Locke in 1834 advocated this system[13], but Robert Stephenson preferred the simpler, and in his view more efficient, method of tipping at the full height[14], and by 1837 Brunel referred to practical difficulties in forming a bank on two lifts (let alone on three or four) which 'render it not a very general practice'[15]; significantly, this remark was in reply to questions on the quantity of earth that could be teamed supposing an embankment to be 'carried forward to its full height in the ordinary mode'.

Data on the rate of construction, in what by 1837 had already become the 'ordinary mode', are available from several sources. Watford embankment on the London & Birmingham line is a good example. 1.4 miles in length with a finished top width of 33 ft (10 m), a maximum height of 45 ft (13.7 m) and 2:1 slopes, the bank had a total volume of nearly 1 million cu.yd. Of this amount rather less than one-half was brought from Watford Heath cutting, a mile to the south, and about 500,000 cu.yd came, in the form of gravel and Chalk, from Watford tunnel approach cutting. Earthwork operations began about August 1834 and reached completion in March 1837. On average therefore 190,000 cu.yd were placed in the embankment in a year; equivalent to 750 cu.yd per day[16].

Details are given by the resident engineer George Buck[17]. He records in July 1835 that the northern part of the bank had then reached a position where the height was 32 ft and the fill, tipped at its angle of repose, gave a width of 48 ft at the bankhead with space for six roads (Fig. 3). Working on a lead of 0.4 miles, a total of 160 men employed on day and night shifts moved 1000 cu.yd in 24 hours, measured as usual over a period of a fortnight which allows for time taken in extending the rails and other minor interruptions. This corresponds to a rate of 6.2 cu.yd per working day per man; a figure which, on the assumption that the workforce on this operation would be halved in the winter, translates into the long-term rates mentioned above.

For Huntonbridge embankment the same contractor, James Copeland, employed 100 men (60 on day and 40 on nights). They teamed 580 cu.yd per day on a 2 mile lead from Watford tunnel north approach cutting, also in gravel and Chalk; again measured over a fortnight in the summer[18]. Thus 6 cu.yd per day per man is an average performance under conditions of a well-managed contract with good banking materials. That the rate achieved at Watford was close to a maximum is confirmed by Robert Stephenson's statement in July 1836 that the greatest quality to date teamed over the end of any embankment on the London & Birmingham works had not exceeded 200,000 cu.yd per year[19].

Two points should be noted: the effective amount of work done is almost double, since for every cubic yard deposited in an embankment on this system nearly the same quantity has been excavated at the same time from a cutting: and, secondly, that the rate per man is averaged for the whole workforce employed in 'getting', 'filling', 'carrying' and 'tipping', as well as those involved with other tasks such as shifting the rails, maintenance of waggons and care of the horses. This average rate should not be confused with the often quoted fact that two navvies could dig and fill into barrows or waggons up to 30 cu.yd of earth per day.

Somewhat higher rates of construction can be deduced, from figures given by the engineer Frederick Swanwick, for two embankments near Wakefield on the North Midland Railway (72 miles in length, from Derby to Leeds)[20]. Both banks were built of sandstone and shale of the Middle Coal Measures from cuttings up to 60 ft deep. Work started about October 1837. By February 1839 a little less than half of the work had been done and 'the most vigorous means' were adopted to complete both contracts in time for opening the line at the end of June 1840; so the earthworks would have been finished by May at the latest. The bank had 1.5:1 slopes and a top width of 35 ft, providing room for four 'roads'. Oakenshaw cutting (contractor Joseph Thornton who, with John Stephenson, had worked on the Olive Mount cutting mentioned above) contained 680,000 cu.yd of which 130,000 cu.yd were run to spoil and the remainder went into the embankment (maximum height 52 ft) on a lead of about 1 mile. Thus 550,000 cu.yd of material were taken from the cutting to the bank at an average rate of 205,000 cu.yd per year; but this volume is 'measured in the solid' and the quantity of fill would be greater, by at least 10 percent, owing to its lower density. Figures for the Altofts embankment (contractor David McIntosh) are very similar: 525,000 cu.yd taken from Normanton cutting on a 1.5 mile lead to a bank with a maximum height of 37 ft in exactly the same period of time.

Information on the rate of constructing clay embankments is available from New Cross on the London & Croydon line and from Willesden on the London & Birmingham. In the former case a mile-long embankment with 2:1 slopes and a maximum height of 25 ft (7.6 m) was built of London Clay from the deep New Cross cutting. By September 1837 the contractor William Hoof was fully organised and during the next six months, to March 1838, working double shifts for part of the time, 81,000 cu.yd were teamed into the bank[21]; thus achieving a rate of 160,000 cu.yd per year. This would seem to be an excellent performance, only 12 percent less than at Watford despite the notoriously difficult nature of the clay as compared to Chalk and gravel. But an identical figure emerges from data given by Stephenson for Willesden embankment[22]. This, too, was built of London Clay, here brought on a 1.75 mile lead from Kensal Green cutting. With a finished top width of 33 ft and 3:1 slopes, the bank in September 1835 had reached a position where its height was 13 ft. The clay was carried to within a few hundred yards of bankhead by a locomotive hauling 12 waggons, and typically making 25 journeys in a single 12 hours shift, when the waggons would be individually drawn and tipped in the usual manner on four roads. During the 8 months before Stephenson recorded the data in July 109,000 cu.yd had been teamed at an average rate of 160,000 cu.yd per year, which corresponds closely with 25 x 12 = 300 waggons a day.

Horse-runs

When all or a considerable part of a cutting went into spoil banks along the top edge, or in some nearby location, much higher rates were achieved since in this circumstance large numbers of men could be deployed along the length of the cut. The technique generally adopted was to use 'horse-runs', as illustrated in a drawing by John Bourne of the nearly completed Tring cutting on the London & Birmingham line in June 1837 (Fig. 4)[23]. The contractor here was Thomas Townshend who also excavated the great Smethwick cut on Telford's Birmingham Canal improvements 1827-30.

Fig 4: Tring cutting, 17 June 1837, from J.C. Bourne, *Drawings of the London and Birmingham Railway* (1839).

The cutting at Tring is 2.5 miles long with 1:1 slopes and a maximum depth of 57 ft (17.5 m) in Lower Chalk. About 60 percent of its 1.3 million cu.yd content was used in forming long, low embankments over the Gault clay vale to the north; the rest being taken to spoil along the top edge of the cut. Work began soon after September 1834 and Stephenson in February 1838 reported that the excavation and embankments on the contract had been ready for opening the line since December 1837, except for 4000 yd of permanent way, completion of which was now delayed by heavy frost[24]; and he added that the embankments, consisting almost entirely of Chalk fill, were 'already well consolidated and little liable to subsidence'. Thus excavation had proceeded at an average rate of 400,000 cu.yd per year.

Details of a large cutting in which the whole content had to be run to spoil are given by J.U. Rastrick[25]. This was in Contract No. 15 on the London & Brighton Railway which included a cutting in Middle-Upper Chalk nearly two miles long, between Clayton and Patcham tunnels, containing 700,000 cu.yd. Within three weeks the contractor Mark Faviell had 300 men at work with 21 horses. Six months later, in January 1840, there were 377 men and 58 horses, and during this period 325,000 cu.yd were excavated. The contract involved four bridges and a short tunnel but these works had not yet started, so the entire workforce had been engaged in excavation; at a rate of 650,000 cu.yd per year or approximately to 6.5 cu.yd per day per man.

Sonning cutting on the Great Western main line provides an extreme example of the concentration of labour. The cutting, also nearly two miles long, was one of the largest on any of the early railways with a volume of about 1.5 million cu.yd. Progress had fallen so far behind schedule by the summer of 1838 that the Company took over, and before the end of the year work was proceeding with exceptional vigour under the direction of the resident engineer John Hammond with Rowland Brotherhood as principal sub-contractor. From February to August 1839 a workforce of 1000 to 1200 men and 200 horses, with 2 locomotives, removed 555,000 cu.yd (chiefly clay) at an average rate of just over 1 million cu.yd per year[26]. Material from the western part of the cut went into a 4-mile embankment towards and past Reading (hence the locomotives) but all the central portion was taken to spoil banks several yards behind the top edge, employing 42 horse-runs up each side working from depths down to 60 ft (18 m) on 2:1 slopes[27]. The cutting was completed by the end of the year in time for permanent way to be laid and opening of the line through to Reading in March 1840.

Side-cutting

The converse process of banking from side-cuttings is exemplified by Swindon embankment on the Cheltenham & Great Western Union Railway (Fig. 5). The bank is 1.8 miles long, averaging 20 ft in height and nowhere exceeding 24 ft (7.3 m), with 2:1 slopes and a volume of about 480,000 cu.yd. It was built of brown and some blue (unweathered) Kimmeridge Clay wheeled up from side-cuttings and deposited in layers from barrows[28]. No large cutting on the line itself existed within a reasonable distance. Work began in the summer of 1839[29]; a few small slips occurred in the following winter, but were easily repaired, and by November 1840, with the bank nearly completed, opening of the line (from Swindon to Cirencester) was planned for January 1841[30]. However, rain delayed the work by two months, and in early spring (probably in March) a slip occurred on the east side; sliding into the side-cut over a length of 330 ft with an initial drop of 8 ft in the bank top. This was eventually stabilised by the simple but tedious method of back-filling the side-cut, and after repairing the bank the line was opened on 31 May 1841.

Assuming that work began in May 1839 and the bank was practically finished in November 1840, the average rate of construction in this 18 month period is about 300,000 cu.yd per year; almost double the rate achieved by teaming at New Cross and Willesden but, of course, no 'useful' cutting had been accomplished.

Embankments and Cuttings on the early Railways

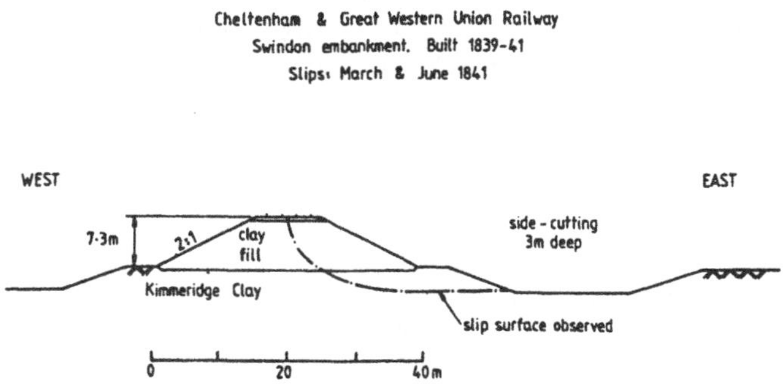

Fig 5: Cheltenham and Great Western Union Railway Swindon embankment. Built 1839–41: slips, March and June 1841.

Embankment Slips

Shortly after opening the line another very similar slip occurred in the Swindon embankment. As a quick first-aid measure Brunel had a row of timber piles driven at the toe of the slope, penetrating 8 ft into undisturbed ground below the level of the bottom of the side-cutting. They were soon sheared off by the relentless movement of the slip. On the suggestion of the resident engineer Charles Richardson pits 6 ft square were then excavated near the toe, 24 ft apart, to a depth of 2 ft below the slip surface – which was clearly seen during this operation as a smooth polished plane – and filled with ragstone brought from a cutting several miles further north. These rubble-filled pits could have been doubled in number but in fact the first set proved effective in stopping the slip, allowing repairs to the bank to be started before completing the more lengthy process of back-filling the side-cut.

The slips in Swindon embankment were essentially foundation failures, though of an unusual kind. Normally, failure occurs simply by the weight of the bank causing shearing and upheaval of the foundation material. A spectacular example is provided by Hanwell embankment on the Great

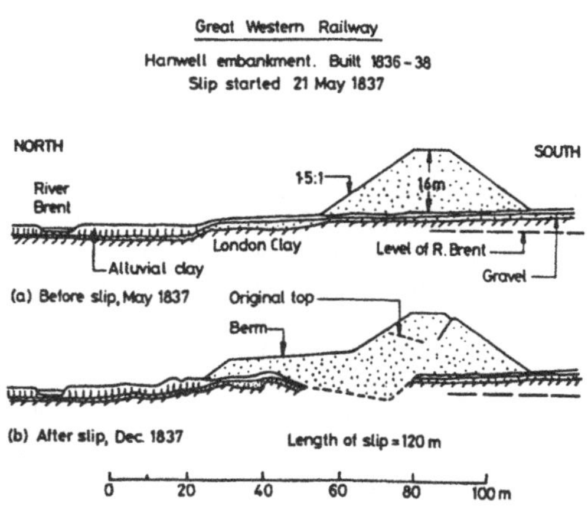

Fig 6: Great Western Railway Hanwell embankment. Built 1836-8: slip started 21 May 1837.

Western Railway, on the west side of Wharncliffe Viaduct, which failed during construction in May 1837[31]. Composed of end-tipped gravel fill 54 ft (16 m) high with 1.5:1 slopes, it is founded on a thin layer of gravel over London Clay (Fig. 6). Gravel was added, and continued to be added, in an effort to maintain a level for further progress, but to no avail; and Brunel then ordered a counterweight berm to be built. By December, when a state of equilibrium had been

Table 3: Slips in Embankments

LINE	EMBANKMENT	HEIGHT m	DESIGN SLOPES	METHOD OF CONSTRUCTION	MATERIAL	DATE OF SLIP	TIME AFTER CONSTRUCTION
London & Birmingham	Bugbrooke	10	2:1	end-tip	Upper Lias	1838	during construction
Great Western	Wotton Bassett	12	2:1	end-tip	Oxford Clay	1841	
London & Brighton	Earlswood	14	2:1	side-cut	Weald Clay	1841	
Eastern Counties	Colchester	15	2:1	end-tip	London Clay	1842	
London & Birmingham	Blisworth	14	2:1	end-tip	Soft mudstone	Jan. 1840	2 years
Great Western	Tockenham	10	2:1	side-cut	Oxford Clay	Jan. 1843	2 years
London & Brighton	Salfords	15	2:1	side-cut	Weald Clay	Nov. 1844	3 years

achieved, the slip extended over a length of 400 ft (120 m), with a subsidence of 30 ft at the top[32]. Distortions of the strata were recorded in exploratory trenches sunk to the depth shown in the section.

Slips within the body of an embankment were nearly always restricted to clay fills (poorly compacted by modern standards) and resulted principally from softening of the clay lumps by absorption of rain water. This process could be quite rapid, leading to failures during construction, but might continue for a year or more after completion of the bank. Examples of both types are given in Table 3.

Obviously the embankment could not be protected from rain, but at least the worst effects could be avoided by not placing clay fill in wet winter conditions. This rule was not always followed, but experience showed that, even with reasonable care, difficulties attended the building of stable clay embankments with heights of more than 10 or 12 m by end-tipping or uncompacted barrow work.

An instructive case record comes from the embankments north and south of Wolverton viaduct on the London & Birmingham Railway[33]. Both had a maximum height of 48 ft (14.5 m). On the north side the materials were mostly glacial sand and gravel with some Boulder Clay. The bank stood well, except for a minor foundation slip in the underlying alluvium. The south embankment, in its early stages, was built with similar materials, brought by locomotive from Denbigh Hall cutting 3 miles away, again without trouble. But as the cutting progressed it yielded 'black, soapy clay' (Oxford Clay) and despite care to mix sand with the clay a large slip developed in the bank. More and more material was added on the top in the hope that a flatter, stable slope would be formed by continued displacement. But as fast as material was placed on top the toe moved out, and after several weeks the bank was still 12 ft below its proper level; even though the slope had now decreased to about 4:1. A temporary timber bridge was then erected to allow end-tipping to continue at bank head. The state of affair in June 1837 is depicted in Bourne's drawing (Fig. 7), about three months after the slip started.[34]

Later in the summer the bridge was removed to try again the effect of adding material, but the response was just the same; so the only remedy was to build a counterweight berm by barrow work. Meanwhile, ahead of the advancing bank, earth was excavated to a depth of 5 ft across nearly the full base-width and used to form mounds on each side as pre-existing small berms, perhaps about 12 ft high with a top width of 20 ft. They proved to be effective, demonstrating that

Embankments and Cuttings on the early Railways

Fig 7: Wolverton embankment, 28 June 1838, from J.C. Bourne, *Drawings of the London and Birmingham Railway* (1839).

it is easier to prevent a slip than to stabilise one after it has occurred. This is due to the fact that the 'residual' strength on a slip surface is much smaller than the strength of the unsheared clay[35]. In February 1838 Stephenson reported that work at Wolverton had for some time been advancing in the most satisfactory manner, and would be completed in eight weeks time, as indeed it was.[36]

The very flat slope required for final stability after a slip in clay fill is illustrated by Colchester embankment on the Eastern Counties Railway. Two slips totalling 380 ft (115 m) in length developed during construction of this end-tipped 50 ft (15 m) high London Clay embankment, in the autumn of 1842. Material was added on the top, as usual, but in this case the process was continued until stability had been achieved, by outward displacement of the toe, with a slope of 6:1[37]; an operation in which 34,000 cu.yd of additional fill was 'swallowed up'[38]. Trestle bridges were then built across the top, still much below its proper level, and the bank left to consolidate until July 1843, when it was completed without further slipping[39].

Slips in Clay Cuttings

Superficial slips were common in clay cuttings (and embankments) either in rainy seasons during construction or in the first or second winter afterwards. They usually required for their repair no more than trimming and a few shallow trench drains. In an altogether different category were deep slips which took place, almost without warning, several years after excavation. Three well-documented cases are summarised in Table 4.

The first of these 'delayed slips' occurred on 2 November 1841 at 8.30 p.m. in the west side of New Cross cutting on the London & Croydon Railway, with an effective depth of 48 ft (14.5 m) in brown London Clay, 3 years after excavation (Fig. 8). It had a length of 360 ft and involved the movement of a mass containing about 50,000 cu.yd of clay. A slip on the opposite side

Table 4: Delayed slips in clay cutting

LINE	CUTTING	EFFECTIVE DEPTH m	SLOPE	MATERIAL	DATE OF SLIP	TIME AFTER EXCAVATION
London & Croydon	New Cross	14.5	2:1	London Clay	2 Nov. 1841	3 years
Great Western	Sonning	13	2:1	Reading Clay	24 Dec. 1841	2.5 years
London & Birmingham	Bugbrooke	14	2:1	Upper Lias Clay	24 Sep. 1842	4.5 years

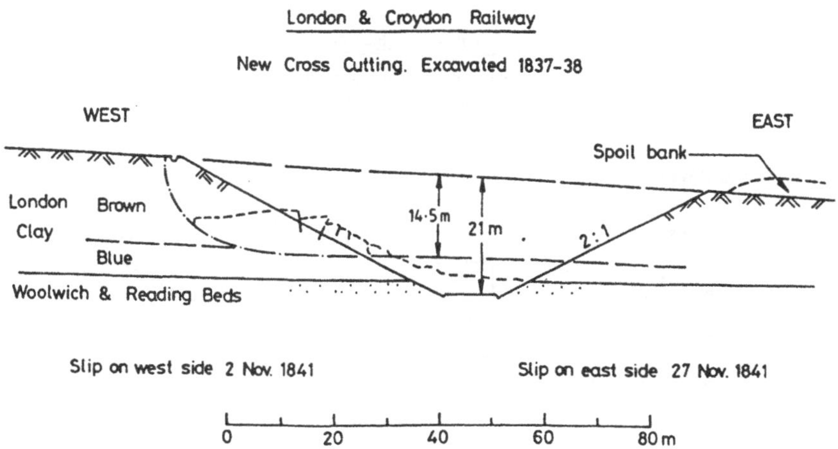

Fig 8: London and Croydon Railway New Cross Cutting. excavated 1837-8.

followed three weeks later, with an extension of the first one in January 1842. The slips, by their magnitude and unexpected nature, attracted much attention in the technical press and formed the principal subject of a paper by Charles (later Sir Charles) Hutton Gregory, the resident engineer under whose direction the remedial works were carried out.[40] Within hours of the first slip a coach service had been organised between New Cross and Forest Hill, the next station down the line, and remarkably the railway was closed only for a total of 12 weeks while dealing with a succession of three of the largest slips ever encountered in British transport history. The slopes were benched back in stages to an overall inclinatin of 4:1, and during these operations the polished nature of the slip surface and its shape were observed (Fig. 8).

On a smaller scale at about the same time (November or December 1841, the exact date has not been found) a slip occurred in Wootton Bassett cutting on the Great Western line: 42 ft (14 m) deep at 2:1 in the Oxford Clay[41]. And this was followed on 24 December by a slip on the same line in Sonning Cutting, which lead to a train de-railment with the loss of eight lives[42]. At the slip the cutting, with 2:1 slopes, had an effective depth of 42 ft (13 m) in clay of the Reading Beds and a total depth of 58 ft[43].

In his report on the New Cross slips Charles Pasley, Inspector-General of Railways, having examined and approved the remedial work then in progress, goes on to say:

> *The original slope of the sides of the deep cutting was two of base to one of height, and they had stood for more than two years without any slip [of significance] until lately, when continued rains have produced this unfavourable effect, which I do not think that any engineer could have anticipated; for it is only our late experience that has developed the disadvantage of deep cuttings and high embankments in certain kinds of clay, even at very moderate slopes. No blame therefore attaches to the original construction of this railway, though the extraordinary slips that have occurred recently will be a lesson to put engineers on their guard for the future in working in such soil.*[44]

Embankments and Cuttings on the early Railways

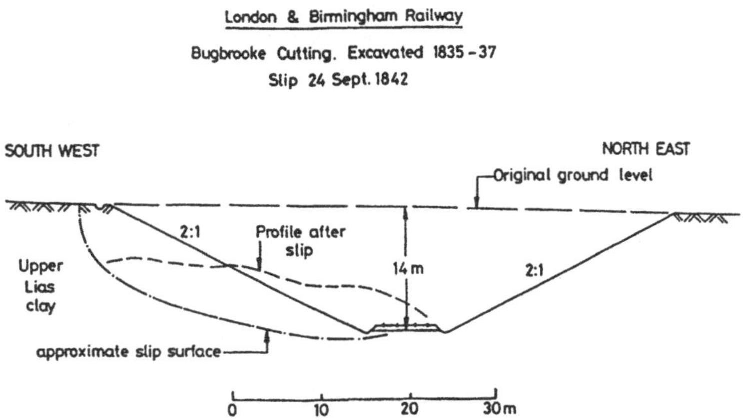

Fig 9: London and Birmingham Railway Bugbrooke Cutting. Excavated 1835-7: slip 24 September 1842.

Thus, while Pasley emphasises the delayed nature of the slips (and he had in mind Sonning as well as New Cross) he attributes their occurrence to heavy rain; an opinion generally held at the time.

In the sense that rain can trigger a slip, this is true. But in his 1844 paper Gregory postulated that additionally a more fundamental process must be involved: deep-seated progressive softening of the clay, connected in some way with the effects of stress release caused by excavation, and he thought that the fissured nature of London Clay played a significant role. This foreshadows the explanation of delayed slips in stiff fissured clays put forward by Terzaghi in 1936[45], from which stemmed a long series of investigations leading, very recently, to a proper understanding of the complex process of pore pressure change and progressive failure involved[46].

The early winter of 1841, however, was indeed exceptionally wet. But in September 1842 a slip occurred on the London & Birmingham line at a time of only moderate rainfall and in a cutting which had remained stable for more than 4 years. This was at Bugbrooke, in Northamptonshire: a cutting 47 ft (14.3 m) deep with 2:1 slopes in Upper Lias clay (Fig. 9) completed before February 1838 and which had survived the winter of 1841-42 (and, like Sonning and New Cross, the scarcely less wet winter of 1839-40).

The circumstances are described by Pasley who now realised that some deep-seated effect must be acting, and he began to wonder if deep cuttings in 'London Clay and other plastic clay formations' were always liable to slip sooner or later.[47] Subsequent experience has shown this view to be not far from the truth. Cuttings with slopes steeper than 3.5:1 in London Clay or 2.5:1 in Upper Lias clay, for example, are prone to slip 10 to 50 years (or even longer) after excavation[48].

While developing his ideas on delayed slips Gregory would no doubt have been influenced by the failure at Bugbrooke, and also by further slips on the Croydon line south of New Cross, in cuttings at Brockley (9 m deep) and Forest Hill (12 m deep): both in brown London Clay.

Remedial Measures

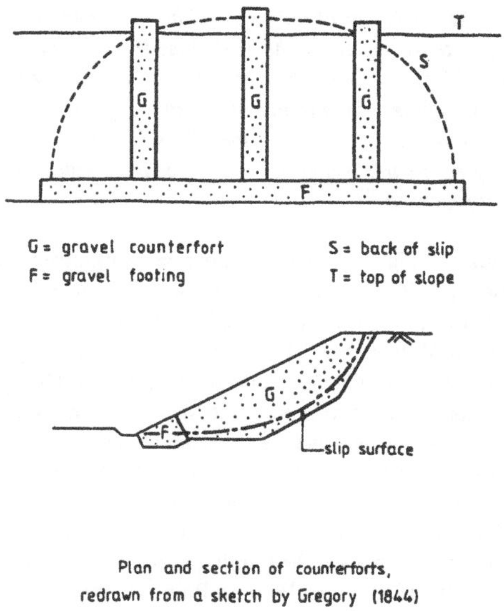

Fig 10: Plan and section of counterforts, redrawn from a sketch by C.H. Gregory, *Min. Proc. Insti. of Civ. Eng.* 3 (1844).

The usual method of stabilising a deep clay slip on the early railways was to flatten the slope, either by trimming or benching back, or in extreme cases by removing the slip mass entirely, as at New Cross. Slope flattening required the removal of considerable volumes of clay (25,000 cu.yd at Wotton Bassett) and the purchase of land outside this existing boundary fence (nearly an acre at Bugbrooke, 1.7 acres at Forest Hill).

Counterforts were an alternative remedial measure, except perhaps for very large slips. They consist of trenches, typically 6 ft wide at 24 ft centres, dug into the slope to a depth of 2 or 3 ft below the slip surface and backfilled with gravel or rubble stone rammed in tight (Fig. 10) and act both as an internal buttress and as a deep drain. Introduced by Robert Stephenson in 1839 to stabilise the Jurassic clay slopes above the limestone in Blisworth cutting[49], counterforts were used by Joseph Locke early in 1842 in the cuttings adjacent to Tapnage tunnel on the Gosport Branch of the London & South Western Railway[50] and later that year by Gregory in Brockley cutting. At Tapnage the slips occurred within a year of excavation; not surprisingly, as the cuttings were made at 1.5:1 in London Clay! In this case the slopes were cut back before installing the counterforts.

It appears that Gregory in 1844 gives the first published description of counterforts, though by that time they were quite well known; as clearly emerges from the discussion on his paper. They became practically a standard remedial measure on the railways, and remained so for the next hundred years.

Note on Steam Power

It has been mentioned above that stationary engines were sometimes used on earthworks. An example is provided by the cutting (600,000 cu.yd) and embankment (430,000 cu.yd) at Leire on the Midland Counties Railway. This was part of a big contract carried out by David McIntosh on the 20 miles length between Leicester and Rugby. In the summer of 1838 on this contract he had nearly 2000 men at work with about 150 horses, and as the engineer Thomas Woodhouse reports:

> *The spirited manner in which Mr Macintosh is working his heaviest cuttings, shows his determination to complete the works in the time specified in his contract, the 1st*

> *of May 1840 ... The excavation at Leir Hill is the most extensive piece of earthwork on the whole line of railway; and here he is erecting a steam-engine to facilitate his operations; he has an inclined plane from the cutting to the embankment (on a lead of 0.75 miles) where the material is to be deposited. The plane descends at an angle just sufficient for the gravity of the waggons to overcome their friction and the resistance offered by the rope. The waggons, when emptied, are drawn back by the engine; by this means, more earth may be removed expeditiously than by any other yet practised.*[50]

McIntosh continued with great vigour, his workforce by the following year having increased to 2600 men and 260 horses, and despite the wet winter of 1839-40 he managed to complete the whole contract, involving upwards of 3 million cu.yd of excavation as well as the Avon viaduct, at the specified time after 2.5 years work; thus averaging at least 240,000 cu.yd per year in the Leire cutting, practically the full depth of 60 ft (18 m) being in Boulder Clay[51]. The embankment, with 2:1 slopes and a maximum height of 40 ft (12 m), was already well consolidated on opening the line in July 1840 and there is no record of any slips in the bank[52].

Other than a fairly restricted application of stationary engines the sole use of steam power for excavation on the early British railways was on the Eastern Counties line when John Braithwaite in 1843 carried out successful trials of the 'American Steam Excavator'. Powered by a 10 h.p. (nominal) engine, this was the fourth machine of its kind made by Carmichael, Fairbanks & Otis and used from 1838 on the Western Railroad (from Worcester, Mass. to Albany, N.Y.). It dug and loaded into waggons about 500 cu.yd of sand per day: equivalent to the work of rather more than 30 men, and at roughly half the cost[53]. Nevertheless, even in America, where labour was less plentiful, steam navvies came into their own only so late as the 1880s with the introduction of far more powerful machinery[54]. In England they were first used on a grand scale on the Manchester Ship Canal (1887-93) where 58 large and 17 smaller steam navvies excavated 54 million cu.yd in 6 years[55].

Conclusions

Cuttings and embankments on the early railways were planned on the principle of designing for normal, expected conditions and dealing with the abnormal when it is encountered. Slips, in the latter category, gave a lot of trouble but mostly in deep clay cuttings and high embankments where previous experience scarcely existed. They should be seen in relation to the great majority of the earthwork which was carried out rapidly and successfully on an unprecedented scale by the methods and with the results described.

Correspondence. Prof. A.W. Skempton. Department of Civil Engineering, Imperial College, London, SW7 2BU.

References

1 L.T.C. Rolt, *The London-Birmingham Motorway* [1960]. For the Great Central Railway see papers by F.W. Bidder and F.D. Fox *Min. Proc. Inst. Civ. Eng*. 142 (1900), pp. 1-48. Data on the early railways come from half-yearly reports supplemented in some cases by information from Francis Whishaw, *The Railways of Great Britain* (1842).

2. Details of the London & Birmingham Railway workforce have not been found, except that the numbers of men employed were upwards of 10,000 in July 1836 and 12,000 in June

A W Skempton

1837; as reported in *Railway Magazine*.

3. Peter Lecount, *The History of The Railway Connecting London & Birmingham* (1839) p. 42. Quantities are given by Arthur Freeling, *The Railway Companion from London to Birmingham, Liverpool & Manchester* (1828) p. 12 and sections of the cutting by F.W. Simms, *Public Works of Great Britain* (1838), pls. 20-35.
4. *Civil Engineer & Architect's Journal* 2 (1839), pp. 224-225. In a note on p. 252 R.B. Dockray says this method was in full operation before he took up his post as resident engineer (in December 1835) on this section of the line.
5. The specification, dated August 1836, is quoted in S.C. Brees *Railway Practice* (1837) pp. 93-108. Macneill worked from 1826 under Telford on the London-Shrewsbury division of the Holyhead road; his specification for the Salamannan embankments may reflect this experience.
6. For example, Robert Stephenson specifies layes not exceeding 3 ft in thickness, each being 'closely packed' previous to succeding layers being laid on, for a viaduct on the Slatley contract, London & Birmingham Railway, 1834. Townshend collection, Inst. Civ. Eng. archives. See also Vignoles' specification of 1836 for the Avon viaduct on the Midland Counties Railway in Brees pp. 85-86.
7. "Resident Assistant Engineer" (J.D. Hughes), *A Practical Inquiry into The Laws of Excavations and Embankments upon Railways* (1840) pl. 1 (see Fig.1) and pp. 163-165. Details of end-tip waggons are given by "O.T." "On Earthworks, Excavation, Cutting and Embankment upon Railways", *Civil Engineer & Architect's Journal* 6 (1843) pp. 265-268 (see Fig.2) and by G. Drysdale Dempsey, *Railways, Papers on the mechanical and engineering operations and structures combined in the making of a railway*, (1846) pl. 9.
8. Dempsey pl. 5 shows an embankment under construction with 4 roads at the bankhead; this seems to have been the most usual arrangement.
9. These figures were widely accepted, and often quoted in, for example, *The Minutes of Evidence of Parliamentary Committeees.*
10. L.T.C. Rolt, *George and Robert Stephenson* (1960) p. 71. For heights of the embankments see *Railways in England 1826 and 1827*, Translation from the German (Berlin, 1829), published for the Newcomen Society (Cambridge, 1971).
11. R.E. Carlson, *The Liverpool & Manchester Railway Project, 1821-1831* (Newton Abbot, 1969) pp. 188-192.
12. Thomas Telford, *Liverpool & Manchester Railway. Report to the Commissioners for the Loan of Exchequer Bills* (1829).
13. Joseph Locke, *Minutes of Evidence*, London & Southampton Bill, 19 June 1834.
14. Robert Stephenson, *ibid.*, 25 June 1834.
15. I.K. Brunel, *Minutes of Evidence*, London & Brighton Bill, 7 March 1837.
16. The Watford contract was let in May 1834 but work on the cuttings and embankments probably started not before August, as the earth waggons were built on site while digging foundation pits for the viaduct and sinking shafts for the tunnel. Completion of the tunnel and embankment is recorded in the April 1837 issue of *Railway Magazine*. Freeling *Railway Companion* gives volumes of the embankment and adjacent cuttings.
17. G.W. Buck, *Minutes of Evidence*, Great Western Bill, 7 July 1835.
18. James Copeland, *ibid.*
19. Robert Stephenson, *Minutes of Evidence*, London & Brighton Bill, 8 July 1836.
20. F. Swanwick, Reports on the North Midland Railway, *Railway Magazine*, (February 1839 and February 1840). This journal in November 1837 reports that operations had commenced in the Wakefield district where considerable numbers of workmen are already employed.
21. Half-yearly reports on the Croydon Railway, including data supplied by the engineer Joseph

Gibbs, *Railway Magazine, (*March and September 1837; March 1837).

22. Robert Stephenson, *Minutes of Evidence*, Great Western Bill, 8 July 1835.
23. John C. Bourne, *Drawings of the London and Birmingham Railway* (1839), pl. 19. An interesting variant was introduced by the contractor James & George Thornton in the Haywards Heath cutting on the London & Brighton line. This operated like a cliff railway with a frame carrying three loaded barrows being hauled up the steep slope on an inclined plane while another, unloaded, descended. See *Railway Magazine* (27 March 1841).
24. This information is given in Robert Stephenson's report, dated 17 Feb 1838, printed in *Railway Times*.
25. J.U. Rastrick, Half-yearly reports on the London & Brighton Railway, *Railway Magazine* (July 1839 and January 1840).
26. Half-yearly reports on the Great Western Railway, *Railway Times* (August 1838, February and August 1839).
27. S.A. Leleux, *Brotherhoods, Engineers* (Dawlish, 1965), p. 11.
28. C.W. Pasley,'Report on the present state of the Swindon embankment', 14 Jan 1842: *Reports of the Railway Department*, 1842, pp. 91-92.
29. Much of the information on Swindon embankment is given by the resident engineer Charles Richardson in a paper published as a separate pamphlet *On Landslips*. (Bristol, 1891).
30. I.K. Brunel. Half-yearly reports on the Cheltenham & Great Western Union Railway, *Railway Magazine* (3 Nov 1840 and 4 May 1841).
31. Joseph Colthurst, discussion on the paper by C.H. Gregory *Min. Proc. Inst. Civ. Eng.* 3, (1844) pp. 163-166.
32. Railway Magazine (January 1838), p. 65.
33. Lecount, p. 37.
34. Bourne, pl. 23.
35. A.W. Skempton,'Residual strength of clays', *Geotechnique* 35 (1985), pp. 3-18.
36. Robert Stephenson, Report on the London & Birmingham Railway., *Railway Times* (17 Feb 1838).
37. Peter Bruff, discussion on the paper by C.H. Gregory *Min. Proc. Inst. Civ. Eng.* 3, (1844) p. 147.
38. John Braithwaite, Report on the Eastern Counties Railway, *Railway Times* (27 Oct 1842).
39. C.W. Pasley, Reports on the Extension of the Eastern Counties Railway from Brentwood to Colchester, 14 March 1843 and 29 July 1843: *Reports of the Railway Department*, 1843, pp. 150-153.
40. C.H. Gregory, "On Railway Cuttings and Embankments; with an account of some slips in London clay, on the line of the London to Croydon Railway" *Min. Proc. Inst. Civ. Eng.* 3, (1844). The paper was presented on 26 March 1844. The lengthy discussion includes several valuable contributions.
41. I.K. Brunel, Half-yearly reports on the Great Western Railway, *Railway Times* (22 Feb 1842).
42. Sir Frederic Smith, 'Report on the accident on the Great Western Railway on the morning of the 24th December 1841'. 25 December 1841: *Reports of the Railway Department*. 1842, pp. 77-79.
43. Joseph Colthurst, discussion on paper by C.H. Gregory *Min Proc Inst. Civ. Eng. 3* (1844) pp. 167-168, including a section of Sonning cutting with a sketch of the slip surface.
44. C.W. Pasley, 'Report on the recent slip on the Croydon Railway, near New Cross', 12 Jan 1842. *Reports of the Railway Department.* 1842, pp. 92-93.
45. K. Terzaghi, 'Stability of slopes in natural clay', *Proc. Int. Conf. Soil Mechanics* (1936) *1* pp. 161-165.

46. P.R. Vaughan. 'Assumption, prediction and reality in geotechnical engineering'. *Geotechnique*, 44 (1994), pp. 573-609.
47. C.W. Pasley, 'Report on the slip of earth in the Bugbrook cutting . . . with observations on embankments and cuttings' 20 Sept 1842. *Reports of the Railway Department*, 1843, pp 225-226.
48. R.J. Chandler & A.W. Skempton,'The design of permanent cuttings slopes in stiff fissured clays', *Geotechnique*, 24 (1974), pp. 437-466.
49. R.B. Dockray, discussion on the paper by Gregory *Min. Proc. Insti. Civil Eng.* 3 (1844) pp. 148-150. He adds that counterforts since their introduction about 5 years ago have been 'extensively' used in repairing slips on the London & Birmingham Railway. Counterforts of solid masonry, acting solely as internal buttresses, were used in 1833 to stabilise the slopes of Panthier earth dam on the Canal de Bourgogne, and similarly in 1835 in Cercey dam, (Alexandre Collin *Recherches expérimentales sur les Glissements des Terrains Argileux*, Paris, 1846). Gravel counterforts in English railway earthworks constituted an independent development; they appear to have been unknown to the earlier canal engineers.
50. C.W. Pasley,'Report on the proposed opening of the Gosport Branch of the London & South Western Railway', 5 Feb. 1842: *Reports of the Railway Department* 1843, pp. 171-172. Locke and the contractor Thomas Brassey ran into all kinds of trouble on the short length of this line between Botley and Fareham: slips in Fontley embankment and Tapnage cutting, and the collapse of Fareham tunnel (in the Reading Beds) when nearing completion in July 1841.
51. T.J. Woodhouse, Progress report on the Midland Counties Railway, *Railway Times* (June 1838).
52. General meeting of the Midland Counties Railway, *Railway Magazine* (Aug. 1840). It should be noted that about 25 percent of the Leire excavation was taken to spoil.
53. *Civil Engineer & Architect's Journal* 6 (1843), pp 268-269. This article includes an engraving of the Steam Excavator. Its bucket held 1.25 cu.yd.
54. Russell Jones, 'A history of civil engineering plant', *New Civil Engineer* 9 Nov. 1978, pp. 37-105.
55. Sir Edward Leader Williams. 'The Manchester Ship Canal', *Min. Proc. Inst. Civ. Eng.* 131 (1897), pp. 14-30. The use of steam power reached its maximum on the Panama Canal where, between 1906 and 1915, some 130 million cu.yd of excavation was accomplished by upwards of 100 steam shovels (as they were called in America) and 160 locomotives. (G.W. Goethals, *The Panama Canal*, New York, 1916). In both these works, however, most of the excavated material was taken to spoil.

16

The railway navvy: a reassessment

David Brooke

In the ninety years preceding the outbreak of the First World War, the construction of over 23,000 'route' miles [1] of line provided work for the navvy in railway building somewhere in the United Kingdom. Certain periods, such as the late 1840s when the 'Railway Mania' collapsed, saw a catastrophic decline in the demand for labour and, by the final decades of the nineteenth century, the navvy was compelled to find an increasing proportion of his employment in the fields of dock and reservoir construction. Nevertheless, no time in that century proved as bleak as was the depression of the early 1930s when seasoned navvies in London competed for the job of excavating the pedestrian underpass from Paddington main line station to the underground system in 1932 [2].

Evidence in the Records of Parliament

In the course of the nineteenth century, the navvy and his masters changed the face of rural Britain to an extent unequalled by any other group except, perhaps, for the diverse interests behind the Enclosure Movement. In view of their importance to the Victorian economy, it is incongruous that so little information on this body of men appears in contemporary official records.

Their existence is, for example, virtually ignored in that voluminous enquiry the Royal Commission on Labour of the 1890s. Earlier, in 1839 and 1840, Parliament had requested railway companies to provide information on the deaths and injuries suffered by labourers at work in building. A few companies, amongst which was the Great Western, responded with detailed reports but others, notably, two major projects of the time, the London & Brighton and Edinburgh & Glasgow, ignored the order. Thereafter, Parliament was interested only in deaths and injuries amongst passengers and company servants [3]. Similarly, in 1847 the Railway Department of the Board of Trade began a potentially valuable line of enquiry by ordering companies to supply statistics on the different types of men engaged in constructing their lines. It specified that they should be listed under several different headings, but the reply from many companies allocated them between only two: 'Labourers' and 'Artificers'. In this case, the latter group often included not only tradesmen but also clerks, inspectors, accountants, timekeepers, etc. [4].

With the exception of some passing references in the 'Criminal Returns' and an enquiry which is mentioned below, this is the sum total of the Parliamentary evidence on this very important labour group. It is random and incomplete and it stands in striking contrast to the several major enquiries into the life style of an equally notorious group, the canal boatmen [5]. Therefore, it is not surprising that most accounts of the origins, working conditions and customs of the navvy rely heavily on the literature produced by middle-class ladies (who usually deplored the fact that navvies did not conform to their standards of morality) [6], on the lurid accounts of their misdeeds which appeared in the press and, above all, on the evidence given to the Select Committee on Railway Labourers of 1846 [7].

The Enquiry of 1846

This committee had its origins in the discovery by John Roberton, a very active member of the Manchester Statistical Society, that the men driving the long Pennine tunnel at Woodhead between Manchester and Sheffield were living in conditions nearly as squalid as those suffered by many of the inhabitants of Manchester. The matter became almost an obsession for Roberton and he harnessed the astonishing energy of Edwin Chadwick to the cause of reform [8].

Pressure from Chadwick led to the appointment of a Select Committee chaired by Edward Pleydell Bouverie which heard evidence from engineers, including Brunel, contractors, policemen, railway missionaries, etc. Men from the group in question played a minor role in the enquiry—only three navvies were invited to speak and they received a mere 171 questions out of more than 3000 put to all witnesses. This situation is in obvious contrast to the structure of other enquiries into working life in the same period. The Royal Commission into Children in Mines of 1842, for example, collected statements from 34 miners in the South Staffordshire and Shropshire coalfields alone [9]. Apparently factory operatives and coal miners had something useful to say on these occasions, but not railway navvies. The most balanced and authentic account of life at Woodhead Tunnel probably appears not in the 1846 enquiry but in a booklet entitled *Strictures* which Thomas Nicholson, the contractor at its eastern end, produced in answer to the charges of the Manchester society [10].

Chadwick was correct in believing that Government inspection and employers' liability in relation to accidents should become features of railway building, but this ran counter to that 'spirit of independence' in the management of railway affairs which Brunel and others wishes to see preserved. The whole matter ended in disappointment for the supporters of the enquiry when Parliament failed to even discuss its proposals mainly because of the intense and prolonged debate over the repeal of the Corn Laws. With the end of the 'Mania' and the decline in the number of men employed in this type of work, the committee's report was quietly forgotten.

The picture of the navvy which emerges from nineteenth-century literature, from the episodes of violence which shaped the opinions of the newspaper-reading public and from the evidence given by the witnesses before the Select Committee of 1846 is of a young man whose ignorance, even in relation to personal matters, was so profound that he often did not know his own name and went by some outlandish sobriquet. In a large number of cases, the navvy came from Ireland and his home on the contracts was frequently a corner of a hut in some insanitary encampment. His progress about the country, so the story goes, was accompanied by theft and drunken assaults on policemen and women. Altogether, the tranquil rural communities which had the misfortune to be visited by navvies experienced only a little less disturbance than had been created by the 'Swing Riots' of 1830.

In the following account, the writer draws upon two other sources of information, the census returns and the records of the criminal courts of south-western England, in

FIG. 1. Image of the navvy: 'spades & trumps' (from *Punch*, 1855).

order to present a somewhat different picture of the origins and behaviour of the railway navvy.

The Workforce in the Census Returns

The British decennial census of population commenced in 1801 and the first to coincide with a time of railway building was that of June 1841. Unfortunately, in this enumeration the population had not to give an exact place of birth and, of further relevance for the listing of those who were building railways, the Registrar General allowed the census takers to give only an estimate of numbers if they encountered outstanding problems in compiling their returns. Thus the official covering the area through which Clayton Tunnel was being driven for the London & Brighton Railway, having found himself in a large encampment of temporary huts, provided only an estimate of the number of people there and, as if to justify his decision, added the comment "Labourers on the Rail Road, Strangers to the Parish—'Here today and gone tomorrow' " [11]. In contrast, complete information was gathered by the enumerator for another large body in the area, those to the north in the Ouse Valley engaged in building Balcombe Viaduct. The next census contains exact places of birth and ages, and the family relationships of the people listed. The same is true of subsequent enumerations through to 1881, which is at present the last census open to the public under the Hundred Year Rule.

As early as 1851, a pattern begins to emerge in relation to the origins of the men engaged in this type of work in England and Wales. Approximately 25 per cent of them in 1841 had been born in the county in which building was taking place and, according to later censuses, about 50 per cent in that county plus those immediately adjacent to it. Lines in the counties of the south-west, where there was little industry to compete for labour, had especially high levels of local participation. For example, 75 per cent of the men working on the Wilts, Somerset & Weymouth in 1851 came from the south-west, and 81 per cent of those building the Devon & Somerset twenty years later [12]. The majority of these people were, almost certainly, agricultural labourers and other groups, such as handloom weavers, who found themselves increasingly under the pressure of competition from machinery. As Charles Robert Hemingway, the head of a leading firm of contractors, testified: "Navvies were the pick of the agricultural districts through which the new railways were being made, both as regards physique and intelligence . . . " [13].

The attractions of railway work for the farm labourer were twofold: firstly, it was better paid. Earnings and wage rates varied considerably according to the experience and physical abilities of individual men, but it appears on evidence derived from court depositions and the proceedings of Petty Sessions that railway labourers in southern Gloucestershire were being offered 11p–14p for a 10 hour day in the early 1840s as compared with 9p on the land. Secondly, many farm workers suffered from underemployment because their work was often of a casual nature. Railway building may not finally have brought the rewards that many hoped for, partly because of the Truck System [14], but it certainly had a better potential than did agriculture.

In these circumstances, contractors rarely had difficulty in finding sufficient workers for the job and thus no formal and frequently-used recruitment machinery appears to have been required. To give an example of what often occurred: hundreds of men gathered in the neighbourhood of Cullompton, Devon, in the spring of 1843 when word spread that work would soon commence on the southern part of the Bristol & Exeter line. News had gone round the district that men would then be needed on the railway and many deserted their existing masters [15].

There were, however, some circumstances that led to shortages of labour. Very difficult pieces of work could produce requirements which exhausted local reserves. Blisworth Cutting in Northamptonshire on the London & Birmingham created such a demand for workers that its contractors posted agents on the roads of the district with instructions to offer any male pedestrian a job [16]. Next, when railway construction occurred in parts of the country that were thinly populated, such as the border counties of Scotland or the northern Pennines, contractors had to rely on itinerant labour. Only about 6 per cent of the workers on the southern sections of the Settle–Carlisle line in 1871 had been born in Yorkshire. In contrast, 17 per cent came from south-eastern England and their presence in north-west Yorkshire and Westmorland almost certainly came as a result of the transfer of labourers from several major contracts in the London area completed by Joseph Firbank in the 1860s to the scene of his works on the Settle–Carlisle [17]. Finally, through the entire history of railway building, men deserted the contracts in the hay and cereal harvest months of June to September. They may simply have wanted a change of working circumstances, and every hand was required on the land at this time of the year. Whatever the motivation, in summer ancient patterns of work and behaviour, allied to instinctive affinities with the land, briefly triumphed over the onward drive of the new technology. For the contractor, the loss of men in summer was doubly unfortunate since this was the most favourable period of the year for his work.

On the basis of this evaluation, therefore, the most important single source of workers for railway construction was generally labour which came from the surrounding countryside. This was simply a consequence of the huge surpluses of men on the land in the mid-nineteenth century and, no doubt, most of these people went back to the fields when work in their district had ceased.

But it cannot be denied that another element existed amongst the men in railway building. Whilst the bulk of the work may have been done by former farm labourers, from the earliest years of the railways there were groups of peripatetic workers who went about the country from one contract to another. In 1841, the census taker for the village of Bletchingley, Surrey, where a tunnel was being driven on the South Eastern Railway, noted in his returns on the workforce: "As to the kind of persons, they are chiefly from Derbyshire, Staffordshire and Yorkshire, and are the most uncultivated set of beings on earth" [18]. In their previous employment they may have been constructing the Manchester & Leeds and North Midland lines. As another example of mobility, when Enoch Naylor was charged before Somerset Quarter Sessions in 1844 with having been a member of a gang which committed a number of robberies in villages in the Black Down Hills, two of his fellow workers at Whiteball Tunnel on the Bristol & Exeter Railway came forward to give evidence in his favour. They had, they said, all left Lancashire together in 1834 and had subsequently worked with him on a series of lines, including the London–Southampton, Paris–Rouen and, shortly before, the Folkestone–Dover, and had "never heard any thing against him before this" [19]. Naylor and his companion, John Green and Jack Champion, were amongst the professional navvies and, no doubt, their training and experience made them individually worth several of the former farm labourers to any contractor.

The sequence of events, it appears, was that the news of railway building attracted agricultural labourers out of the fields and they constituted the bulk of the labour in most parts of England and Wales. Construction gangs were reinforced, however, by a leavening of men, the 'true' navvies, whose life was dedicated to heavy manual labour. These people moved around the country choosing their own projects or as part of a rather select group which followed a particular contractor.

As for the Irish, the nationality that is sometimes given the credit for having built the entire system, they were at work on railways throughout England even before the great outflow from Ireland which followed the famine of the 1840s. But in England they played a significant role only in those parts of the country where at least one of two conditions held: where industry was a counter attraction for resident labour, and where there was heavy immigration at the time, such as Lancashire and the west Midlands. Only approximately 2 per cent of the men on the Great Western in Wiltshire in 1841, for example, came from Ireland, and 3 per cent of those building the South Eastern in the neighbourhood of Tunbridge Wells ten years later. In contrast, 23 per cent and 26 per cent of the labour forces of the Shrewsbury & Hereford (1851) and West Lancashire (1881), respectively, had been born in Ireland. However, in Scotland, which experienced considerably more Irish immigration than did England or Wales, they were the leading railway builders of the country from the Borders to the Grampians; in northern Scotland their place was taken by Highlanders who were experiencing the same problems in relation to employment as were English farm workers [20].

Tradesmen

The majority of the work in railway building, such as excavating cuttings and making embankments, could be carried out by a reasonably fit labourer after only a little training. But any corps of railwaymen also contained workers with a wide variety of skills. Amongst the main trades which sustained the industry for almost a hundred years were masons, carpenters, blacksmiths, quarrymen, brickmakers and layers, miners, wheelwrights, millwrights, engine men, etc. On most lines, skilled workers probably accounted for 10–20 per cent of all labour [21]. Contractors rewarded these men with wages that were about 5p per day more than those received by labourers. By the 1880s they had been joined by at least one new skill, steel erecting, principally for bridges. Labourers also had to learn to use a new material. Concrete made a very early appearance with its use for a bridge on the District line in London in the 1860s [22], but it was first employed on an extensive scale in the building of the West Highland Railway thirty years later.

Tunnel miners were a rather special group of workers because of the hazards and unpredictable nature of their chosen occupation. Many railway miners may have gained their first experience of this craft in coal mining, but by the 1870s they were being joined by others in the mining fraternity, notably by men from the declining Cornish copper industry. The customary method of tunnelling used in Britain, the 'English System', began with the sinking of working and ventilation shafts. Miners then cut a narrow heading at the height of the crown of the proposed tunnel and this was later enlarged downwards. A complicated system of poles and boards supported the new bore [23]. Pneumatic drills were not given a sustained and mass trial in British tunnelling until the building of the Severn Tunnel in the 1880s—that is at least ten years after they had been successfully used in Europe for driving the Alpine Frejus Tunnel.

It is also late in the century before some of the slog of removing masses of earth, clay and loose rock is reduced by the introduction of mechanical excavators, the 'Steam Navvy'. Until then the only mechanical source of power, apart from cranes, which could regularly be found on contracts was the locomotive. At an early date in the construction of Britain's railways, Robert Stephenson and Brunel encouraged contractors to buy them for use on the London & Birmingham and Great Western main lines, chiefly for hauling trains of wagons between cuttings and embankments. Nevertheless, taking the whole railway building scene, for most of the century it was mainly human and horse power, shovels, pick axes and common gunpowder.

Casualties and Crime

The misdeeds of the navvy are given a prominent place in most accounts of railway building and, according to many authorities at the time, he clearly belonged to the 'dangerous classes' of Victorian society. Navvies were apparently capable of the most outrageous behaviour: they raffled the corpses of dead comrades, encouraged prostitutes to live in their insanitary hovels, gratuitously attacked policemen and Irishmen, spread smallpox and typhus wherever they went and even destroyed the last vestiges of the gravestone of Robin Hood, the legendary protector of the oppressed [24]. After reading these catalogues of immorality and violence, one is left to wonder how such atrocious folk could have finished a single mile of line, let alone several thousands. It would be equally wrong to suppose that they were a docile group. Railway building could be a savage business and it sometimes produced a savage response from the men.

Deaths and serious injuries occurred frequently on the contracts. Thomas Nicholson in his *Strictures* listed 26 deaths at Woodhead [25] and at least 21 occurred at Box, usually through men falling down shafts. Altogether, 65 deaths and 57 very serious injuries are recorded as having happened on the Great Western between Maidenhead and Bristol in the years 1838–41. Even the very easy route for building between Swindon and Kemble, near Cirencester, saw nine deaths in 1840 alone [26]. This is the minimum number of casualties since the extant records on the matter are far from complete.

It is clear from the reports of coroners that contractors and their agents could have done more to prevent accidents; sometimes not even the most commonsense precautions were taken. When three men died in an explosion at Wickwar Tunnel (near Bristol on the line to Gloucester) in November 1841, it transpired that J. W. Nowell, the tunnel contractor, stored his gunpowder in the blacksmith's shop and it had, almost certainly, been ignited by a spark [27]. Prosecutions of contractors for negligence were a rare event. After a drunken Irishman had fallen to his death in an unfenced cutting near Chippenham in 1839, the coroner ordered the Surveyor of Highways for Wiltshire to proceed against the contractor [28]. This usually resulted in only a fine by Petty Sessions. To those outside navvy circles, accidents were, more often than not, the fault of the men themselves. As the editor of the *Wiltshire Gazette* commented after a series of deaths at Box: "There is great carelessness in the workforce" [29].

The impression amongst the men that many contractors felt no sense of responsibility towards them can only have been reinforced by the knowledge that they could be left destitute and without the wages which they had earned. In May 1838, hundreds of men gathered in the centre of Reading after William Ranger, contractor at Sonning cutting, had become insolvent. He owed them two weeks' wages and so they were in an ugly mood. Trouble was averted only because the Mayor and Town Clerk offered to provide £1000 out of their own pockets [30]. Things did not often end so well. When Thomas Townshend, a contractor on the Bristol & Exeter, failed in 1842, a mob of navvies attacked the house of his agent, Robert Dunn, smashed up his furniture, threatened to kill Mrs Dunn and drank several casks of beer. Although they complained before Somerset Assizes that at the time of the offence they were starving, having not eaten for four days, the court sent the ring leaders to Australia for fifteen years [31].

Episodes such as this must not, however, be taken as proof that the navvy and the law were in perpetual conflict. On scores of other occasions, as the records of the Petty Sessions courts of the south west show, the men had sufficient confidence in the law and its machinery to take their wage claims before magistrates and they rarely failed to get a satisfactory award against their employers.

At the same time, it would be wrong to ignore the fact that the general public—or perhaps, more accurately, those in the middle classes who were their self-elected spokesmen—felt apprehensive at the news that railway builders were coming to their area. This may in part have stemmed from the absence in early Victorian society of a satisfactory rural police system. Many towns and cities possessed police forces by 1840, but the enforcement of the law in the countryside, where most railway building took place, was largely the responsibility of unpaid and part-time parish constables. There is a good deal of evidence to show that these men varied considerably in their prowess and dedication. The Rural Police Act of 1839 allowed county authorities to establish paid, uniformed and full-time forces, but few counties immediately adopted the Act because of opposition on financial grounds from ratepayers. Thus, until well into the second half of the century, when England and Wales had become truly policed societies, in many parts of the country during episodes of civil disorder there was virtually no agency of the law between the parish constables and the military—and the authorities were as reluctant to deploy the military then as they are today [32]. It is hardly surprising, therefore, that the indigenous populations of small towns and villages felt alarm about the impending arrival of hundreds, perhaps even thousands, of lusty navvies in their vicinity.

FIG. 2. Reality of the navvy: labourers with a contractor's locomotive in the 1870s (Wiltshire Record Office).

The majority of the episodes of disorder which occurred during railway building had their roots in one or both of two circumstances. Firstly, the antipathy which the English, Scots and Welsh felt towards Irish navvies. Trouble involving the Irish usually happened when they were a substantial proportion of a workforce and it arose chiefly because of the conviction that the Irish were prepared to work for lower rates of pay. Contractors claimed that they did not pay the Irish less than other nationalities, but it is difficult to get at the truth of the matter and dislike of the Irish was fanned by those occasions on which they were used as strikebreakers.

The second leading cause of the majority of the trouble involving navvies was excessive drinking; this is what lay behind the majority of the assault, drunk and disorderly, etc., charges in the courts. For example, in June 1838 the house of the constable of Twyford, Berkshire, was besieged by a mob of about 100 navvies after he had attempted to arrest a man who was seen stealing a parcel from the Reading Mail Coach. One of the leaders of the crowd, 'Lankey' John Johnson ('out of Lancashire') in a statement before Berkshire Quarter Sessions pleaded: "I was mad drunk and when I have had any Beer I do not know what I am doing" [33]. It did not save him from a sentence of six months hard labour, but the courts heard the same excuse on scores of other occasions. Whilst it is true that navvies were renowed for their heavy drinking, this has to be put into the context of an age in which drunkenness and its associated violence were common. It was part of the accepted and regular pattern of entertainment in town and country for an evening of heavy drinking to be concluded by a series of pugilistic contests in the streets, and not infrequently things got out of hand. The most serious disturbance which occurred in the neighbourhood of Box Tunnel in the years 1838–41, when most of the construction took place, was caused not by navvies but by a drunken mob from Box village. In September 1839, a crowd of 50–60 people descended on Kingsdown Common, where the largest annual fair in the district was being held, and destroyed stalls and tents, stole money and drink and attacked anyone who did not escape. A navvy who was violently assaulted later said in his deposition to Wiltshire Quarter Sessions: "I was beaten so much that I could not walk . . . I crawled away and went home" [34].

The early Victorian countryside was thus a crude, rough and boisterous place and, on the strength of a review of over 8000 cases heard by the Quarter Sessions courts of the counties Gloucester, Somerset, Wiltshire, Berkshire and Devon in the years 1837–45, it can be said that early railway building in the south-west made no significant contribution to the incidence of indictable crime. Even in the north of England and Scotland, where the majority of the worst attacks on the Irish occurred, its impact on levels of crime may have been only minimal. According to one well-placed observer, Frederic Hill, Chief Inspector of Prisons for Scotland and Northumberland and a former Government railway inspector: "The whole number of railway labourers committed to prison has been but small, and with few exceptions . . . the offences have been of a minor description, chiefly drunken quarrels" [35].

In conclusion, it appears that the battalions who built the railway network consisted of two types of men. Firstly, there was the itinerant element, the sort of men whose lives were dedicated to work on a series of civil engineering projects; they may have been a minority on most contracts but their experience gave them an importance out of all proportion to their numbers. Secondly, there were the masses of 'common' labourers, people from agriculture and declining industries, who grasped the opportunity to make a little money whilst construction was under way in their district. This dominant group was not, as popular legend claims, composed of social outcasts whose background and customs divided them from the inhabitants of the town and villages near which they were working. For the thousands of householders who provided them with lodgings during their brief stay in a locality, they were fellow country people driven by necessity into trying their luck in another form of employment. Their proclivities were those of the general working population. Thus, they showed an inclination towards excessive drinking in times of full employment, and a dislike of the policeman who interfered with their old and often brutal pastimes, and yet they were prepared to use the machinery of the law to prosecute those who stole a tool or an article of clothing from them.

Finally, the navvy, like other members of the working classes in a time which had no official retiring age or state pension, worked as long as his physical strength allowed even if this meant going on into his sixties. They were, according to the censuses, an ageing group; 32 per cent of the men on construction work in England in 1881 were aged 40 years and over, as compared with 13 per cent in 1841. This may indicate that by the end of the century, in a time of greater labour mobility, navvying had lost some of its attractions as a form of employment for young men [36].

The last word should be left to Elizabeth Garnett, secretary of the Navvy Mission Society and one of the few people to devote their lives to the welfare of these workers:

> "Certainly no men in all the world", she wrote in 1879, "so improve their country as Navvies do England. Their work will last for ages, and if the world remains so long, people will come hundreds of years hence to look at and to wonder at what they have done". [37]

References

[1] The use of the alternative measurement of line distance, 'track' miles, gives an even greater total.
[2] D. Sullivan, *Navvyman* (1983), p. 228.
[3] Parliamentary Papers, 1841, Vol. XXV.
[4] Parliamentary Papers, 1847, Vol. LXIII, and annually to 1860, Vol. LXI.
[5] Amongst these enquiries are the Select Committee on Canals, 1883, the Select Committee on Canal Boats, 1884, and the Royal Commission on Labour (Group B), 1893–4.
[6] See, for example, A. R. Tregelles, *The Ways of the Line: A Monograph on Excavators* (London & Edinburgh, 1858).
[7] Parliamentary Papers, 1846, Vol. XIII.
[8] An excellent account of the background to this investigation is contained in R. A. Lewis, 'Edwin Chadwick and the Railway Labourers', *Economic History Review*, New Series, 3, no. 1, (1950), pp. 107–118.
[9] Royal Commission on Children in Mines, Parliamentary Papers, 1842, Vol. XVI.
[10] T. Nicholson, *Strictures on a Pamphlet Published at the Request of the Manchester Statistical Society* (Manchester, 1846).
[11] P.R.O. 1841 Census, Poynings, HO 107/1112.
[12] For further information on statistics obtained from the census returns for England & Wales and Scotland, see D. Brooke, *The Railway Navvy* (Newton Abbot, 1983), pp. 175–92; also, D. Brooke, 'Railway Navvies on the Pennines', *Journal of Transport History*, N.S. III, no. 1 (February 1975), pp. 41–53, and 'The Railway Navvy of the 1881 Census', *Quarterly Journal of Social Affairs*, 2 (4) (1986), pp. 363–77.
[13] G. Y. Hemingway, 'The Hemingway Masons and Contractors from Dewsbury and their Associates' (typescript, undated), p. 675, Dewsbury Central Library.

D. Brooke 45

[14] The system used by many contractors under which men were paid not in coin of the realm but in tickets; these could be exchanged for goods at the contractor's Tommy Shop. The general opinion was that this depreciated the value of earnings.
[15] *Taunton Courier*, 26 April 1843.
[16] P.R.O. London & Birmingham Railway: Engineer's Reports, 1 Sept. 1836, and Birmingham Committee Minutes, 22 July 1836, RAIL 384/68 and 103.
[17] Brooke, *Railway Navvy*, p. 189.
[18] P.R.O. 1841 Census, Reigate, HO 107/1076.
[19] Somerset County Record Office, Taunton, Depositions in the cases of Enoch Naylor and Ann Starr, Somerset Epiphany Adjourned Quarter Sessions, 1844.
[20] Brooke, *Railway Navvy*, pp. 28–9.
[21] ibid., pp. 53–4.
[22] B. Morgan, *Railways: Civil Engineering* (1973), p. 149.
[23] A succinct description of this method is contained in G. E. Sandström, *The History of Tunnelling* (1963), p. 114.
[24] P. Valentine Harris, *The Truth About Robin Hood* (Mansfield, 1978), p. 36.
[25] Nicholson, *Strictures*, pp. 4–8.
[26] Based on reports for various dates in the *Devizes & Wiltshire Gazette, Reading Mercury, Bath Chronicle* and *Gloucester Journal*, and Returns on Railway Accidents, Parliamentary Papers, 1841, Vol. XXV.
[27] Gloucestershire County Record Office, Gloucester, Inquest Book of J. G. Ball, 7 Nov. 1841, CO 3/2.
[28] *Devizes & Wiltshire Gazette*, 10 Oct. 1839.
[29] ibid., 1 Nov. 1838.
[30] *Berkshire Chronicle*, 2 June 1838.
[31] *Taunton Courier*, 6 April 1842.
[32] Although at this time men from the Metropolitan force, the first experts in crowd control, were "borrowed" on several occasions by authorities in the north to suppress disorder caused by navvies and opposition to the New Poor Law.
[33] Deposition in the case of John Johnson, Berks. Midsummer Quarter Sessions, 1838, Berks. County Record Office, Reading.
[34] Depositions Nos. 62–70, Wilts. Michaelmas Quarter Sessions, 1839, Wilts. County Record Office, Trowbridge.
[35] Twelfth Report on the Prisons of Scotland, Parliamentary Papers, 1847, Vol. XXIX.
[36] This shift is not related to movements in the age structure of the entire population, and it had occurred amongst both itinerant and locally-born navvies.
[37] Quoted in Sullivan, *Navvyman*, p. 89.

17

Railway contractors and the finance of railway development in Britain

Harold Pollins

I

The contractor is an elusive figure in the history of British railways. There is the important study of his activities by Jenks[1] and several useful biographical works,[2] but those who write on railway history are not usually concerned with the company as a business unit and the constructional side of the story is seldom discussed in financial terms.[3] Yet the contractor was an imposing and well-known figure. Charles Thomas Lucas was described as 'one of the foremost men of the nineteenth century'; the firm of Lucas & Aird is said to have been 'one of the largest employers of labour in the country'.[4] William Dargan was regarded as 'a gentleman who has constructed Irish railways and employed Irish labour to a larger extent than any other Irishman ever did, and who has thus acquired a noble and well-earned fortune'.[5] Contemporary railway periodicals, not noted for their inhibitions or selectivity, attacked contractors constantly—often more than directors or officials. In the mid Victorian period the contractor moved up the social scale easily and quickly, and the prototype career (portrayed by Trollope in *Dr Thorne*) of a fortune, knighthood, election to

This article was first published in the *Journal of Transport History*, Vol III (1957–8), 41–51, 103–10, and appears in this volume in a slightly revised form.

I should like to thank the following for their criticism and advice: Dr Charles Feinstein, Professor W. E. Minchinton, Dr E. A. Course, Mr E. W. Cooney, Mr Charles E. Lee.

[1] L. H. Jenks, *The Migration of British Capital to 1875* (London & New York, 1927). He is mainly concerned with overseas railways, but has a great deal to say on British railway development.

[2] A. Helps, *Life and Labours of Mr Brassey* (1872); F. MacDermott, *The Life and Work of Joseph Firbank* (1887); Henry Peto, *Sir Morton Peto* (1893); G. A. Hobson, *The Life of Sir James Falshaw* (1905); J. A. Spender, *Weetman Pearson* (1930); I. Thomas, *Top Sawyer; a biography of David Davies of Llandinam* (1938); R. K. Middlemas, *The Master Builders* (1963); D. Young, *The Member for Mexico* (1966).

[3] This is the case too of *The British Civil Engineering Contracting Industry* (Federation of Civil Engineering Contractors, 1956), but it has some useful points on finance.

[4] *Proceedings* of the *Institution of Civil Engineers*, Vol 124 (1895–6), 439–40.

[5] *Illustrated London News*, 21 May 1853, 393. His contracting life began in the early 1830s on the Dublin & Kingstown Railway (O. J. Vignoles, *Life of Charles Blacker Vignoles* (1889), p 182). Another important, and now forgotten, contractor was Peter Thompson, 'for several years . . . known as one of the largest railway and general contractors in the country', *Herapath*, 20 May 1876, 563.

Parliament, and fall from grace was no mere caricature.[1] Even sober bodies, such as the Institution of Civil Engineers, could find room in their publications for caustic remarks on some of the earlier contractors:

> The nature of the relation which subsisted between the engineer of a public company and the contractors who undertook the execution of the works was, at the commencement of Mr Brassey's career, very different from that which subsequently obtained. The early contractors were, for the most part, men of strong natural abilities, insight into the cost and method of executing work amounting to instinct, low tastes, violent habits, and grasping tenacity of purpose. A contract being once made, it seemed to be regarded as natural that the contractor should set his wits to work to make the most of it. This was to be done, on the one hand, by grinding his labourers under the pressure of the truck system and the 'tommy shop', and on the other hand by 'scamping' his work. . . . As a rule, vexatiously minute inspectors were open to bribes. They gave trouble unless they were bought off.

Brassey, it noticed, did a great deal to ameliorate these conditions, as did other large contractors, notably Peto.[2] It was not common for the *Railway Times* to say anything laudatory about anybody in the railway world, but in a discussion on the East London Railway in 1865 it stated that 'the contract with Messrs Brassey fully assures completion of the works within the capital defined in the act of Parliament'.[3] In railway history that is high praise.

It would be absurd to suggest that the contractors were the most important figures in the railway world, but in the period after the great mania of the 1840s any history of railway construction which ignored them would be ludicrously incomplete.[4] There were well-known 'railway politicians' (concerned with negotiations between companies about working arrangements and also construction) and railway operators—James Allport of the Midland Railway, for example. But there is no doubt that in the period between the fall of George Hudson, the 'Railway King', in 1849, and the emergence of Sir Edward Watkin (who came into prominence in the late 1860s) it was the contractors who were the centre of attraction.

Their activities after 1850 are usually associated with the overseas railways. As the *Railway Times* put it:

[1] There is a short description of the kind of rich, vulgar contractor who started as a navvy in J. Francis, *A History of the English Railway* (1851), Vol I, 271–2.

[2] *Proc Inst Civ Eng*, Vol 33 (1871–2), 248–9 (Brassey). For complementary remarks on Peto, see *Select Committee on Railway Labourers*, PP 1846, XIII, Report, p iv; also J. Francis, op cit, I, 266–73.

[3] *Railway Times*, 17 June 1865, 790. See also Joseph Devey, *Life of Joseph Locke* (1862). Chapter XI, 'The Railway Contractor', is largely in praise of Brassey.

[4] A virtue of such books as Charles E. Lee, *Metropolitan District Railway* (South Godstone, 1956); C. P. Gasquoine, *Story of the Cambrian* (Wrexham, 1922); and C. G. Maggs, *The Midland & South Western Junction Railway* (Newton Abbot, 1967) is that the activities of the contractors are given due place.

> No one can have failed to observe the active movements of our wealthy and enterprising contractors, gentlemen who have made large fortunes within a short time, and who now seem bent on losing them as quickly. Having completed as they consider their work at home, and having given us an almost perfect system of railways here, they have started on a kind of knight-errantry to supply railway deficiencies all over the world.[1]

Herapath's Railway Journal, which often took a less jaundiced view of railway events than did the *Railway Times*, could be more sympathetic. During the Crimean War it praised the railway contractors in these terms:

> Our great railway contractors will figure in history as having performed a conspicuous part in the destruction of the Russian enemy, as well as in the construction of railways all over the world—both operations being most beneficial to the interests of society at large.[2]

Their work in Britain was also important. They were associated with numerous public works, and their financial guarantee made possible the Great Exhibition of 1851.[3] In particular the 'contractors' lines' are worth investigating; these were the railway companies in which the contractors' activities amounted to a great deal more than merely building the line.

The railway contractors were not new in the 1850s; railways had almost always been built by them from the earliest days, as had, indeed, the canals before them. Railway companies might resort to direct construction in an emergency—the most famous example being the Kilsby tunnel on the London & Birmingham Railway in the 1830s, when the engineering difficulties proved too great for the contractor.[4] And the great contractors emerged early; Peto and Brassey were engaged in extensive overseas contracts in the 1840s.[5] Nevertheless their greatest significance in railway finance came after the period of the great mania of the 1840s.

By the end of the mania just over one-third of the total route-mileage ever built had been constructed; by 1870 the figure was about three-quarters, and in the succeeding fifty years another 5,000 miles were built. A great deal was obviously accomplished in the 1850s and 1860s—about 8,000 route miles were opened[6]—and it was done in a period of acute diffi-

[1] *Railway Times*, 27 November 1852, 1257.

[2] *Herapath*, 2 December 1854, 1239. Also M. Robbins, 'The Balaklava Railway', *Journal of Transport History*, Vol I (1953).

[3] C. R. Fay, *Palace of Industry, 1851* (1951), p 7. William Dargan virtually financed the Dublin Industrial Exhibition of 1853: *Illustrated London News*, 21 May 1853, 393.

[4] S. Smiles, *Life of George Stephenson* (3rd ed, 1857), pp 329–31. There is a useful description of early contracting methods in *Railway Times*, 20 January 1844, 63–4, and 23 February 1844, 228. Much information is contained in *SC on Railway Labourers*, PP 1846, XIII, and Jenks, op cit, pp 134–6.

[5] In May 1846, at the peak of railway construction, Peto was employing, directly or indirectly, 9,000 men. (*SC on Railway Labourers*, PP 1846, XIII, QQ 1228; 30).

[6] There is a useful analysis of the growth of railway mileage in G. A. Wade, 'Twenty-two thousand miles of railway', *Railway Magazine*, October 1904, 290–7.

culty for the railways. Broadly, for a generation after the mania British railway companies, old or newly established, found it hard to raise money.

Some of the reasons were peculiar to the railways: the excesses of the 1840s, and the low rate of dividend with share prices below par for much of the 1850s, even in the case of the major companies, did not endear them to the investing public, despite the continuing expansion of traffic. The high proportion of capital with fixed interest and fixed dividends reduced the amount of net profit available to holders of ordinary shares. Nor was confidence enhanced by the continued depredations of railway servants—most companies at some stage or other had to report absconding officials. Some companies exhibited peculiar finances and committees of investigation were appointed from time to time. The affairs of at least one company—the West Hartlepool Harbour & Railway Company—were the subject of a parliamentary inquiry.[1]

Particularly noteworthy was the prohibition in 1847 of the payment of interest on calls. This may well have been a major cause of the shortage of capital, even though it was evaded by, amongst other methods, contractors paying interest.[2] The long period between the promotion of a railway company and the first dividend paid meant that shareholders might wait for years before any net profit was divided. (Existing shareholders often insisted that the capital of extensions and branches should be separate and should not participate at once in the general profits of the company.) Finally, the most profitable routes had already been built.[3]

The unpopularity of British railways to investors came just as other opportunities became available. The extension of limited liability to industry and commerce, and, more especially, the growth of overseas investment, notably in railways, provided important competitors in the money market.[4] In the 1850s and 1860s the railway journals commented continually on the slowness with which investors took up shares in British railway companies. The older companies were particularly affected. In 1863, a year of improving trade in the upswing of the trade cycle, the chairman of the London & North Western Railway, Richard Moon, said:

> You will find that capital is going out to India, and all over the world seeking for employment in railways. What is the state of things in England? There is not one of the great companies in this country who can raise sixpence without preference or guaranteed shares. There are no proprietors willing to come forward to make a railway. They are

[1] *SC on the West Hartlepool Harbour and Railway Company*, PP (Lords) 1863, XXXIV (113).

[2] *SC on Standing Order 167*, PP 1882, XIII (196), Report, p iii, Q 263.

[3] A useful study of the financial problems of railway promotion and construction in the 1850s is contained in J. R. T. Hughes, *Fluctuations in Trade, Industry and Finance; a study of British Economic Development, 1850–1860* (Oxford, 1960), Chapter 7.

[4] *Railway Times*, 19 April 1856, 491, 'French and English Investments'.

made by contractors, engineers, and speculators, who live on the fears of the companies.[1]

Capital was needed in the first instance to meet the needs of growing traffic (thereby defeating the brave attempts to close capital accounts).[2] In any event they were forced to engage in new capital works. For the promotion of new lines was bound to affect the 'territory' of an existing company; if it ignored them or opposed them in Parliament, a competing company might take them up. Thus 'the fears of the companies' were real enough, for though new lines might be promoted in the 'traditional' manner, with local meetings at which subscriptions might be obtained,[3] it became normal practice for new companies to try to get the support of a main line company. A proposed line, strategically placed between two existing routes, would have no difficulty in either getting assistance from one of them or being taken over.

As well as these ways of financing new construction—old companies building to meet the demands of traffic or for self-preservation; new companies assisted by old companies—there was a third way: new companies associated with contractors.[4] It is not easy to find examples of new companies obtaining their funds entirely from the public;[5] nor of the major, established companies being concerned in these contractors' lines.[6]

II

The normal arrangement between company and contractor was that the contractor was paid monthly on the certificate of the railway company's engineer. Usually 10 per cent of the amount would be retained by the company as a guarantee fund. A contract between the London & North Western Railway and Thomas Brassey and Alexander Ogilvie stated:

[1] London & North Western Railway, *Reports and Accounts*, year ending 31 December 1862 (BTHR, RAC 1/234), Chairman's speech.

[2] H. Pollins, 'Aspects of Railway Accounting before 1868', in A. C. Littleton and B. S. Yamey (eds), *Studies in the History of Accounting* (1956). (Reprinted in this volume, p 138 ff.)

[3] G. H. Evans, *British Corporation Finance, 1775–1850* (Baltimore, 1936); H. Pollins, 'The Marketing of Railway Shares in the first half of the Nineteenth Century', *Economic History Review*, 2nd ser, VII (1954).

[4] T. Coates, *Railway Deposits Discussed* (1864), p 8. This is one of the few contemporary publications to praise the contractors' lines.

[5] Studies of some of the small railway companies bear this out: D. S. Barrie & C. R. Clinker, *The Somerset & Dorset Railway* (South Godstone, 1948); J. G. Spence, *The Caterham Railway* (South Godstone, 1952); D. S. Barrie, *The Rhymney Railway* (South Godstone, 1952).

[6] A minor, isolated example in 1851. The old-established Eastern Counties Railway discussed building a line by paying the contractor in bonds: ECR Board Minutes, 25 September 1851 (BTHR, EC 1/10, p 74).

> The Company shall and will pay to the Contractors for and in respect of the said Railway and Works . . . the sum of Three hundred and sixty nine thousand five hundred and ninety one pounds sixteen shillings. The said sum shall be paid monthly in such proportions as shall be certified by the Engineer as being in his judgment the proportionate value of the works executed during each month subject to a deduction of Ten pounds per cent to be held and retained in hand by the Company as a guarantee . . . against any default or neglect of the Contractors.[1]

In the case of the contractors' lines this was changed. A deduction was still made but the method of payment was different: the company's securities were taken by the contractor instead of cash. An abortive contract between the Swindon, Marlborough & Andover Railway and William Wright stated that the contract price of £399,045 was to be paid as to one-third in debentures and two-thirds in shares. (This was modified later to £100,000 in debentures and the remainder in shares.)[2]

It is true that even before 1850 contractors and other creditors would take shares and bonds instead of cash. In 1848 the Great Northern Railway paid Thomas Brassey £22,000 by a mortgage bond, and in the same year the North Staffordshire Railway paid him £45,000 in bonds.[3] But the records of the companies show that these methods were irregular and infrequent, and companies would as soon rely on promissory notes, if they were short of money, or temporary loans.[4] A hint of the future was the offer by the contracting firm of Charles Fox & Co to invest £150,000 in a proposed Cornwall Railway;[5] and the tendency of contractors to lend money or to be owed large sums, in the late 1840s, was another indication of things to come.[6]

When a contractor was paid in securities the railway company was obviously solving its financial problems. For the contractors a possible motive was the need to keep their permanent staff and equipment occupied. By the 1850s some had large organisations; in 1860 Peto, Brassey & Betts were described as 'the largest contractors in Europe'.[7] Of Thomas Brassey it was said:

> The capital that Mr Brassey dealt with was never idle. As soon as any part of it ceased to be wanted for any one great work, it was required for another, which either had to be commenced or was entering into a

[1] Contract for the Runcorn & Aston branch, 1 May 1863, cl 36 (BTHR, LNW 3/28).
[2] Contracts dated 27 July 1875 (cl 18); 19 October 1875 (BTHR, MPR 34/7).
[3] Great Northern Railway Finance Committee, 2 August 1848 (Minute 108) (BTHR, GN 1/219); North Staffordshire Railway, Board Minutes, 25 November 1848 (Minute 349) (BTHR, NS 1/4).
[4] G. H. Evans, op cit, *passim*.
[5] Devon & Cornwall Railway, Provisional Committee, 30 October 1843 (BTHR, COR 1/1). This is the company referred to in E. T. MacDermot, *History of the Great Western Railway* (2 vols, 1931), II, 261–2.
[6] H.P., *Peto*, pp 20–1; *Bradshaw's Railway Shareholders' Manual*, 1854, 109.
[7] *Engineer*, 20 January 1860, 35.

> phase of full activity. It must be recollected, that a main object with Mr Brassey was to furnish sufficient work for all that staff of skilled agents and for those large bodies of workmen whom he had collected around him.[1]

With railway companies and investors not keen to go in for new works, contractors had to participate in the promotion of schemes if their capital equipment was to continue to be fully employed. It is difficult, however, to decide who exactly were the promoters of many of these lines. In the boom of the 1860s there was a great deal of talk about lawyers' and engineers' lines,[2] as there had been in the mania of the 1840s. And, as ever, there were numerous schemes which began in provincial towns, with local notabilities trying to get railway accommodation. But although the evidence is not very clear—the minutes of companies are often silent on these matters—it seems likely that contractors at least helped to promote. Matthew Sheild and G. F. Bruce, two contractors who became bankrupt in the late 1860s, were described as promoters of railway schemes.[3] It is not always clear, however, because there was a great deal of talk of contractors' nominees sitting on railway boards,[4] the absence of a contractor's name in the early proceedings of a company does not, therefore, preclude the possibility of his inspiring it. And although it was said on several occasions that they did not promote lines,[5] the fact that the subscription contracts of two Irish companies were signed only by contractors provides strong, if not conclusive, evidence of promotion.[6]

Railway companies were statutory companies, obtaining their powers and legal status from Parliament. Part of the necessary procedure was the submission of a subscription contract containing the names and other details of potential shareholders who had paid a deposit and agreed to take shares if the company obtained its Act. The House of Commons abolished this requirement in 1858 on the grounds that such subscription contracts

[1] Helps, *Brassey*, p 154. However, another view refers to 'the habit of contractors to wind up the services of the entire staff at the end of a contract, and to engage new staffs for each contract as it came along'. Spender, *Pearson*, p 38.

[2] *The Economist*, 14 January 1865, 36.

[3] *Railway Times*, 26 January 1867, 89.

[4] Richard Samuel France, the contractor for the Mold & Denbigh Junction Railway, stated this quite explicitly:

> It is usual, that when a Contractor undertakes the entire responsibility of finding the whole of the money, and carrying out the works, he has the option of putting some of his own friends on the Board in order to take care of his own interests.

Letters and correspondence between Lord Redesdale and Mr France upon the subject of the Mold and Denbigh Junction Railway, PP (Lords) 1867, XIX (20), p 8.

For an attack on R. S. France and the Welsh contractors' lines, see *Railway Times*, 23 February 1867, 203–4.

[5] *SC on Railway and Canal Legislation*, PP 1857–8, XIV, Q 1294. *SC on Standing Orders (Parliamentary Deposits)*, PP 1864, X, Q 666 ff. *SC on Standing Order 167*, PP 1882, XIII, Q 170.

[6] The companies were the Cork, Midleton & Youghal, and Dublin & Wexford, both in 1853 (BTHR, SUB 3/1).

were not very reliable in providing a guarantee that the company had a solid body of financial support. Doubts as to their validity had been common for many years before they were discontinued,[1] but while they continued to be required the companies had to take seriously the question of 'filling up' the contract. In addition a deposit was submitted—usually in government stock—as a security, while the Bill was being considered and the line being built. Thus a company's initial financial and administrative needs might be sufficiently acute for its progress to be held up. The nature of the parliamentary timetable was such that a delay was disastrous. The documents and funds had to be submitted by a certain date, otherwise the Bill might not be examined that year and the company would have to wait 12 months, by which time the shareholders might well have lost heart or another company might have got in first. This all accounts for the feverish activity in October and November each year on the part of companies preparing to go to Parliament (or watching for possible competing schemes).

In the 1840s it had been common for directors of existing companies to sign the subscription contract on behalf of their own branches or for nominally independent lines.[2] This continued to be done in the 1850s; but in the case of new companies the contractors played a large part too. The largest individual subscriber to the West End & Crystal Palace Railway in 1853 was Edward Ladd Betts (£60,000 out of a total of £223,750). William Dargan subscribed over one-third to an Irish line. And Brassey, Peto & Betts subscribed to others. Four contractors subscribed £33,500 out of £72,070 to a Mid-Kent extension line in 1857,[3] and there are numerous other examples.[4] Contractors, too, supplied money for the parliamentary deposit. In 1861 one-half of the deposit for the Bill of the Metropolitan Railway's extension to Moorgate Street was provided in this way.[5]

The fact that contractors signed the subscription contract and financed the deposit is not in itself important—the records of many companies show that the deposit was in fact often borrowed.[6] What is significant is that the signing of the subscription contract became part of the negotia-

[1] O. C. Williams, *Historical Development of Private Bill Procedure and Standing Orders in the House of Commons* (2 Vols, 1948–9).

[2] Pollins, 'Marketing of Railway Shares', loc cit, 235.

[3] *Railway Times*, 11 June 1853, 601–2; 7 May 1853, 481; 10 January 1857, 32–3. This journal printed many subscriptions, or extracts from them, in the 1850s, as did *Bradshaw's Railway Shareholders' Manual.*

[4] See, for example, the two volumes of subscription contracts, BTHR, SUB 3/1 and 3/2.

[5] *Railway Times*, 31 August 1861, 1098. This journal argued that the contractors in their turn borrowed money for the deposit: ibid, 26 March 1864, 469.

[6] Robert Baxter, a parliamentary agent, said the deposit was invariably borrowed. *SC on Standing Order 184, Section 2*, PP (Lords) 1866, XXVIII (168), Q 775.

Many of the deposits for the London tube railway Bills in the early twentieth century were borrowed: Underground Electric Railways, Board Minutes, 5 August 1902 (Minute 71); 29 March 1904 (346); 10 January 1905 (432); 12 September 1905 (493) (BTHR, UER 1/1).

tions between company and contractor. Examples are to be found in the case of two companies, which may be regarded as case-studies.

The Rhymney Railway was formed during the boom of the early 1850s. The board had the support of the neighbouring Taff Vale Railway, the Rhymney Iron Company, and the trustees of the Marquess of Bute. Nevertheless the company found itself compelled to approach George Wythes, the contractor, 'to fill up the Subscription list in case of need'. He agreed to take the contract for the line and also 2,500 shares, but in the end he signed the subscription contract for £15,000. A year later an extension line was proposed and again Wythes signed the contract for £15,000, paid the deposit, and obtained the contract.[1]

The East Kent Railway (the name was changed in 1859 to the London, Chatham & Dover) was also a result of the boom in the early fifties. The minutes of the Provisional Committee, which begin in October 1852, show quite clearly that the Railway was leaning quite heavily on Sir Charles Fox (of Fox, Henderson & Co). After some vague references to negotiations the company agreed that out of the 30,000 £25 shares[2] Fox and his friends should take 14,500 and pay the deposit of £2 10s 0d on them, provided that 5,000 were subscribed by individuals on application to the company and the deposit paid. In return Fox would have the contract. Fox agreed to find £36,250 for the parliamentary deposit. And out of the authorised capital of £700,000 the subscription contract was signed to the amount of £544,375, of which only £135,375 was subscribed locally (by 334 local shareholders). Fox himself signed for £31,875.[3]

It follows from all this that such competition as there was between contractors was not over the price of the tender, but rather over the ability of the contractors to provide funds in some way.[4] There were contemporary allegations of a price ring among contractors;[5] there was certainly an association of railway contractors which held quarterly meetings.[6] In this situation the likelihood of capital accounts being inflated was clearly very great.

[1] Barry, *Rhymney Railway*, pp 52–3. Rhymney Railway, Board Minutes, 14, 22, 27, 31 December 1853, and correspondence with Wythes in BTHR, RHY 1/1, p 393. He also signed the subscription contract for a later branch line: ibid, 30 December 1856.

[2] After various amendments to the prospectus—detailed in the minutes—the capital was fixed at £700,000 (ie 28,000 shares): *Railway Record*, 4 December 1852, 786.

[3] Provisional Committee, 22 November 1852 (BTHR, LCD 1/5). See also *Railway Record*, 11 December 1852, 790, where it was stated that two thirds of the capital had been taken by 'capitalists'; *Reports & Accounts*, 31 December 1854 (speech by Fox); *Railway & Travel Monthly*, November 1919, 295; *Railway Times*, 15 January 1853, 62–3.

[4] Swindon, Marlborough & Andover Railway, Letters etc relating to tenders . . . (BTHR, MPR 34/2). In this file there is a list of tenders from various contractors. Those which made no offer about finance were rejected.

[5] *Herapath*, 17 June 1848, 644; 4 November 1848, 1161; 25 May 1850, 508–9; 18 September 1852, 1017. These news items also refer generally to the absence of competition and to the tendency to tender for the whole of a line.

[6] *Railway Times*, 6 April 1839, 295.

III

It is not possible to give a comprehensive list of contractors' lines. It is apparent that many were promoted in the 1850s and 1860s, especially in the boom of 1864–6; many of the lines promoted in East Anglia, associated with the name of Peto, were of this character.[1] Some of the companies which were analysed in a series of articles in the *Railway Times* in 1869 under the heading of 'Insolvencies' and 'Abortions' were contractors' lines, and it is interesting to observe that *The Economist* estimated that two-thirds of the capital authorised in the session 1864–6 was for schemes which could be characterised as contractors' lines.[2] Robert Baxter, a parliamentary agent, said in 1864 that 'nine railways out of ten are made with the aid of contractors'.[3] Not all of them were small; some of the most important ones were the London, Chatham & Dover, the Metropolitan District and the various companies that made up the Cambrian Railways.

While contractors might have promoted companies, and have taken part in the parliamentary proceedings, the essential characteristic of contractors' lines was that payment for construction was in the railway company's securities. The activities of the contractors would often go beyond construction. It was not uncommon for contractors to arrange to buy the land, pay the railway company's engineer and solicitors, and guarantee the interest on the shares during construction. The contract[4] between the Swansea Vale & Neath & Brecon Junction Railway and John Dickson of 1864 had these clauses:

> The Contractor as the Agent of the Company will as soon as can be enter into and carry into effect the requisite Contracts for the purchase by the Company of all Lands Buildings and other property which shall be requisite either temporarily or permanently for the Railway and the Sidings Stations and other works and conveniences thereof. (clause 9)
>
> The amount of the purchase money or compensation in every case of voluntary agreement shall be determined exclusively by the Contractor and in all cases of reference to Arbitration or to a Jury the lawful and reasonable requirements or directions of the Contractor as to the proceedings thereof shall be observed by the Company and their Solicitor. (clause 11)

Similarly in the case of Thomas Savin and the Brecon & Merthyr Railway. In 1860 Savin made a proposition about the construction of the line.

> Mr Savin undertook to construct their lines for so much a mile, out of which he was to provide for all incidental expenses such as the payment

[1] See the account in *PETOVIA, being a review of the scheme for a railway from Pitsea to Colchester . . . by a tooth of the dragon* (1857).

[2] *Economist* Commercial History and Review of 1866, 9 March 1867, p 31.

[3] *SC on Standing Orders (Parliamentary Deposits)*, PP 1864, X, Q 645.

[4] Contract dated 29 September 1864 (BTHR, MPR 261).

of the Company's Engineer and Solicitor, and all charges for land, or parliamentary expenses, he also undertook to lease the line for 5 per cent for ten years after opening and to pay the same rate of interest during construction.[1]

The paying of interest during construction was a fairly normal feature of contractors' lines. In 1853, for example, Sir Charles Fox agreed to pay interest at 4 per cent during construction,[2] and a great deal of evidence testifies to the continuance of this practice into the eighties and nineties.

It may seem strange, after the discussion so far about contractors taking shares, that they should have agreed to pay interest on them. The point is that in many cases shares were issued to the public in order that the more normal relation between company and contractor should supervene. And when the contractor took shares his prospects of getting rid of them, or borrowing on their security, were correspondingly improved by the offer of guaranteed interest. Indeed one contract, between the Swindon, Marlborough & Andover Railway, and Watson, Smith & Watson stated explicitly that the contractors should take the contract for shares and debentures, or cash if it were available. That some cash would be forthcoming was guaranteed in the contract, because the directors and solicitors arranged to take a certain number of shares. The eventual result was that a large proportion of the contract was paid in cash, promissory notes, and Lloyd's bonds.[3] Similarly, the Central London Railway arranged to pay its contractors partly in bonds; the company had 1,911 shareholders, and the shares were paid up.[4]

Nevertheless there is no doubt that in many cases contractors were paid wholly or largely in shares and debentures. The shares were usually taken at a discount; this was well known at the time. In 1863 the secretary of the London & North Western Railway gave evidence before a select committee of the Lords:

> Earl of Airlie: You know, as a matter of fact, that the contractor does take shares in the case of small Companies who are in difficulties at a lower rate than their nominal amount?
>
> Charles Stewart: Yes; but those shares must be considered as paid up in full.[5]

In one case, it is true, the contract specified that the shares were to be taken

[1] 'Conybeare and the Brecon & Merthyr Railway' (BTHR, MPR 4/32). Conybeare was the engineer to the company.

[2] East Kent Railway, Provisional Committee, 24 August 1853. Also *Railway Record*, 27 August 1853, 580.

[3] The contract is dated 15 September 1879. It is in a file of papers (BTHR, MPR 34/1) which contains a list of payments made to the contractors specifying the various methods of payment.

[4] *Railway News*, 24 February 1900, 307.

[5] *SC on Railway Companies' Borrowing Powers*, PP (Lords) 1863, XXXIV (175), Q 361.

by the contractor at par. But since the contractor increased his prices the eventual result came to the same thing as an issue at a discount. The contract was between the Swindon, Marlborough & Andover Railway and William Wright. The contractor at first offered to take the contract for £344,797 if he were paid in cash, or £407,828 if paid in shares and debentures. The contract price was eventually fixed at £399,045, to be taken in shares and debentures. The specification shows that the contract price was increased by a flat 17½ per cent to bring it up to £399,045.[1] This seems to be a different situation from that described by the engineer to the Brecon & Merthyr Railway in 1866. In the course of a letter to the company's solicitors, the engineer, H. Conybeare, wrote:

> In the case of Contract lines for a lump sum per mile, it is not usual that anything should be paid to the Contractor, except on the Engineer's certificates, and out of the paper given by the Board on such certificates, the Contractor finds money to pay all charges whatsoever, law, land, Parliamentary, and engineering; it is so on the Irish Contractors' lines, to which I am engineer; it is so on the London, Chatham, and Dover, where the price certified for earthwork is 2s 11½d per cubic yard, instead of the usual rate (for earthwork 'pur et simple)' of 1s per yard; the reason being that all other expenses, land, law, Parliamentary, and engineering are charged on works and paid for by Contractors, out of proceeds of Engineer's certificates.[2]

John Fowler, the engineer, generalising on this matter said: 'I think the Committee may take it as a broad fact, that in most cases where the contractors find the greater part of the capital, and takes (*sic*) payment in shares, the contract price will be higher than if the company could go into the open market, and pay for the works in cash.'[3]

These details provide further evidence of the difficulties involved in examining railway accounts to try to establish capital expenditure. But more important is the question: how did the system work? The contractors needed money for their men and for their material. They often had to pay for the land in cash. To some extent, no doubt, contractors obtained long credits from their suppliers. But they still required a certain amount of cash for their work. As has been shown, some companies were able to issue shares and so obtain money from the public; some contractors were engaged in many different works (some of which would be paid for in cash) and one contract might well have assisted another. In addition, the contractors often had large investments in other industries—George Wythes, for example, invested heavily in mining and blast furnaces.[4] The income

[1] William Wright to Swindon, Marlborough & Andover Railway, 2 June 1875. (BTHR, MPR 34/2 and 34/7). See Maggs, *M & SWJR*, for a detailed history of this company.

[2] 'Memorandum on Engineers Certificates', BTHR, MPR 4/32.

[3] *SC on Standing Order 184*, PP (Lords) 1866, XVIII, Q 218.

[4] *Proc Inst Civ Eng*, Vol 74 (1882–3), 294–7.

from these activities could be used to tide them over difficult periods.[1] But it is noteworthy that in the case of William Wright, in 1875, a backer was obtained, Raisbeck Robinson (Temple Chambers, 32 Fleet Street), who wrote to Wright that he was 'prepared to finance this Railway for you on your obtaining the Contract'.[2] But Wright was a small man, who gets into this narrative through the accidental survival of some documents, and he may not be a typical case.

The shares and debentures issued to contractors could be used as security for borrowing, or be sold to the public. In fact contractors often held on to the shares, perhaps because they could not get a reasonable price for them; the reports of bankruptcy proceedings (there are many of them) in the journals invariably refer to the fact that the contractor had in his possession large numbers of worthless shares. And John Dickson, the contractor for the Neath & Brecon Railway, said in 1867, a few months before he became bankrupt, that he was the largest shareholder in the railway.[3] But it is not easy to decide how much of the contractors' shareholding was held as *bona fide* investment, and how much they were simply unable to dispose of.

The great expansion of contractors' lines in the 1860s required sources of finance greater than could be arranged in this *ad hoc* way. These came with the emergence of credit and finance companies. Their history has been traced by Mr King,[4] and it is sufficient to say here that it was those companies above all which permitted the great expansion in the number of contractors' lines in the boom of the 1860s. Lord Redesdale, who (as chairman of committees in the House of Lords) was an acute observer of railway matters and a participant in them, stated: 'It was in 1864 that the difficulties began to arise, the railway contractors then experiencing pressing difficulty in obtaining money for their undertaking, and certain "finance companies" were formed for the avowed purpose of providing the capital which would enable contractors to carry out their schemes.' He even went so far as to say: 'Last year all the schemes which were brought before Parliament by Bills more or less looked to these finance companies for assistance.' He argued that the fall of Overend, Gurney & Co in 1866 (which precipitated the crisis of that year) was occasioned to a large extent by the 'advances made by them to railway companies or to contractors on railway securities'.[5] Little enough is known about the detailed workings of these

[1] Noted by John Fowler, *SC on Standing Order 184*, Q 85.

[2] Robinson to Wright, 1 July 1875 (BTHR, MPR 34/2).

[3] Minutes of Evidence on the Neath & Brecon Railway Bill (Lords, 28 May 1883), QQ 2–3 (BTHR, MPR 261). But George Wythes did not remain a shareholder in the Rhymney Railway. See Rhymney Railway Share Registers (BTHR, RHY 2/1).

[4] W. T. C. King, *History of the London Discount Market* (1936), especially Chapter 7. Also Jenks, op cit, Chapter 8.

[5] *Hansard*, CLXXXIII (1866), col 858. See for example the prospectus of the Contract Corporation Ltd. (*Herapath*, 19 December 1863, 1345) which refers to the onerous terms on which contractors had to raise money.

finance companies, but there is no doubt of the extent of their activities. Even such an important contractor as Peto came into contact with them.[1]

Another innovation, which had particular reference to the contractors' lines, was the use of Lloyd's bonds, which were invented 'to meet the cases of young companies struggling with monetary difficulties'.[2] They were not necessarily initiated with a view to their use by contractors' lines; any railway company which had difficulties in getting funds to continue work could use them. It might be that the capital powers were exhausted, the bankers would not advance funds and the directors could not, or would not, give their personal security for a loan. In this situation a Lloyd's bond would be issued. They were, however, taken up to a great extent by contractors, who received them from the railway companies instead of cash or shares.

The Railway Act of 1844 (s 19) had made illegal the issue of loan notes—a commonly-used instrument of the late 1830s and early 1840s to secure debts; they were little more than IOUs, with no mention of security or maturity date on the document.[3] Loan notes had provided a useful method, adopted by many railway companies, to tide them over a period when it was impossible, for some reason, to issue more capital (their powers may have been exhausted, for example). The way to get round this prohibition on loan notes was to make use of Lloyd's bonds. They were not a debenture or loan note but 'in the form of an acknowledgment, under the seal of the company, of a debt incurred and actually due by the company to a contractor or other person for work done, goods supplied (or otherwise, as the case may be) with a covenant for payment of principal and interest at a future date'.[4] That is to say the companies did not raise money on these bonds; they merely secured a debt.

They were named after the barrister with whom they originated, John Horatio Lloyd, who gave a detailed account of them to the Select Committee of the House of Lords on Railway Companies' Borrowing Powers (1864). Lloyd explained that the bonds were useful in a situation where a railway company had exhausted its funds. The contractor in such a situation would be able to continue work if the company gave him 'an acknowledgment, under seal'. This the contractor would take to a banker 'or some capitalist' and obtain an advance, providing him with the funds to continue the work. The rate of interest on them was higher than that on a debenture —usually 5 or 6 per cent; they were issued by small, new companies, not

[1] *Railway Times*, 9 June 1866, 708; *Herapath*, 16 June 1866, 670. *Railway Times*, 16 November 1867, 1158. Further discussion of railways and finance companies is in *Railway Times*, 18 September 1875, 904–5.

[2] *Railway News*, 9 January 1864, 33.

[3] G. H. Evans, op cit, p 72; F. Clifford, *History of Private Bill Legislation* (2 Vols, 1885–7), I, 127–8; *Railway News*, 9 January 1864, 33; *Stock Exchange Official Intelligence*, Vol 20 (1902), 1935.

[4] H. Godefroi and J. Shortt, *Law of Railway Companies* (1869), p 39.

the large ones such as the London & North Western or the Great Western. And they were eventually discharged by applying to Parliament for powers to issue capital or were paid out of revenue.[1] The advantages of these instruments were summarised by a contemporary writer.[2] The contractor was enabled to go into the market and obtain cash upon the bond 'at a much better rate than he would be able to procure the same upon his bare personal security' and it allowed him 'where the work was paid for in this way by sections, to undertake considerably more extensive contracts than his capital would otherwise justify him in doing. The railway company also would be able to carry on their works with a reduced capital, having time to pay cash for its construction until profits came in'. And of course companies could use them without bothering about the statutory one-half of the share capital being paid up before exercising their borrowing powers.[3]

Lloyd thought there were several millions in Lloyd's bonds in circulation, and the Select Committee agreed that there was no limit to the gross amount that could be used. Their most important advantage was that they were not illegal. (This did not stop the *Railway Times* from describing them as 'a peculiar but inevitable fraud'.)[4] Lloyd's bonds were not a feature of the 1860s and 1870s only. In 1900, during the negotiations about the electrification of the Metropolitan and Metropolitan District Railways, one method of payment being considered was by these bonds.[5]

It is not surprising that this kind of financial dealing often led into difficulties.[6] While it would be too gross an exaggeration to say that the history of contracting is a history of bankruptcy, it would be reasonable to suggest that most contractors found themselves embarrassed at one time or another. The most spectacular case was that of Peto in 1866—associated with the fall of Overend, Gurney, and with the London, Chatham & Dover Railway. At that period Thomas Brassey is said to have lost £1 million[7] and, as well as a host of minor people, John Watson & Co, Smith Knight & Co, and Thomas Savin were affected even before the crisis of May 1866.[8] The typical reason was that their railway shares, which they held in great numbers, were worth very little; their value was often insufficient to cover their debts. This is in addition to the more usual kind of problem. Peto in the 1830s had to borrow to carry on the work of the Hanwell Viaduct be-

[1] *SC on Railway Companies' Borrowing Powers*, PP (Lords), 1864, XXVII (127); reprinted in PP 1864, XI; QQ 117; 180–1; 207; 235.

[2] H. J. Tarrant, *Lloyd's Bonds; their nature and uses* (1867), p 8.

[3] *SC on Borrowing Powers*, PP 1864, XI, Q 117.

[4] Ibid, Q 117; Report, p iv; *Railway Times*, 12 August 1865, 1039.

[5] *Railway News*, 24 November 1900, 746.

[6] Jenks, op cit, p 259, refers to the 'many railways, within the British Isles and without, that contractors had built upon the credit of their smiles'.

[7] *Engineering*, 16 December 1870, 443. But he was able to offer Peto half a million pounds in that year and take over his contracts: *Railway News*, 17 December 1870, 714; *Railway Gazette*, 4 February 1938, 215.

[8] *Railway Times*, 26 March 1864, 464; 28 October 1865, 1391; *Herapath*, 23 December 1865, 1429; 31 March 1866, 383–4.

cause of a dispute with the Great Western Railway over the contract price.[1] And in the 1870s Peto was in trouble again with the Cornwall Minerals Railway. The private diary of William Pease, steward to George Fortescue, Boconnoc, Lostwithiel, gives an interesting sidelight on this.

> *13 February 1875*. Sir Morton Peto came this day and gave me particulars as to the state of his affairs. The Minerals Ry contract has ruined him. He has not a shilling of his own and owes thousands of pounds to those who lent him money to enable him to proceed with the contract. The Railway Compy has taken all his plant and he owes a large sum for rail iron. He does not know what course his creditors will take, but there is nothing more for them. In order to make the payments which has been made, his plate was sold and also his wife's jewels. He said the estimates for his contract were far too little in consequence of the great rise in wages and in the price of material. The land cost *three* times as much as was calculated on, and rail iron £5 per ton beyond what it can now be had for and many of the labourers had 7/6d. per day.[2]

IV

This article has been concerned with only one aspect of the contractors' activities in the construction of British railways in the second half of the nineteenth century. It has shown their transformation from a prime concern with engineering to an equal if not greater interest in finance. It is not surprising to find Peto, in the early 1860s, helping the London, Chatham & Dover Railway to liquidate over £1¼ million of Lloyd's bonds.[3]

In the course of this article I have not referred to the contractors' activities as railway operators; that would be worthy of a separate study, as would a discussion of their organisation (including the significance of subcontracting and truck). Nor must it be thought that the railways were the only industry in which contractors were financially important. Some of the urban horse-tramways were built with the financial aid of contractors.[4] Without their financial aid railway building would have been on a smaller

[1] *Proc Inst Civ Eng*, Vol 99 (1889–90), 401.

[2] I am grateful to Mr C. R. Clinker for providing me with this extract. It should be pointed out that the crisis of 1866 had already virtually finished Peto's career; compared with the London, Chatham & Dover Railway, with which Peto's downfall in 1866 was associated, the Cornwall Minerals Railway was a small affair.

[3] *Herapath*, 27 October 1866, 1181–2. This was explained in the course of a long speech to his constituents at Bristol (Peto was an MP) in which he defended himself against the attacks made on him by a Committee of Investigation of the LCDR. Similarly, William Dargan was a director of the International Financial Society Ltd, formed to make advances on landed estates: *Railway News*, 23 January 1864, 96. Jenks, op cit, p 252, generalises about 'the readiness with which railway contractors . . . lent the prestige of their names to the directorates of finance companies'.

[4] William Jaques, the manager of the South London Tramways Company, said that the contractors paid the dividends on his company's shares during construction: Minutes of Evidence on the South London Tramways Bill (6 June 1883, Commons), QQ 1693–6 (BTHR, PYB 1/1078).

scale—if, that is, it had been left to the existing companies or to new companies financed by the public, or to landowners. Not only did the contractors build new lines but they probably stimulated older companies to build in order to protect their territories. How desirable it all was is difficult to decide. Obviously many of the lines were necessary and would have been built at some time or other—some of the London Transport railways for example. But many had an unhappy history, which included the payment of small dividends or none at all, and even bankruptcy. A full assessment would need to go beyond the profit and loss of the particular line, and take in any net additions to the profits of the company or of the network, as well as general social and economic benefits.[1]

The history of railway construction, as it is provided in many standard works, peters out at about 1850, despite the impressive mileage built after that date. Whatever the reasons for this—and there is a crying need for the work done by H. G. Lewin to be continued after 1852—this article has shown some of the financial questions involved. Clearly the 1850s were a watershed in constructional finance; the problems were new and so were the solutions. To that extent the traditional story of railway finance in Britain needs modification.

[1] See G. J. Ponsonby, 'The Closing of Branch Lines and Intermediate Stations', *British Transport Review*, Vol I, 1951, and the recent applications of cost-benefit analysis.

18
The origin of American railroad technology, 1825–1840

Darwin H. Stapleton

More than a dozen American civil engineers visited Britain from 1825 to 1840 to learn more about their profession, and in doing so acquired the knowledge and skill to build railroads. All of these men had previous training which gave them the technical background to assess and understand the important elements of the modern railroads being constructed in Britain. Some of them were apprentices of important early civil engineers, some of them had learned the skills of surveying and construction "on-the-job" as assistant engineers on various engineering projects, and a few others had started with academic training at the United States Military Academy or under Alden Partidge. Partidge's Norwich (Vermont) Academy was the earliest private school to teach civil engineering.[1]

The visits to Britain which these men made had no common characteristic, except that the visits demonstrated a fascination with technical innovation. Some men, like Moncure Robinson, John Edgar Thomson, and Edward Miller, went out of their own desire to become acquainted with engineering improvements in Britain, and Europe in general. Others, like Jonathan Knight, William G. McNeill, and George W. Whistler, were commissioned by organizations to make the trip.[2] All of them returned to be advocates of railroad construction, and each became the chief engineer of one or more major American railroads.

They acquired the necessary technical knowledge in Britain through meetings with British engineers and close observation of the newest railroads. For example, William Strickland became a close friend of Jesse Hartley, a British engineer, through his visit of 1825, and Horatio Allen found both Hartley and George Stephenson "perfectly willing to converse on whatever topic I wished to introduce."[3] Edward Miller and Wirt Robinson inspected the Liverpool and Manchester Railroad during their visits,[4] and Knight, McNeill, and Whistler also saw the Stockton and Darlington.[5] Apparently there was nothing to keep American engineers with adequate credentials from seeing all they wanted to see and from asking about all that they wanted to learn. As a result the American engineers developed knowledge of railroads in three areas: (1) locomotives and inclined planes, the two "new" elements in railroads, (2) the uses of materials—especially stone, wood and iron—in construction, and (3) the principles of laying out routes feasible for railroad travel.[6]

The American civil engineers who returned with this special knowledge were readily hired by railroad corporations, and they dominated the market for chief engineers in the early years. Many of the early railroads built by the visitors were imitations of the most impressive of the

British railroads, the Liverpool and Manchester. The Baltimore and Ohio engineers particularly attempted to copy its solidity of construction, and its use of locomotives and inclined planes.[7] The Columbia and Philadelphia, Boston and Lowell, and Camden and Amboy had similar designs.[8] In order to follow the British pattern most of the locomotives and all of the rails for American railroads in the first years were imported from Britain.[9] The visiting engineers transferred virtually an entire technological "package" to the United States.

Had the skill to build railroads been confined to the group of American civil engineers who had visited Britain, however, the rapid growth of railroad mileage which the United States experienced in the 1830's would never have occurred. Instead a large number of men were taught the necessary skill and given the needed information through the "on-the-job" training system which existed on practically all American railroad projects. Typically a railroad had an engineering corps, at the top of which was a chief engineer. (Rarely the man at the apex held the title of consulting engineer.) Frequently the chief engineer was a man who was in charge of several lines at the same time, and did not devote his full attention to the work. If that was the case there was usually a trusted and experienced associate on location who was called the "principal assistant engineer," or "resident engineer." Occasionally there was more than one man of this rank, and each had responsibility for a different section of a lengthy work, or for different technical aspects, such as motive power, inclined planes, or track. Under these men there were a large number of assistant engineers with more or less experience who were learning the trade. When the railroad was still in the stage of survey and location there were further subordinate ranks, called at various times and places such names as draftsman, rodman, leveler, targetman, surveyor, chainman, and axeman. Many men began at these ranks for little or no pay and, if they proved adept, worked their way up the heirarchy. Sometimes they succeeded to the office of chief engineer, but more frequently were hired by another company because of their experience.[10]

In the early years of railroad construction the engineering corps as an institution was closely related to the tradition of apprenticeship. A man who was a recognized railroad engineer attracted to him young men with mathematical talents or surveying training who wished to learn the profession. If they were acceptable (family relationship often helped to make them so) they were assigned to a routine task and paid a small salary. The engineer usually paid these men out of his own income, which partly explains the rather large official salaries of some early engineers.[11]

Subordinates who did their work well and were liked and trusted were advanced in rank as men above them in the hierarchy left for other work. Those under engineers who were in charge of several railroads at once were often shifted from one railroad to another as they were needed, frequently with an advance in position. It appears that an engineer was occasionally able to have one of his protégés appointed to a position which he was unable to accept personally, such as chief engineer of a railroad for which he had done the preliminary survey.[13]

In order to demonstrate the dominance of American railroads up to 1840 by civil engineers who had visited Britain ("visitors"), and the men with whom they worked on railroads ("associates"), the following tables show known visitors and associates (Tables Ia and Ib), and the total mileage of American railroads for which they were responsible (Table II). The tables are based largely on printed works, such as the *Dictionary of American Biography*, *National Cyclopedia of American Biography*, and the *American Railroad Journal*, although some manuscript sources were consulted.

Several problems have been encountered in compiling the tables. The most frustrating is the lack of sufficient biographical information on a few major railroad engineers, especially John Wilson, John Randel, Jr., and Elisha Johnson. Each of these men was chief engineer of more than one railroad,[14] and probably trained several assistants,[15] but nothing is known of the origin of their railroad skills. Perhaps they were visitors or associates themselves, or they may have been skilled enough prior to the building of the earliest railroads, like the Baltimore and Ohio, that they were able to learn a great deal by observing them and talking with the visitors who were in charge. If so, they form the exception to the general course of the transmission of railroad technology. At any rate, they are a significant omission from the tables of visitors and associates.

It has been difficult to find out about all the men who may have served under any one engineer. In each case there must have been many; very active men like William G. McNeill and Moncure Robinson probably had several dozen men in their corps of engineers from 1830 to 1840. Those men are not well represented in biographical encyclopedias. Sketches which exist often fail to corroborate other evidence of an association, and frequently generalize about when and where they held a position as chief engineer (if they did). Still, a list has been developed which indicates the wide influence of the visitors through the associates which they trained.

It has also been necessary to list two engineers as associates who officially were not subordinate to the visitor with whom they were associated. John B. Jervis was chief engineer of the Delaware and Hudson Canal Company from 1827 to 1830 while the Gravity Railroad, a feeder to the canal, was being built. However, Horatio Allen, one of his resident engineers who had visited Britain, directed the construction of the Gravity. Similarly, although Stephen Long was a co-engineer with Knight, McNeill and Whistler, little of the construction of the Baltimore and Ohio was done until the latter three returned from their British visit. It must be concluded that both Jervis and Long derived their knowledge of railroad construction from their association with the visitors.

Even with these difficulties, which make any compilation at this time incomplete and an understatement, the amount of railroad mileage which is attributable to the combination of visitors and associates is very impressive. Using several sources (of varying quality, but the only ones available[15]) which list railroad mileage completed in the years 1835-1840, it is possible to attribute at least 63 per cent of the railroad mileage in each year to visitors and their associates. They clearly dominated early railroad construction in the United States.

The table of railroad mileages shows an interesting feature of the transfer of railroad technology. The high percentages of attributable railroad mileage in New England and the South are the result of four visitors dominating the two regions. William G. McNeill, John Fessenden, and George W. Whistler had been on the Baltimore and Ohio, traveled to Britain together, and in 1831-1832 moved on to New England railroads such as the Boston and Worcester, Boston and Providence, and the Providence and Stonington. They remained in that region throughout the rest of the decade. Moncure Robinson and his associates Walter Gwynn, John H. Hopkins, William S. Campbell, and James Herron accounted for a large portion of railroad mileage in the South during the 1830's. Robinson had a virtual monopoly of Virginia railroads, and Gwynn was chief engineer of four railroads in North Carolina.

There is no apparent cause of this regional dominance, except that in purely statistical terms the lower mileage of the two regions compared to the Middle Atlantic/Midwest makes it easier for fewer engineers to have a greater impact. Moreover, New England and the South had comparatively few branch railroads and mostly major lines, so it was easier for a few men to hold most of the chief engineer positions. On the other hand, there may have been something peculiar in the two regions to which these few engineers were best able to adapt railroad technology, and thus make it a success. Thomas Parke Hughes has suggested that regional differences in the diffusion of technology should be studied in order that "the problem of diffusion might be more clearly defined."[17]

The decrease in attributable mileage in the five-year period is probably the result of having less information available on associates than visitors. While the visitors dominated railroad construction in 1835, only five years later there had been a huge increase in the mileage of railroads constructed, much of which was done under known associates. Further research may make it possible to increase the attributable mileage.

This assessment of the role of American civil engineers who visited Britain in the years 1825-1840 leaves no doubt of the origin of American railroad technology. Although the visitors had previous training in the basic skills of civil engineering in the United States, it was through their discussions with British engineers and their observations of British railroads that they learned the skills and obtained the knowledge necessary for railroad construction. On their return to the United States they became the chief engineers of the early railroads, and they instructed other Americans in railroad technology through the system of on-the-job training which prevailed on those railroads. Some of those associates went on to become chief engineers of other railroads.

The transfer of railroad technology to the United States by American civil engineers who visited Britain is one of the clearest cases of the transfer of technology to the United States, since the visitors who carried it out are not obscure, and the railroad was a distinctly new element in the American scene. It is possible to trace the source of the technology, the means of its transfer, and its diffusion. It is most important that this story be understood, as it lays a firm basis for railroad history in the United States.

Footnotes For Text

[1]See Table Ia; Daniel H. Calhoun, *The American Civil Engineer* (Cambridge, Mass.: M.I.T. Press, 1960) Chapter II.

[2]"Moncure Robinson," "John Edgar Thomson," *Dictionary of American Biography* (New York: Scribners, 1927-1936); Solomon W. Roberts, "Obituary Notice of Edward Miller," *Proceedings of the American Philosophical Society*, XII (1871-1872), 582; Robert E. Carlson, "British Railroads, and Engineers and the Beginnings of American Railroad Development," *Business History Review*, XXXIV (Summer, 1960), 147-148.

[3]Agnes A. Gilchrist, *William Strickland: Architect and Engineer*, 1788-1854 (New York: Da Capo Press, 1969, enl. ed.), 6, 12; Carlson, 140, 145.

[4]Roberts, *loc. cit.*; 30 December 1831, W.D. Lewis diary, Historical Society of Delaware, Wilmington Delaware; Eugene S. Ferguson, ed., *Early Engineering Reminiscences (1815-1840) of George Escol Sellers* (Washington, D.C.: Smithsonian Institution, 1965), 134.

[5]Carlson, 148.

[6]For an example of how one visitor understood British railroad technology, see: Moncure Robinson, "Report on the survey of canal and railway routes between the waters of Delaware and Susquehanna," *Hazard's Register of Pennsylvania* (Philadelphia, Pa.), III (1829), 54-56.

[7]Jonathan Knight, 17 February 1832, to P.E. Thomas, *American Railroad Journal* (New York, N.Y.), I (1832), 228-229; 14 November 1831, W.D. Lewis diary; W. Hasell Wilson, *Reminiscences of a Railroad Engineer* (Philadelphia: Railway World, 1896), 13. The inclined plane at Parr's Ridge on the B&O was never supplied with a stationary engine: *Eleventh Annual Report...Baltimore and Ohio Rail Road Company* (Baltimore: Lucas & Deaver, 1837), 19.

[8]J. Knight and Benj. H. Latrobe, *Report Upon the Plan of Construction of Several of the Principal Rail Roads in the Northern and Middle States* (Baltimore: Lucas &

Deaver, 1838), 11, 17, 46; "Columbia and Philadelphia Railroad. John Barber, Superintendent," *Pennsylvania House Journal* (Harrisburg, Pa.), 1831-1832, II, doc. 22.

[9]John White, *American Locomotives* (Baltimore: Johns Hopkins Press, 1968), 5, implies that British locomotives dominated the American market to about 1835. American railroads routinely purchased rails from Britain, and through the 1840's imports were the major source of new rails. For estimates, see: Robert W. Fogel, *Railroads and American Economic Growth* (Baltimore: Johns Hopkins Press, 1964), 194 [table].

[10]Calhoun, 24-37, 47-50. However, this description of the engineering hierarchy is based on my impressions after reading a variety of sources and is not derived directly from Calhoun.

[11]*Ibid.*, 47-48. See: Moncure Robinson, 26 October 1828, to John Robinson, Moncure Robinson Papers, The College of William and Mary in Virginia, Williamsburg, Virginia. In this letter Robinson states that for a railroad survey he had "been compelled to employ...some 16 to 18 hands all of whom as well as my assistants have been paid by me."

[12]E.g., John H. Hopkins, Principal Assistant Engineer on the Richmond, Fredericksburg and Potomac Railroad, was appointed Chief Engineer on the Louisa Railroad after Moncure Robinson, the Chief Engineer of the R,F&P, conducted the survey. See: Charles W. Turner, "The Louisa Railroad, 1836-1850," *North Carolina Historical Review*, XXIV (January, 1947), 37, 39.

[13]John Wilson was engineer of the Columbia and Philadelphia Railroad and the West Chester (Pa.) Railroad. John Randel, Jr., was engineer of the New Castle and Frenchtown (Del.) Railroad and the Ithaca and Owego (N.Y.) Railroad. Elisha Johnson was Superintendent for construction of the Rochester Railroad, and engineer of the Buffalo and Black Rock Railroad, and the Tonawanda Railroad. This list may not be complete.

[14]E.g., 24 April 1830, Minute Book of the New Castle and Frenchtown Turnpike and Railroad Company, Acc. 722 (microfilm), Eleutherian Mills Historical Library, Greenville, Delaware. In this minute it was recorded that Randel asked for compensation for "his Assistants."

[15]The major problem is that each source has known errors in its list of completed mileage. *Mitchell's Compendium* and Stevenson's work probably err on the low side, because they seem not to have noticed many short and branch lines. The census of 1880 gathered its evidence from reports of companies then in existence, thus, at the very least, not including those railroads which were built up to 1839 but failed and were never reconstituted. Tanner's compilation clearly includes a large amount of mileage which was projected or surveyed, but not under construction in 1840.

[16]Thomas Parke Hughes, "Comment on Papers by Brittain and Robinson," *Journal of Economic History*, XXXIV (March, 1974), 129. A recent article by David Jeremy emphasizes the regional variation in transfer. See: David J. Jeremy, "British Textile Technology Transmission to the United States: The Philadelphia Regional Experience, 1770-1820," *Business History Review* (Spring, 1973), 24-52, esp. 52.

TABLE Ia
AMERICAN CIVIL ENGINEERS WHO WERE VISITORS TO BRITAIN BEFORE 1840

Name	Pt[1]	Year(s) of Visit	Railroads as Chief Engineer[2]	Source(s)[3]
Horatio Allen	2	1828	Gravity Railroad;[4] South Carolina	a
John Fessenden	2,3	1828-1829	Boston & Worcester; Eastern	b,c
Erskine Hazard	2	1826	Mauch Chunk[5]	d,e
Samuel Kneass	1	1825	Mine Hill & Schuylkill Haven; Lexington & Ohio; Philadelphia & Trenton; Feliciana; Philadelphia, Wilmington & Baltimore[6]	a
Jonathan Knight	2	1828-1829	Baltimore & Ohio	a
William McNeill	2,3	1828-1829	Baltimore & Susquehanna; Paterson & Hudson; Boston & Providence; Providence & Stonington; Taunton & New Bedford; Long Island; Western; N.Y., Boston & Providence	a,b,f
Edward Miller	2	1831	Allegheny Portage (Prin. Asst. Engr.);[7] Little Schuylkill & Susquehanna; Sunbury & Erie	g

TABLE Ia

Name	Pt[1]	Year(s) of Visit	Railroads as Chief Engineer[2]	Source(s)[3]
Solomon W. Roberts	2	1836-1838	Little Schuylkill & Susquehanna	a
Moncure Robinson	2	1825-1827	Chesterfield; Little Schuylkill; Petersburg; Danville & Pottsville; Winchester & Potomac; Richmond, Fredericksburg & Potomac; Philadelphia & Reading; Richmond & Petersburg	a,h
William Strickland	1	1825	Philadelphia, Germantown & Norristown; Delaware & Maryland; Wilmington & Susquehanna	j
Wirt Robinson	2	1831-1833	Philadelphia & Reading	h
Robert Stevens	2	1830-1831	Camden & Amboy[8]	a,i
Charles Storrow	1	1829-1830	Boston & Lowell	a
John Edgar Thomson	2	c.1830-1832	Georgia	a
George Whistler	2,3	1828-1829	Baltimore & Susquehanna; Paterson & Hudson; Providence & Stonington; N.Y., Boston & Providence	a,b,k

TABLE Ib
ASSOCIATES OF VISITORS

Name	Pt[1]	Associate of	Railroads as Chief Engineer[9]	Source(s)[3]
Joshua Barney	2,3	McNeill, 1832-1836	Predecessors of Boston & Maine	b,l
Uriah Boyden	2	Storrow, c. 1832-1836	Nashua & Lowell	a
William Campbell	4	M. Robinson, 1835	Alabama, Florida & Georgia	m
William Cook	2,3	Knight, McNeill, Whistler, 1829-1830	Camden & Amboy; Port Gibson & Grand Gulf; Philadelphia & Trenton	b,p
Walter Gwynn	2,3	Knight, 1829-1830; M. Robinson, 1830-1833	Portsmouth & Roanoke; Roanoke, Danville & Junction; Halifax & Weldon; Wilmington & Weldon	b,h,p
James Herron	2	M. Robinson, 1836	Raleigh & Gaston (Prin. Asst. Engr.)[10]	n
Robert Higham	5	Young, c. 1834-1836[11]	Auburn & Rochester	r
John H. Hopkins	2	M. Robinson, 1835-1837	Louisa	h
John B. Jervis	2	Allen, 1828-1829	Mohawk & Hudson, Schenectady & Saratoga	a
Edwin F. Johnson	2,4	Young, c.1833-1834[11]	Auburn & Syracuse, New York & Erie	q

TABLE Ib

Name	Pt[1]	Associate of	Railroads as Chief Engineer[9]	Source(s)[3]
Benjamin Latrobe (younger)	2	Knight, 1830-1835	Baltimore & Port Deposit	a
Stephen Long	2	Knight, Whistler, McNeill, 1829-1830	Atlantic & Great Western	a
Henry Ranney	2,4	Knight, 1831-1832	Lexington & Kentucky; New Orleans & Nashville	q
W. Milnor Roberts	2	Hazard, 1827-1829; Miller, 1831-1835	Harrisburg, Portsmouth, Mt. Joy & Lancaster; Cumberland Valley	a,o
Charles B. Stuart	5	Knight, 1832	Syracuse & Utica	q
John C. Trautwine	1	Strickland, 1828-1835	Hiwassee	a
Issac Trimble	2,3	McNeill, 1832-1835	Baltimore & Susquehanna; York & Wrightsville	b
William C. Young	2,3	Knight, McNeill, Whistler, 1830	Utica & Schenectady	b

TABLES Ia AND Ib—FOOTNOTES

[1]Pt=training as a civil engineer prior to visit or association. Code to numerals: 1=apprenticeship to an engineer; 2-state or corporation hierarchies ("on-the-job" training); 3=United States Military Academy; 4=Partridge's Academy; 5=unknown.

[2]Railroads on which the subject is known to have been chief engineer or an active consulting engineer before 1840 are listed. A few of the visitors were chief engineers after 1840.

[3]Key to sources:

a. Article in *Dictionary of American Biography* (New York: Scribners, 1927-1936), 20 vols.
b. Article in George W. Cullum, *Biographical Register... U.S. Military Academy (Boston:* Houghton, Mifflin & Co., 1891), I.
c. Francis B.C. Bradlee, *The Eastern Railroad* (Salem, Mass.: Essex Institute, 1922), 4, 5, 10-12, 17.
d. W. Milnor Roberts, "Reminiscences and Experiences...," *Transactions of the American Society of Civil Engineers,* VII (1878), 197.
e. Alfred Mathews and Austin N. Hungerford, *History of the Counties of Lehigh and Carbon* (Philadelphia: Everts & Richards, 1884), 670-671; Erskine Hazard, "Observations upon Rail-roads," *The Franklin Journal and American Mechanics' Magazine,* III (1827), 275-277.
f. C. Palmer, 19 April 1838, to W.D. Lewis, W.D. Lewis Correspondence, New York Public Library, New York, N.Y. (Providence & Stonington).
g. Solomon W. Roberts, "Obituary Notice of Edward Miller," *Proceedings of the American Philosophical Society,* XII (1871-1872), 581-586.
h. Darwin H. Stapleton, "Moncure Robinson; Railroad Engineer, 1828-1840," in Barbara E. Benson, ed., *Benjamin Henry Latrobe & Moncure Robinson* (Greenville, Del.: Eleutherian Mills Historical Library, 1975), 33-60.
i. J. Elfreth Watkins, *The Camden and Amboy Railroad* (Washington, D.C.: Gedney & Roberts, [1891], 30.)
j. Agnes A. Gilchrist, *William Strickland* (New York: Da Capo Press, 1969, enl. ed.), 1-12; Jack C. Potter, "The Philadelphia, Wilmington and Baltimore Railroad, 1831-1840," (Unpub. M.A. thesis, University of Delaware, 1960), 69, 99, 106; John V. Hare, *History of the Reading* (Philadelphia: John Henry Strock, 1966, reprint), 311-313, 317.
k. C. Palmer, 2 January 1838, to W.D. Lewis, W.D. Lewis Correspondence, New York Public Library, New York, N.Y.
l. Francis B.C. Bradlee, *The Boston and Maine Railroad* (Salem, Mass.: Essex Institute, 1921), 2-5.
m. *Hazard's Register of Pennsylvania,* XVI (1835), 402; Michel Chevalier, *Histoire et Description des Voies de Communication aux Etats-Unis* (Paris: Librairie de Charles Gosselin, 1841), II, 433.
n. Entries for 3 May 1836 and 9 December 1837, Wilson Miies Cary Fairfax diary, Virginia Historical Society, Richmond, Virginia.
o. Roberts (source d), 197-208.
p. Forest G. Hill, *Roads, Rails and Waterways* (Norman, Okl.: University of Oklahoma Press, 1957), 105, 105n.
q. Article in *National Cyclopedia of American Biography* (New York: 1898-), 50 vols.
r. Frank Walker Stevens, *The Beginnings of the New York Central* (New York: G.P. Putnam's Sons, 1926), 185.

[4]Horatio Allen was the Resident Engineer for the Delaware and Hudson Canal Company and was in charge of constructing the Gravity Railroad.

[5]Erskine Hazard and Josiah White constructed the Mauch Chunk Railroad in 1826. Although a short, lightly-constructed railroad, it was one of the first in the United

States, and was visited by many people interested in them. Little is known of Erskine Hazard, except that he was an associate of White's for many years and in consequence he was involved in iron manufacture and canal construction, as well as railroads.

[6]Originally Kneass was chief engineer of the Philadelphia and Delaware County Railroad, which joined with two other railroads to become the Philadelphia, Wilmington and Baltimore in 1838. Kneass then became chief engineer of the new corporation.

[7]Sylvester Welch was engineer. He had formerly worked on the Erie and Lehigh canals. Edward Miller was specifically in charge of the inclined planes, but provided valuable information on other British practice to Welch. See: "Allegheny Portage Railroad. Report of Sylvester Welch, Engineer," *Pennsylvania House Journal*, 1832-1833, Appendix to vol. II, doc. 15.

[8]Although William Cook was appointed chief engineer in 1831 to succeed Stevens, it seems clear that Stevens was the technical genius behind the Camden and Amboy. See: Franz von Gerstner, *Der Innern Communicationen der Vereinigten Staaten* (Vienna: L. Forster's Artistische Anstalt, 1843), II, 83.

[9]Railroads on which the individual was chief engineer (up to 1840) after his association as a visitor. None of the associates were chief engineers before their period of association.

[10]Charles F.M. Garnett was chief engineer of the Raleigh & Gaston; no biographical data on him have been found.

[11]Although Young was an associate himself, two engineers who were trained under him are included on this table. It is interesting that the diffusion of railroad technology was in its third generation of American engineers in less than a decade.

TABLE II

RAILROAD MILEAGE ATTRIBUTABLE TO A VISITOR TO BRITAIN OR AN ASSOCIATE

1. *1835.* Source: *Mitchell's Compendium* (Philadelphia: Mitchell & Hinman, 1835), p. 10.

	New England[3]	Middle Atlantic/Midwest[4]	South[5]	Total
PA[1]	92%	80%	87%	83%
T[2]	(50 mi.)	(564 mi.)	(355 mi.)	(969 mi.)

2. *1837.* Source: David Stevenson, *Sketch of the Civil Engineering of North America* (London: John Weale, 1838), pp. 275-276.

	New England	Middle Atlantic/Midwest	South	Total
PA	98%	78%	82%	84%
T	(203 mi.)	(935 mi.)	(472 mi.)	1,610 mi.)

3. *1839.* Source: Census Office, Department of Interior, *Report on the Agencies of Transportation in the United States [Tenth Census, 1880, IV]* (Washington, D.C.: Government Printing Office, 1883), 322, 326, 330, 334, 338, 342, 346, 350, 354, 358, 362, 378.

	New England	Middle Atlantic/Midwest	South	Total
PA	89%	67%	96%	77%
T	(357 mi.)	(1,400 mi.)	(508 mi.)	(2,265 mi.)

BULLETIN 139

4. *1840.* Source: H.S. Tanner, *A Description of the Canals and Rail Roads of the United States* (New York: T.R. Tanner & J.D. Disturnell, 1840), 223-234.

	New England	*Middle Atlantic/Midwest*	*South*	*Total*
PA	70%	56%	67%	63%
T	(611 mi.)	(2,324 mi.)	(2,184 mi.)	(5,119 mi.)

TABLE II—FOOTNOTES

[1]PA=percentage of regional or United States mileages attributable to visitors or associates.

[2]T=total regional and United States mileage (in parentheses).

[3]Maine, N.H., Vt., Mass., R.I., Conn.

[4]N.Y., N.J., Pa., Ohio, Mich., Indiana, Ill., Del., Md., D.C.

[5]Va., Ky., Tenn., Miss., Ala., Ga., Fla., N.C., S.C., La., Ark.

The China Basin

One of the outlays of a railroad company which does not add to the number of miles, although it materially augments the cost, is that of terminals. These involve an amount little understood by those outside of railroad circles. An illustration of such outlay is found in a little improvement in the terminal facilities of the Santa Fe at San Francisco, recently described in The Railway Age, where in order to enlarge its yards it is obliged to fill in what is known as China Basin, at an expense of $2,500,000. It is a very common impression that a railroad can be built for about $7,500 per mile, but even admitting that to be the fact (which it is not) the mere laying of track is only the beginning of the expense. A railroad even when sufficiently constructed to permit the operation of trains, is very far from complete. In fact, a railroad is never finished, outlays similar in kind to the one above referred to being constantly required.

Railway Age Nov. 15, 1901

A House on the Tracks

At St. Louis a few nights ago a special train on the Missouri Pacific, running between St. Louis and Kirkwood, encountered an obstruction in the shape of a house on the track, which it completely demolished, besides damaging the engine. It appears that an ingenious house-mover, having a contract to move a building, consulted the time-tables and chose an hour late at night, when he was certain there would be no regular trains to interfere with his work, to cross the railroad tracks. To guard against any possible danger, however, he had employed a flagman to stay on guard and flag any train that might happen along. Having done this, he told the occupants of the house to quietly go to bed and he proceeded with his moving. At about 11:20 o'clock, when he had just fairly gotten the building on the track, the special came along, and the flagman having deserted his post, it dashed into the building with result above stated. Strange to say, none of the occupants were injured, although the house was a complete wreck.

Railway Age Oct. 30, 1884

19
Tracks and Timber
John H. White

The railroad dependence on wood as the basic material for track construction is the subject of this paper. Capital-poor American lines turned to this abundant material in their earliest years and quickly evolved a remarkably cheap method of construction that permitted a rapid expansion of the system. Pile, strap rail, and pole roads were obsolete even before the end of the nineteenth century, but the wooden crosstie continued to be a fundamental part of the track structure.

Our Wooden Roads

The railroads were in the vanguard of major wood consumers and their appetite for forest products was so great that one writer labeled them the "insatiable Juggernaut of the vegetable world." During the first half of the nineteenth century everything about the American railway was wooden. The track structure, stations, engine houses, bridges, snow sheds, water tanks, coaling stations, freight and passenger cars, locomotive fuel—even the brake shoes and rail joints. Ironically, railroads were frequently called iron roads during the last century, when in fact they were actually wooden roads, for rather little of the total physical plant, beyond the locomotive itself, was a product of the forge. *Our Iron Roads*[1] was the title of an early history of British railways and it is entirely accurate and plausible, for the railways of that country were indeed iron roads—timber played a minor part in their construction, except for cars, and even in them metal underframes were adopted at an early date.

Britain was recognized as having the best railway system in the world, durable, swift and technically as perfect as the engineering of the day could produce. Why then didn't America copy these model iron roads, why did we instead build wooden roads? To explain that curious deviation we must say something on the beginnings of inland transportation in North America.

Early America, so rich in natural assets, was so very poor in all the other trappings thought necessary to a great western civilization. We had little capital, a small population, few skilled workers, and still fewer manufactories. But our greatest lack was a means of travel between east and west. Mountains blocked an easy passage and the main rivers all seemed to flow the wrong direction. But while we lacked the basic tools for economic growth, we suffered no shortage of spokesmen with ready solutions for providing the means necessary to open the inland empire. In a society eager for land development, speculation and industrialization, one finds fertile territory for promoters and visionaries. A review of some of the schemes for internal improvements might prompt one to accept Alfred Nobel's statement that "Second to agriculture, humbug is the biggest industry of our age." Maybe so, but somehow one is convinced that a certain amount of humbug is essential to any great enterprise. Practical men can always find fault with every new scheme, and reasons to reconsider or go slow, if at all, into a major undertaking. And all of this to the good, so long as it doesn't entirely stifle progress. Yet great works start with a dreamer; impractical, disturbing, even annoying men, who somehow point the way to the future. American history is full of such men; technical history has had its visionaries as well. Some are well known, like James Watt, George Westinghouse, or Oliver Evans, but others might be less familiar; John Stevens, Thomas Earle and Robert Mills.

[1]Fredrick S. Williams, *Our Iron Roads*, 6th ed. (London: 1885).

Figure 1. John Stevens' elevated wooden railroad proposal, 1812.

They also were great dreamers. They believed that a key for opening America's imprisoned interior was available in the construction of cheap wooden railways.

Wooden Pile Railroads

The railways had been available since the first days of Independence, and a few short, industrial lines were in existence. As early as 1808, Benjamin Latrobe reported to the U. S. Treasury on prospects of such lines for opening the western lands.

But the canal and highway were considered more sensible solutions. John Stevens, a wealthy businessman of Hoboken, disagreed and in 1812 published a pamphlet on the advantages of the railway over existing forms of land travel.[2] Of more importance to our discussion, he proposed the all-wooden railway (Figure 1). To the British engineer, the scheme was a giant step backwards to the now discredited tramways of Gothic times. The modern engineer knew that stone blocks and solid iron rails were the only proper way to build a permanent way. Stevens rightly contended that America could not afford such luxury and that we needed a cheap, provisional railway. Choice timber was abundant; iron was scarce and dear; stone was expensive to gather and dress. A wooden railway could be quickly constructed by unskilled labor, a great advantage in a country short on patience and artisans. Stevens carried his idea for an all-wooden railway to its ultimate limit. He would avoid cuts and fills by building the entire line on an elevated structure; low to the ground on the level and as high as necessary over ravines and valleys. He envisioned a continuous wooden trestle. The rails would be wooden, but he admitted "shoeing" with iron would greatly increase their life. Erected, a mile of track could be built for an estimated $2,500; 1/8-inch iron strap rails would add another $1,500 to the cost. Stevens suggested a cast iron plate for a cheaper rail cap. In all, Stevens' prices were wonderfully optimistic but the idea caught the fancy of many men who were seeking a radical and expeditious solution to the transport dilemma. Stevens fanned their enthusiasm by noting that with his system a line from Albany to Lake Erie could be constructed in one or two seasons, whereas a road or canal might take a decade to complete. Stevens tried unsuccessfully over the next several years to promote the construction of a railroad on his pillar plan.[3] Distribution of his pamphlets and repeated efforts to secure state support amounted to a continuous propaganda campaign that circulated the railway idea far and wide.

Others, possibly inspired by Stevens' proposal, began to offer variations on the same theme. Always the cheapness of wooden construction was central to the arguments for long distance railway lines. In 1826 the architect Robert Mills proposed a wooden monorail to connect Washington and New Orleans.[4] He claimed it could be built for $1,300 a mile, if all wooden, or $2,200 if fitted with an iron strap rail. Mills was a great dreamer, perhaps the greatest, to think that a horse-powered, wooden monorail, 1,200 miles in length, would prove successful. Like Stevens, Mills could entice no one to support his project. However, a Quebec merchant named James George had better luck.[5] He suggested a cheap form of wooden railway suitable for ordinary road wagons. Notches cut in parallel logs would form a smooth roadway; the notched rails would be held in gauge by crossties. An experimental line was actually built in about 1826.

By this time the railway was recognized as more than a utopian projection. Commercial lines were opening in

[2] John Stevens, *Documents Tending to Prove the Superior Advantages of Railways and Steam-Carriages over Canal Navigation,* (New York: 1812). (Reprinted in 1936 by the Railway and Locomotive Historical Society.)

[3] A. D. Turnbull, *John Stevens,* (New York: 1928).

[4] "Three Papers on Rail-Roads." Pamphlet reprinted from *The American Farmer,* Washington, 1826.

[5] Railway and Locomotive Historical Society *Bulletin 78,* October, 1949, p. 53.

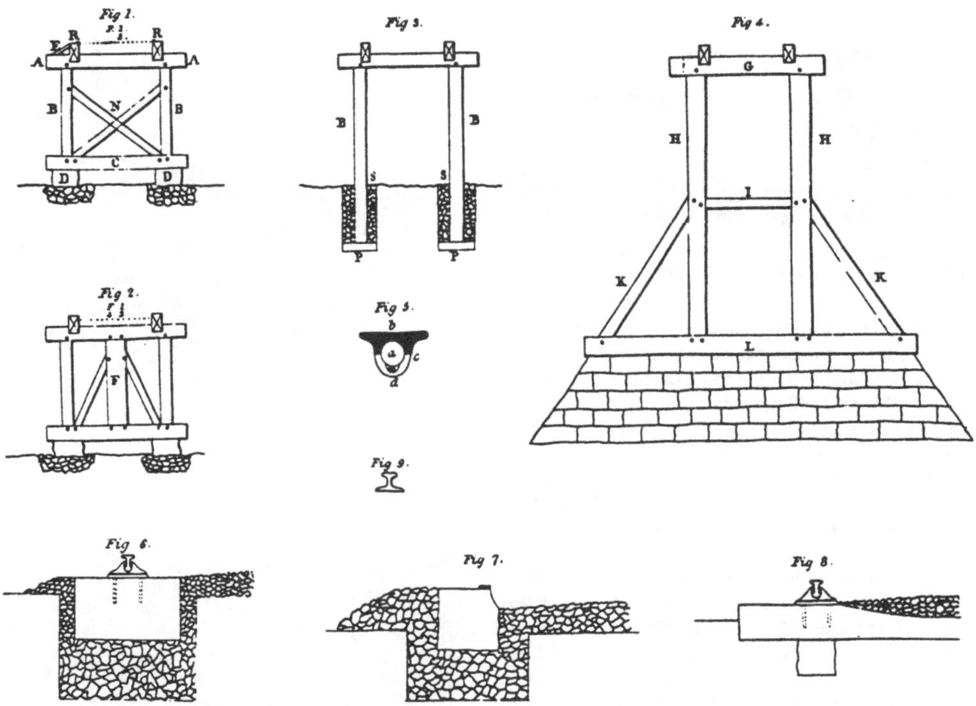

Figure 2. Delaware & Hudson R. R. trestle track, 1828c. (N. Wood, *Treatise on Railroads*, 1832 edition, Plate IX.)

Great Britain, yet the sophisticated and costly design employed convinced many that perhaps the railway really wasn't practical for the United States after all. However, the opening of two industrial lines in eastern Pennsylvania in the 1820s proved the wooden railway was a workable design. Neither the Mauch Chunk nor the Delaware and Hudson railways (Figure 2) were built to a very high level of engineering but their iron-capped wooden rails provided a full size operating model that dispelled in part the skeptics' worst criticisms. The D. and H. line was built largely on the elevated wooden structure because of the rolling terrain it crossed in northeastern Pennsylvania and was accepted by admirers of Stevens as a vindication of his pillar railway. The idea might have quietly died had it not been for the appointment of Horatio Allen, an assistant engineer on the D. & H., as chief engineer of the South Carolina Railroad. Allen was satisfied with the trestle plan and thought it particularly well suited for the cotton states where good lumber was in ready supply. The main advantage remained the avoidance of the major cost of grading but the citizens of Charleston were also informed that the trestle railroad would be safe from floods, and require no fencing because man and beast might safely pass under the structure. And so between 1830 and 1833 an incredible serpentine wooden structure, often ten feet above the ground, was built across the state.[6] Roughly 3.25 million board feet of lumber was used in building the 136-mile line. Cost per mile was about $7,000, roughly twice the estimate, but the road had been opened with remarkable speed. The line fell into disrepair with equal speed; just months after the opening it was realized that the trestle plan was a colossal mistake. The supposedly indestructible yellow pine decayed rapidly; the Kyan and Earle preservation solutions had some effect but not enough to save the sagging, rotting

[6]S. M. Derrick *Centennial History of the South Carolina Railroad*, (Columbia, S. C.: 1930).

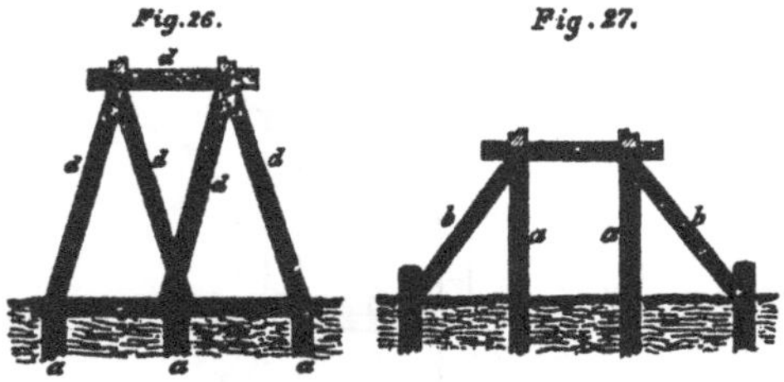

Figure 3. South Carolina R. R. trestle track. (David Stevenson, *Civil Engineering of North America*, 1838, p. 249.)

structure. The elevated tracks also proved dangerous; what might be a minor derailment on a level line turned into a major disaster on the South Carolina R. R. (Figure 3). In the days of unsound axles, wheels and rails, derailments were all too common. Even Allen agreed that it was imperative to rebuild the line as a conventional railroad. By 1835, 75 miles of the ramshackle structure was filled in with an earth embankment. A lack of funds delayed reconstruction of the entire line until 1839. The cost, an additional $463,000.

The failure of the South Carolina Railroad did not persuade others not to try the stilt railway. At least three other trestle railroads were built: the Syracuse and Utica, the New York and Erie, and the Ohio Railroad. Only the Syracuse line saw any operation. The Erie expired after over a hundred miles of trestle were finished–reorganized and started construction afresh several years later along more conventional lines.

The Ohio Railroad was one of the more ambitious of several ill-starred western railroads projected in the Jackson era.[7] It was to run along the Lake Erie shore connecting Toledo with Pennsylvania. Construction was semi-automated with a traveling steam pile driver and saw mill. The machine drove the piles, cut them off to the desired height, and sawed lumber for the trestle work and rails. It could complete about 330 feet of track a day. The eight-man crew was housed in a traveling boarding house that was towed by "the locomotive sawmill." Similar rigs were used in construction of the Syracuse and Utica. Advanced as the building techniques of the line may have been, it was a financial failure. The project collapsed in 1843, the state lost one quarter of a million dollars. What was meant to be the new Appian Way became a ghostly silhouete against the northern Ohio skyline. Some portions of the 63-mile long viaduct stood until the 1890s.

[7]*Magazine of Western History*, Vol. 13 (1891), p. 742.

Figure 4. B. & O. R. R. strap rail, 1830. (J. G. Pangborn, *World's Railways*, 1896.)

Figure 5. Virginia and Tennessee R. R. strap railroad track and locomotive, *Roanoke*. Scene about 1855. (Photo: Norfolk & Western Railway.)

Strap Rail Track

While rather few railroads were built on the exotic, all wooden schemes proposed during the introductory period of steam transport in America, the strap rail system enjoyed a wide popularity during the third decade of the nineteenth century (Figures 4 and 5). Engineers recognized the superiority of the solid British plan, and indeed a number of early U. S. lines were faithfully patterned on the English design–the Boston and Lowell was such an exact copy of the Liverpool and Manchester that it might have even fooled the Stephensons; but in general, stone blocks, cast iron chairs and wrought iron edge rails were beyond the meager capital of most American railway projectors. They were forced to settle on a much cheaper style of construction. A mile of British railroad, counting all costs of equipment, land, track and bridges, totalled about $180,000; while in America the cost was only $20,000 to $30,000.[8] A basic saving was realized by the small amount of iron used for the Yankee line. We used 25 tons of iron per mile, against the British average of 91 tons a mile. Both plans produced a working railway but the contrast was like champagne and a very poor white table wine. But the cheap vintage won the American market, its great appeal was obvious. It quenched the thirst for railway mileage and the conoisseur was left to worry about the refinements at some later time. Most lines were designed in the counting house and not the drafting room and what would produce the most for the smallest investment was considered the best bargain. And so two-thirds of the American railway opened by 1840 were of the strap rail plan.[9]

A typical flat bar line was built in the following way: An iron strap, generally ½-inch thick by 2½ inches wide was spiked to a longitudinal timber, say 4 by 6 inches, thus forming the rail. The wooden rail or stringer was fastened to wooden crossties by spikes, wooden wedges or in some cases, iron brackets. A longitudinal sub timber or mud sill was placed under the ties parallel to the rails. The structure might sound rather substantial, but it was inadequate for even the ten-ton locomotives of the day. John B. Jervis thought the strap railways were a very poor excuse for so important a public work and admonished those who advocated them for doing a very grave disservice to the "science of rail roads."[10] Yet Jervis himself oversaw the construction of several strap rail lines.

The composite track structure was barely adequate when new but a few seasons in the weather left it a hopeless ruin. Had the timbers been placed on crushed rock ballast and the right-of-way ditched, the rains might have drained away, but such niceties were costly and so the tracks were generally laid on the bare ground. The mud sill was in fact buried to hold the track in place and earth was piled against the upper members. No better condition could be found for rapid decomposition—moisture and the lack of air and light encouraged fungi and insects. Spring rains washed more mud over the crumbling timbers, weeds and deferred maintenance encouraged further decay.

The cheap strap railroad produced a fearful hangover of high maintenance. After four years of operation the Philadelphia and Columbia commented regretfully on its foolhardy decision for a flat bar track by saying in its report of 1837 that its wretched line was "a fruitful source of injury to cars and engines. . ." In a similar fashion the B & O lamented that the plate rail sections of its line cost well over twice as much to maintain as its edge rail divisions.[11] The Little Miami's Chief Engineer cursed the constant attention and heavy outlay necessary to keep his road even in passable condition.

Maintenance of schedules became a joke on most such lines, but their reputation for safety was no laughing matter. The rolling action of the locomotives and cars caused the bars to curl up at their ends often to such a height that they were picked up by the next train. In some cases the train was derailed—such an incident on the Portsmouth and Roanoke Railroad resulted in two deaths and eight injuries. On other occasions the engine would pass over safely but the snake head would spring up to enter the underside of a car, like a giant lance. To eliminate such unwelcome visitors, some roads sheathed the cars' underframes with heavy planks. A line in Michigan went all out to reassure its passengers by attaching iron plate beneath its cars.

The defects of strap rail construction were well established by 1840 and many lines had already begun the painful process of rebuilding on a better plan. This general knowledge led to no immediate elimination of the old form. *The American Railroad Journal* might speak bravely of the disappearance of the "ancient order of flat bars" in its 1842 issue, but it was bravado for the most part because the Panic of 1837 was reaching its lowest trough and all the betterments were irrevocably delayed by the hard times. Most roads simply made do until the business revival later in the decade.

The false economy of the system continued to lure a few new adherents. The Cleveland and Pittsburgh admitted the failings of the scheme but went ahead in 1846 with a flat bar line. A year later the second hand rail market became glutted with used strap rail as the New York State lines began to rebuild with T rail. The New York legislature outlawed strap rails in 1847 and gave the public carriers within its borders three years to convert. Two poorly financed midwestern lines, the Aurora Branch and the Galena and Chicago Union rail-

[8] J. H. White, Jr., *American Locomotives 1830-1880,* (Baltimore: 1968).

[9] F. A. Von Gerstner, *Die Innern Communication der Vereinigten Staaten von North-America,* (Vienna: 1842). This work offers the most detailed engineering account of the first American railways. In his introduction he concluded that two-thirds of the lines were of the strap rail style.

[10] Railway and Locomotive Historical Society *Bulletin 55,* May, 1941, p. 30.

[11] B & O Annual Report 1845, p. 18.

roads, are known to have been constructed with cast off New York State rails. After mid century rather little new strap rail construction was considered for main line service and it is likely that most such trackage was replaced by the beginning of the next decade. Thus ended what was probably the greatest wooden nickel in American engineering practice.

Trackage for Street and Industrial Railways

Following the demise of strap rail construction for main line service, the wooden railway was removed from public notice but it did not entirely disappear. Short lines, industrial tramways, logging railroads and even the city street railways found it acceptable for their needs. Again the principal attraction was economy. The total mileage and technological importance of these lines is surely negligible but they are an interesting curiosity and thus form an engaging, if not an important chapter in the history of the American railway.

During the 1850s one large American city after another began to adopt horse-powered urban railways. This special railway form had originated in New York some twenty years before. An obvious parallel can be drawn between this movement and the beginnings of the steam railroad. Like his predecessors, the ambition of the street railway promoter was greater than his means. He wanted to build long routes quickly before a competitor moved in but he couldn't raise the cash for first class construction. Nor was it really thought necessary. The tiny four-wheel cars weighed only a few tons and rarely traveled faster than a walk. In addition, some prudent managers were none too sure just how good a thing the street railway might be so favored the smallest investment possible. The old strap railway was unquestionably the cheapest way to go and its constructional details, the first horse car tracks differed little from the pioneer steam railroads. To be sure, the strap rail was buried in the street.[12] Small boulders or granite blocks were loosely set in around it to form a paving that was easy to remove for repairs. Primitive and impermanent, it did the job rather well and until the coming of the electric trolleys in the late 1880s, it remained in general favor except on the heaviest traffic lines. The greater weight and speed of the electrics quickly doomed strap rail for street railway service.

If the wooden railway was discredited for public railroads, including street car lines, it seemed to have a future with industrial railways. These lines were often temporary and were intended for the job at hand. They need not be elaborate or highly furnished and they must not be expensive. Second hand rail and well-worn rolling stock were typical of such operations. However, during the late 1860s the Clifton Iron Company thought even greater economies could be realized by reviving the wooden railway. They required a means of transport between their furnace near Ogdensburg, New York, and the mines some miles away in the Adirondacks. The firm's engineer, J. B. Hulburt, constructed an all-wooden standard guage railway. The rails, hard maple scantlings 4 by 6 inches, were fastened to crossties by wooden wedges. Roughly 21,000 board feet of maple plus 1,760 ties were required for a mile of track. Very light locomotives, 10 to 14 tons and a speed limit of 6 to 8 mph, would permit the passage of a 30-ton train. The line was expected to carry 50 to 100 thousand tons of ore yearly. Grading was minimal, with cuts and fills like those made for a country road. At first Hulburt claimed the line could be built for $1,000 a mile, this was later upped to $1,500 and finally to $1,760 a mile, less equipment and other contingencies. But even at the final price quoted the Clifton railway was remarkably cheap and many accounts of it appeared in the trade press.[13]

These reports prompted the Canadian government to examine Hulburt's system and some enthusiasm for the idea was soon abroad in the eastern parts of the country for cheap wooden railways with maple rails. At least one line was built for service as an ordinary common carrier between Quebec and St. Johns. The twenty-five-mile Gosford wooden railway opened in 1870. Heavier wooden rails (7 by 14 inches) and closer tie spacing were used to support heavier locomotives of 21 tons and faster schedules. Reckless speeds up to 25 mph were now thought possible. The railroad apparently operated with some success but when a new extension was opened in the early 1880s it was built with conventional metal rails. However, the wooden railway was not discredited, not even for common carrier service. No idea in engineering is ever put to rest. No matter how defective or stupid the idea is, someone will uncover it and knowingly or unknowingly press it forward as a profound, new solution. Some humbugs, like the monorail or the

[12]Alexander Easton, *A Practical Treatise on Street or Horse Power Railways*, (Philadelphia: 1859).

[13]*American Railway Times*, February 6, 1869, p. 43. *The American Railroad Journal* carried several articles on the same subject between 1868 and 1870.

three-wheel automobile, are particularly persistent because a very simplistic and convincing case can be made for their supposed advantages. The wooden railway is a less notable but parallel humbug that was advocated as recently as 1925. In that year Arthur W. Arnold opened an all-wooden railway in Oregon.[14] The eighteen-mile line ran out of Junction City. It was powered by converted automobiles or trucks. The rails were built-up from 2- x 6-inch planks. Like his predecessors, Arnold envisioned great economies for his wooden railway. Many came to ride. The incredulous went away believing, but the railway soon disappeared behind a screen of weeds and saplings.

If the wooden railway was not suitable for ordinary service, it might logically be deemed appropriate for logging railways. And for a time it was a widely approved plan for such lines. As logging operations worked away from streams, it became a tremendous problem to move logs to market overland. Oxen and wagons were inefficient in good weather and all but inoperative in bad weather. But a cheap, lightly built railway could do the job well. Unlike a road, there was no lost investment. When the timber was gone, the track was taken up and moved to the next stand. If an all-wooden track was used, it was simply abandoned.

One of the most imaginative and curious of the early logging railroads was a line operated in Sonoma County, California, in the 1880s.[15] It was said to be the only railroad graded with a crosscut saw rather than a shovel because the line was built directly on the tree stumps. When a low spot was reached, the trees were cut off to level agreeing with either bank, thus forming a trestle. The ties were fastened directly to the stumps, some of which stood 75 feet high. By the time the roots rotted out, the loggers had moved on to fresh timber.

Perhaps the cheapest and worst of all possible logging railroads was the so-called pole road. It's hard to imagine a more primitive and basic system of railway construction, if indeed it can properly be called a railway. (Figure 6). Pole roads consisted of two reasonably straight logs or poles, roughly a foot in diameter, laid parallel on the ground. Nothing more, no crossties or spikes or other appurtenances were needed (Figure 7). The weight of the logs was enough to keep them in gauge. As a precaution it was deemed best to tamp some dirt around them but after

Figure 6. Pole road, sixteenth-century Transylvania silver mine. (Scientific American, December 9, 1893.)

a few trips the weight of the passing trains would mash the poles securely in place. Because the logs are tapered it was also necessary to trim the end of the larger pieces at a joint but little else was necessary. Like everything about the pole road, switching from one track to another was simplicity itself. One end of the poles was lifted and swung over in line with the second track. To keep the locomotive and cars on the track, very broad, cupshaped wheels like gigantic spools or pulleys straddled the top contour of the pole rails. They were in effect double flanged wheels. The locomotive wheels were often notched or grooved for improved traction.

The locomotives, if they can be called that, were actually overgrown steam, farm tractors. A popular design was patented by W. E. Cole of Montgomery, Alabama. The chain driven Cole engine was produced by several builders. In 1888 the Richmond Locomotive Works offered Cole engines in 10½-, 14½- and 15½-ton sizes. The lighest engine could handle trains on grades of 150 feet to the mile. Six car trains and five-mile-an-hour speed limits were the rule. Even at this modest pace, we can imagine the teeth shaking ride afforded by the bumpy pole road.

The raison d'etre for the pole road was its cheapness. It could be built of native material that required no finishing. What could be simpler for a timberman than to chop down some trees, trim off the branches, and roll them in place. It could be done for $75 to $250 a mile. Not even Hulburt's wooden road could come close to this bargain among railways. However, the economy was a fraud, for wooden rails rotted out in four or five years. Iron rails lasted at least twice that long and permitted heavier, faster, and safer operations. Old iron also had a

[14]Railway and Locomotive Historical Society *Bulletin 59*, October 1942, p. 43.

[15]*American Railroad Journal*, October, 1884, p. 198.

Figure 7. Southern lumber railroad using pole road, climax locomotive and car. Scene about 1890. (Photo: T. T. Taber III.)

salvage value when retired, wood did not. The maintenance of wooden lines was very high, particularly if much traffic developed. As early as 1886 the *Railroad Gazette* stated that "no man is rich enough to run a wooden tramway." During the next decade the wooden logging lines and pole roads became increasingly rare. Lumbering became more the province of national corporations who saw greater long-term economy in operation of standard railroads. The wooden railroad disappeared even from its haven in the most remote forest.[16]

The Wooden Crosstie

An uncritical review of the facts might suggest that the railroads' involvement with wood technology ended with the nineteenth century. Wood-burning locomotives vanished by the middle eighties; strap rail tracks were already gone by a quarter of a century. Wood freight cars were made obsolete by steel in the late 1890s and passenger cars followed in another decade. Timber trestles, though by no means gone, were considered suitable only for secondary bridges. Wooden stations are now little more than Victorian Relics. Yet wood remains essential to the everyday workings of the modern railway and it continues as a basic element in the structure of some 340,000 miles of track. We are speaking, of course, about the ubiquitous wooden crosstie. Today there are approximately 800 million ties supporting our tracks; about one million new ones are installed each year. Each tie measures roughly 9 inches square by 9 feet long.

Why has the crosstie survived when all other wooden elements of the railway have vanished? To answer that question we must first say something about what is expected of the track structure which guides and supports the train. If you have ever closely observed a passing train, you may have noticed that the track rises and falls. This movement is no cause for alarm, it is precisely what the track must do to absorb the energy of a ten thousand-ton train moving at perhaps 50 to 60 mph. A rigid structure would have to be incredibly massive and precise to absorb this combination of forces. A semi-elastic structure can handle them readily and at the same time afford a smoother, safer path that need only be kept within reasonable limits of gauge and alignment. The giant, girder-like rail can bend and flex to a degree, of course, but the real cushion or shock absorber in the track structure is the tie. A wooden beam remains the cheapest, simplest form of tie and hence the survival of this seeming antique among railway appliances.

The wooden crosstie, or "cross stick" as it was occasionally called, is as old as the medieval mine tramway.

[16]Pole roads were still used in Russia as late as 1942 according to Lee's *Evolution of Railways*, p. 103.

It was discarded by the early British railway builders who felt stone block sleepers (i.e. ties) made a more durable foundation for the rail.[17] It could not rot out and was thus suited to what they chose to call the "permanent way" in the belief that once constructed their iron roads would endure for the ages. Several early American lines accepted the British argument and dutifully laid out their lines with stone block ties. The defects of such lines were soon apparent. The Camden and Amboy switched to wooden ties when inmates at Sing Sing Prison could not swing their hammers fast enough to keep up with the demand for more blocks. Other roads ran out of cash; blocks cost $1.50, wooden ties only $.20. An even more serious failing was the rigidity of the track, there was no cushion, and the maintenance accounts showed it. The Boston and Lowell discovered that a system so suitable to England's mild climate was incompatible with New England's winter season. Its stone block track was heaved up by the first hard frost and became a near wreck by the end of the cold weather; permanent way indeed! Most lines, including the British, experienced difficulty in holding the stone block tracks in gauge as train speeds increased. Most resorted to intersplicing wooden ties or iron tie rods.

Meanwhile in New Jersey, the Camden and Amboy Railroad had inadvertently stumbled upon the perfect form of railroad track. They had set out to modify the British plan ever so slightly by eliminating the rail chair, an openmouthed cast iron fixture fastened to stone blocks. The rail, shaped like a capital I in cross section, was dropped into the chair and held in place by wedges. The Camden and Amboy's chief engineer, Robert Stevens, saw a way to eliminate the chair by producing a flat-bottomed rail which thus formed its own base or chair and could be fastened directly to the block. Stevens thus invented the common T rail of today.[18] He did not intend that it be attached to anything so crude as a wooden crosstie but the shortage of stone block

Figure 8. Standard American track, T-rail and wooden crossties. Repair crew near Murfreeesboro, Tennessee, 1864. (Photo: Library of Congress.)

forced him to do so as a temporary expedient. The flat-bottom rail was quickly attached to wooden ties by hook headed spikes (Figure 8). The classic pattern of American track construction was born. It went together more quickly than any other form of track and yet was comparatively cheap, durable and flexible.

The excellence of the scheme was apparently lost on all who first saw it. Stevens finished as much of the line as possible with stone blocks. Within a few years other roads began to use the T-wood tie form of track, seeing it as the best of all possible combinations of the provisional and permanent. The Boston and Worcester built its second track on this plan in 1834. The B&O's Washington branch was so constructed during the following year. The Reading chose it over all others. Late in 1837 Jonathan Knight and Benjamin Latrobe, Jr. made a survey of the northern and middle state railways; they concluded that Stevens's plan had no rival and that it was being adopted by all forward thinking engineers.[19] It was the national standard within a few years and soon spread overseas. Even Britain, so proud of its traditional rail chair trackage, has come to adopt the American

[17]Until recent years C. E. Lee's *The Evolution of Railways*, London, 1943, was the only comprehensive survey of primitive railway construction. Since then two excellent book length studies have appeared: *Early Wooden Railways* by M. J. T. Lewis (London, 1970), and *Stone Blocks and Iron Rails* by Bertram Baxter (Newton Abbot, 1966). The Newcomen Society also has published many papers on ancient tramway construction in its *Transactions*.

[18]J. E. Watkins, *Ceremonies Upon the Completion of the Monument Erected by the Pennsylvania Railroad Co. at Bordentown, New Jersey*, (Washington, D. C: 1891).

[19]Jonathan Knight and Benj. H. Latrobe, *Report Upon the Plan of Construction of Several of the Principal Rail Roads in the Northern and Middle States...*, (Baltimore: 1838).

system since the end of World War II.

These developments were largely the concerns of railway technicians and surely attracted little general notice. The amount of timber consumption by the tie market appears to have gone unnoticed in a national economy glutted with wood. Even the unprecedented growth of this market after the Civil War as new railroad construction skyrocketed, seems to have attracted rather little attention, even though the statistics were ever more alarming: 1870, 39 million ties; 1890, 89 million; 1910, 129 million, or about a quarter of the nation's total wood production.[20] By the end of the century public opinion had been aroused to the dangers of an impending wood famine. Much of this sentiment had been stirred up by the U. S. Forest Service, a small but vocal federal agency which directed much of its campaign against the railroads' wasteful and proliferating use of crossties.

According to Sherry Olson's *The Depletion Myth*, the problem was first attacked by an effort to grow more trees. The railroads responded in a positive fashion by setting aside vast tracts of land for tree plantations. The Delaware and Hudson maintained 150,000 acres in the Adirondacks for this purpose. The Pennsylvania planted 3½ million black locust trees. Other roads planted fast-growing soft woods. All of this with an eye toward future tie supply.

By 1910 the railroad industry was disenchanted with the advice it was receiving from the Forest Service. The tree plantations were by and large a failure. Of greater importance, the normal sources were supplying an abundance of ties. Prices remained stable. The predicted shortage never materialized. The supply of choice building timber and some hardwoods declined but railroad ties were produced from the lowest grade of trees, second only to pulp wood. Many ties came from farmers' woodlots—a source overlooked by the Forest Service. The fast growing rate of the South and Pacific Northwest replenished the supply well beyond the projections which were based on European figures. The tie shortage was a much discussed but totally unreal phantom.

Even though the supply remained adequate, the railroads became more sensitive to the conservation problem. The price of ties did creep upwards as did the cost of tie replacement with rising wages. Longer tie life became of prime importance as these costs continued to increase.

The most radical alternative pursued was the effort to devise a substitute for wooden ties. Metal ties had been used with some success in Europe. Cast iron pot ties had been tried in England as early as 1846. A full width wrought iron tie was introduced in Holland in 1862 and large-scale use was gradually effected.[21] By 1900 half of Switzerland's ties were pressed steel. Germany had 26 million steel sleepers in service six years later. American railway managers listened patiently to the virtues of metal ties. A few test installations were made in the 1880s. The New York Central's general manager, John M. Toucey, tested his own plan with some satisfaction. The patent office had approved over 250 specifications for such devices by 1888. In 1889 the Department of Agriculture issued a special report on metal ties.[22] However, behind all of this activity there was little real interest, for so long as a good white oak tie could be purchased for 65 or 70 cents who would pay 2 or 3 dollars for a metal substitute? Only the U. S. Steel-controlled Bessemer and Lake Erie showed much enthusiasm for pressed steel ties. Beginning in 1904 the big steel subsidiary installed well over one million metal ties. They proved costly to maintain and presented difficult insulation problems to the signal department.[23] The idea was all but abandoned by 1940.

Other wood substitutes were suggested at an early date. One British inventor patented a ceramic tie in 1837. A more practical imitation stone material was found in concrete. Such ties were reportedly tested in this country as early as 1893. The Lake Shore and Michigan Southern installed 6,000 concrete ties between 1902 and 1906.[24] Old T rail was used for the reinforcing member. Several years later Henry Ford became enamored with concrete roadbeds but like most of his railroad ventures it proved a dismal failure.

After World War II Europeans perfected a superior pre-cast tie that incorporated a practical rubber pad seat. This style of concrete tie has superseded the metal tie and is now being seriously considered for use here. By 1969 wood ties were up to $6.50. Train weights were

[20]Sherry H. Olson, *The Depletion Myth*, Harvard Univ. Press, 1971.

[21]*Railway Review*, January 12, 1895, p. 24.

[22]Department of Agriculture Forestry Division Bulletin 3, "The Use of Metal Track on Railways. . ." by E. E. R. Tratman. Washington, 1889.

[23]Roy C. Beaver *The Bessemer and Lake Erie Railroad*, (San Marino, Calif.: 1969), p. 72.

[24]*Railway Review*, August 8, 1906, p. 627.

Figure 9. The wooden railway in all its elements–wood burning locomotive, wood crossties and trestle, and wooden cars. Scene: Nevada County Narrow Gauge R. R., 1876. (Photo: G. M. Best.)

beginning to inflict increasing damage to wooden ties beyond the protection tie plates can offer. At least one U. S. railroad has adopted concrete ties and it is likely others will follow.

If metal and concrete are dismissed as a minor chapter in the story of the American railway tie, preservation must be introduced as a very major chapter. The progress made in preservation has played a major role in the continued success of the wooden crosstie. For generations white oak was preferred for railroad ties. It was mechanically strong, resisted rot up to eight years and was notable for tightly gripping spikes. Other woods were reluctantly accepted. Even today roughly half the new ties are oak.

While the supply was good and prices low, there seemed little to be gained from tie preservation. Wood preservation was an ancient art studied and applied by ship builders. British patents on the subject date back to 1737.[25] American interests cannot likely be traced back to the same period, but they were under consideration when the railway age opened. Both the Camden and Amboy and Ohio railroads used salt for the preservation of pilings. A British chemist, J. H. Kyan, perfected a chloride of mercury solution in 1832 that found some favor in America. The creosoting and Burnettizing (chloride of zinc) processes both came into being six years later and like Kyanizing were tried at an early date. In the 1850s the Reading opened a creosoting plant in Pottsville, Pennsylvania, that employed pressure tanks and was laid out not too differently from a small plant of the present day.[26] In the first years of the following decade the Philadelphia, Wilmington and Baltimore had over 30,000 preserved ties in service. One could go on to list other railroad users of preserved wooden ties but that exercise would prove only that many lines were willing to experiment with preservation at an early date.

---

[25] *Railroad Gazette*, May 21, 1880, p. 257.

[26] Z. Colburn and A. L. Holley, *The Permanent Way and Coal-Burning Locomotive Boilers* . . . , (London: 1858), plate 10½.

The techniques were well understood but so long as wood was plentiful and labor cheap, it was not economically feasible to pursue a large-scale preservation program. About the only exceptions to this rule in America were certain prairie or southwestern railroads which had no native forests to draw upon. The cost of imported ties was enough to make the expense of preservation (17 cents a tie) an attractive alternative. The Rock Island embarked on a large-scale treated tie program in 1886.[27] By 1892 they had 1.7 million treated ties in service. At the same time the neighboring Santa Fe had 1.3 million. Large as these figures might appear, one must remember that they represented less than 1 percent of the total American tie population. As the fears of a wood famine grew, so did the acceptance of the preserved tie. In 1900 about 5 percent of all ties were treated, within fifteen years the figure was 30 percent. Rapid progress was made after World War I as labor and material costs rose. By the late 1920s over three quarters of our ties were of the preserved variety. By this time almost no class one railroad was installing any untreated ties and so within a few years all main line trackage was carried by some form of preserved tie.

Creosote is the most familiar and perhaps the best of the tie preservatives.[28] A coal tar derivative, it is nearly insoluable in water and thus acts as a weather proofing. More valuable is its toxic effect on fungi, the great destroyer of wood. Painting, dipping and spraying provided only a thin and ineffectual surface coating. Because ties are in constant and nearly complete contact with the ground, they require more protection if any service life is to be realized. A nearly complete chemical saturation is required which can best be done in a pressure tank. Ties are placed in an air-tight cylinder, often 6 feet in diameter by 50 feet long. A vacuum is effected causing the sap to gush out. Then creosote is pumped into the open pores of the wood. Each standard 90-pound tie will absorb 30 pounds of creosote. A tie so pickled will give 25 years of service. The pressurized treating idea was apparently suggested in the 1830s by Burnett and, as already noted, was tried in this country as early as the 1850s. Once the process became accepted for tie treatment, most railroads built and operated their own facilities, but in recent years they found it preferable to purchase them from large-scale commercial producers.

This article briefly summarizes the railroads' once nearly total dependence on wood (Figure 9) and their gradual conversion to other fuels and materials. Today only the crosstie links them to the wooden past and even this bond now seems threatened by an advancing technology.

---

[27]See Note 21.

[28]Much valuable data on wood ties and their preservation is given in the *Railway Engineering and Maintenance Cyclopedia*—first published in 1921; it has since appeared at four- and five-year intervals.

Name and Place Index

Please note: Page numbers which appear in italics are references to tables or illustrations.

Subject Index